The Montana Cooperative Extension Service: A History 1893—1974

by
Merrill G. Burlingame
and Edward J. Bell, Jr.

Library of Congress Number 84-62142

ISBN 0-934318-45-X

Published by Montana State University

DEDICATED TO

The thousands of Montanans of all ages

and to

The staff: national, state and county

who have worked together to improve the quality of living for all of the people of Montana.

INTRODUCTION

At the observance of the 75th anniversary of the founding of Montana State University in 1968, the contributions of the Montana Agricultural Experiment Station and the Cooperative Extension Service were emphasized along with the academic achievements of the institution. A strong feeling developed for the need to begin recording these contributions while a number of former important leaders could contribute.

Extension Director Torlief S. Aasheim encouraged the project, and we received hearty support in our many requests for help. We were fortunate in being able to discuss with former leaders, including Jack Taylor, Bob Tootell, N. E. Beers, Harriette E. Cushman, Frances Smith Patten, Bess McClelland, Geraldine Fenn, Clifford W. "Brick" Vaughn and others, the highlights of their periods of service.

References will be found to several earlier histories in manuscript form available in the Montana State University Archives: Fred S. Cooley, "History of the Montana Extension Service, 1914-1924;" Charles E. Potter, "Twenty Years of 4-H Club Work in Montana, 1914-1934;" Ralph Mercer, "The Montana Extension Service, Fifty Years, 1914-1964;" Harriette E. Cushman, "Built for the Land," not completed. The major part of this study comes from the annual and special reports of the Extension personnel, also available in the University Archives.

Our beginning date is 1893 with the opening of Montana State College when the staff of the Agricultural Experiment Station immediately began working with the rural people of Montana. The closing date is 1974, the year Director Aasheim retired.

During the 16-year term of Director Aasheim, 1958-1974, many new fields of Extension activity were added. Late in his administration, the earlier strongly agriculturally-based programs were broadened into people-oriented programs including both rural and urban areas. Many of these "new directions" are still developing under the direction of Dr. Carl Hoffman, vice president for Extension. The story of these programs can best be told at a later time.

We are indebted to more people than we can list. In addition to those mentioned, the history could not have been completed without the support and active participation of former Director Aasheim. We also wish to acknowledge the assistance of the University Archivists, of Donald J. Luebbe, John C. Bower, Robert F. Rasmusson, Louise Stone Shunk, Vivienne Kintz and Jim Sargeant; also to Mrs. John P. (Margaret) Parker for skillful editing and typing, and to MSU President William Tietz for his encouragement.

We hope that this account of the dramatic advancement in the quality of living of the people of Montana, with the assistance of the Extension Service, will stimulate continued achievement in future years.

Merrill G. Burlingame and Edward J. Bell, Jr.

THE EXTENSION BUILDING

The building which has housed the state staff of the Extension Service since 1923 was the first major structure to be erected on the Montana State College campus. The first president, Augustus M. Ryon, spent considerable time during 1893-1894 designing and supervising the construction of what was described as "a very fine brick veneered building." Federal Experiment Station funds were used for the construction costs of about $4,000.

James M. Hamilton, a member of the State Board of Education and later president of the college, reported to the board, "It is difficult to see how a building of its dimensions and construction can be erected for the money. It is well built, well arranged, and is adequate for the purpose for which it is intended."

Limited space in campus buildings for many years resulted in the Extension building being used for various purposes. Upon completion, the basement was used for engineering shops, the first floor contained staff offices, the director and his family lived on the second floor, and the third floor was occupied by the Veterinary Department.

A greenhouse was built in 1903, adjoining the building to the east and south. This was enlarged in 1911 and continued to be used until late in the 1930s. Several science departments used rooms in the structure and in 1907 the agricultural work of both the College and Experiment Station was carried on in the basement and the first floor. In 1907, however, the Biology Department occupied much of the building, remaining there until 1923. Numerous room changes were made during that time.

In 1923 the Extension Service took over the entire building and again internal changes were made. In 1965, a 24-by-42 annex was added to the east, and in 1967, a sturdy, roofed concrete entrance replaced the former worn, narrow, wooden steps.

The annex enlarged the basement, making room for an expanded multilith and publications operation. A program of renovation took place on the other floors involving improved lighting, painting, carpeting and storage cabinets which changed the character of the building into pleasant, although still crowded, quarters for the director and many of his state staff. For 60 years the building has been the nerve center for the Montana Extension Service.

The Extension building. Constructed in 1894.

TABLE OF CONTENTS

APPENDIX

ILLUSTRATIONS

Photographs courtesy of the Montana State University Archives and the U.S. Soil Conservation Service.

The Montana Cooperative Extension Service: A History 1893—1974

I. The Grass Roots Beginnings of the Montana Extension Service

This is the story of the Montana Extension Service and of its work in steadily making Montana agriculture a more successful and stable enterprise and of making the Montana farm and ranch home one of the most pleasant in the nation.

The story really begins with the first record of Montana written by Lewis and Clark on their memorable expedition of 1804-1806. They carefully noted the many kinds of grasses, fish, fruits, coal veins, forests, and the unbelievable clarity of the rapid streams. According to their records, they observed the "extensive and beautiful plains and meadows" of the Gallatin Valley, and on the Yellowstone River they repeatedly saw "several gangs of elk from 100 to 200 in a gang, great numbers of antelopes," and great flocks of geese and ducks, and described a land readily adaptable to agriculture.

They voiced surprisingly few objections to the high plains and mountain country. Occasionally they found "the country rugged, the hills high . . . the soil poor and sterile, sandy near the tops of the hills." Lewis made a pointed first statement of what has become a familiar theme: "It is astonishing," he said, "what a quantity of water it takes to saturate the soil of this country."

The most thorough early report of Montana's resources, and the most favorable, was that of the Isaac I. Stevens Expedition of 1853, commissioned by the Secretary of War to search for a northern route for a transcontinental railroad. On the Marias River, the expedition members observed, "The grass was so luxuriant as to form a heavy turf, and the soil a rich loam." They praised "the broad and fertile Deer Lodge prairie." In the Bitterroot Valley they found that "the soil is a rich, dark loam, mingled with sand and gravel, in which wheat, oats and potatoes have been found to grow exceedingly well, and is naturally covered with luxuriant grass, supplying inexhaustible pasture, over which already several thousand cattle and horses are roaming." Looking back over the high plains and mountain valleys before they plunged into the massive mountain chain, they summarized optimistically, "The observer . . . will be astonished at the conclusions which he will reach in regard to the agricultural advantages of this country."

The Extension story in Montana began to take shape 40 years later, following intermittent "boom and bust" periods of mining development, the "hard winter of 1886-1887" and widespread beginnings of agricultural development. On February 16, 1893, the Agricultural College of the State of Montana (renamed Montana State University in 1965) was created by legislative act. A major division of the College was the Agricultural Experiment Station, established by the federal government under the Hatch Act of 1887. The census of 1890 reported a population of 142,924 people in Montana supported by 5,603 farms and ranches. The state boasted 1,200,000 head of cattle and 2,900,000 sheep, largely on the open ranges east of the mountains, and 230,000 horses in use by the stockmen. One million two hundred thousand bushels of wheat grown on 48,000 acres of land and 320,000 bushels of barley on 10,000 acres pointed toward the future. When the Experiment Station opened, the farmers and ranchers on their far-flung holdings had many questions to ask.

The Experiment Station largely carried the heavy load of receiving the questions, searching for the answers, and taking the information to the people until the Extension Service was formed by the Smith-Lever Act of May 8, 1914. The beginnings of Extension activities go back long before 1914, however.

Beginnings in the United States

The United States has been, of course, singularly fortunate in that its physical features adapted so readily to agricultural progress. Its national history has not yet given suitable recognition to the importance of soil fertility, favorable climatic factors, the varied nature of agricultural production and the diligence of its people in agricultural pursuits.

The United States has also profited in having had two important founding fathers who were vigorous agricultural innovators. George Washington was a scientific farmer who made a strong impact upon the agricultural practices of his time by his inquiring mind and experimental approach. He imported blooded stock from England and Europe, for instance, and regularly tested new crop varieties. He also practiced careful farm management which is reflected in his detailed records of planting and harvest, and of profit and loss. Thomas Jefferson was even more diligent in inquiry, experiment and record making. The visitor to Mount Vernon and Monticello to this day is forcibly impressed that these were experiment stations and demonstration farms.

The American people have been especially blessed by their early determination to secure an education for everyone, as evidenced by The General Education Act of Massachusetts of 1789. This led, no doubt, to the varied educational activities requested of the state governments and the classical universities of the day. Columbia University, for example, established a professorship of natural history, chemistry and agriculture in 1792. A State Board of Agriculture was created in New York in 1819, and a New York

State Agricultural Society in 1832. The Massachusetts Society for Promoting Agriculture was chartered by the state in 1792, and by 1813 the *Massachusetts Agricultural Journal* was being published. In 1819 the important *American Farmer* began publication in Baltimore. Yale and Harvard Universities, the Rensselaer Institute and Amherst College, notably, also gave encouragement to the acquisition and dissemination of agricultural information. New Hampshire created a State Board of Agriculture in 1820, and Ohio's State Board of Agricultural Control began in 1846. Rural people responded by forming supporting organizations. By 1860 some 941 local agricultural societies had been formed and were listed by a national organization, the United States Agricultural Society.

Farm leaders moved quickly to provide suitable educational methods and materials. The public lecture or lyceum movement, which developed into the Chautauqua Society in the 1850s, suggested a similar approach. President Edward Hitchcock of Amherst queried in the newly formed Massachusetts Board of Agriculture in 1853, "Why should we not have farmers' institutes as well as teachers' institutes?" He explained his idea as consisting of "an ambulatory agricultural school where the young, especially, would learn very rapidly from the best masters." The Ohio State Board of Agriculture promoted farmers' institutes as early as 1846.

A result of this interest in studying agricultural procedures was the increasing awareness that our basic knowledge in the fields of agriculture and industrial technology was pitifully limited. Jonathan B. Turner, in Illinois, proposed what would have been the first agricultural college as early as 1852. He saw the movement ripen rapidly until in 1862, the Morrill Act was passed providing for the land grant colleges. Also in 1862, the Department of Agriculture was established, and the Homestead Act revolutionized rural land holding in the United States.

The land grant college was the breakthrough which took the American public educational system from the "classical" European pattern of emphasis upon law, religion, medicine and letters, and broadened the idea of a university to a point where agriculture and mechanic arts could be included. At last John Milton's concept in 1644 of a "complete and generous education is that which fits a man to perform justly, skillfully and magnanimously all the offices both private and public of peace and war," was applicable.

Seaman A. Knapp is usually credited with being the originator of the "cooperative" idea of bringing the colleges and the national, state and local governments together with the farmer and his problems. After distinguished work at Iowa State College, Knapp went to Lake Charles, Louisiana, in 1886, where he undertook the management of a huge program to improve the culture of rice and cotton, as well as the standard of living of the people of the area. He secured an appropriation of $40,000 from the U.S. Congress to assist the state and local governments in the project and thus developed an early prototype of experiment station and extension service cooperation.

The logical next step, then, was the Hatch Act of 1887 establishing Agricultural Experiment Stations at each of the land grant colleges.

The new personnel and facilities were cordially received and immediately besieged for assistance beyond their limited capabilities. This demand encouraged Kenyon L. Butterfield, president of the Rhode Island College of Agriculture at Kingston, to suggest the next step. He became chairman in 1905 of a Committee on Extension in the Association of American Colleges and Experiment Stations, where he promoted his favorite "idea of systematic, long-continued and thorough instruction to the farmers the year through." He foresaw that "this work will not only be dignified by a standing in the college coordinate with research and the teaching of students, but it will rank as a distinct department with a faculty of men whose chief business is to teach people who cannot come to college."

The first county agent with duties comparable to those of future agents was W. C. Stallings, appointed November 12, 1906, in Smith County, Texas. A county agent in the northern states was not named until John H. Barron took a position at Binghamton in Broome County, New York, March 1, 1911. Rumors were afloat over the entire country during this period that the Congress would soon provide support for extension programs on the same cooperative basis as that of the Hatch Experiment Station Act. In each state activity was pointed toward developing a program with legislative support which would match the hoped-for federal assistance.

Beginnings in Montana

Montana was just beginning to consider itself an agricultural state in 1893 when the Agricultural College was established. The emphasis in the past had been on mining. The fertile, well-watered valleys of the west had been used, however, to provide agricultural products for the burgeoning mining camps.

Nelson Story had driven cattle the entire length of the Texas Trail in 1866 to supply the gold camps. The range cattle had mushroomed, assisted by the marketing convenience provided by the arrival of the Utah Northern Railway in 1880, and the Northern Pacific in 1883. The devastating blizzards of 1886-1887 called a crashing halt to the get-rich-quick philosophy of the English, Scotch and Eastern "stockmen" and other enterprising speculators who proposed "to trap it, to shoot it, to mine it and get out." Their withdrawal was easily accomplished since resident managers who had formerly rendered highly favorable financial statements now reported 125 percent losses. The industry returned to home ownership and was rapidly re-established; the population increase included many hardy farmers. Montana became a state in 1889.

The surprising accomplishments of Montana agriculturists in 1893 can be illustrated in the awards made at the Columbian World's Fair in Chicago in that year. W. N. Sutherlin, manager of the Montana exhibit, reported jubilantly:*

> There were 11 awards given Montana agricultural products. The first was on oats, the entire exhibit being superior to the other exhibits, weighing from 41 to 48½ pounds to the measured bushel. There were two awards on hard wheat—one to W. Caldwell, Bozeman, and one to J. E. Dyer, Missoula. There were four awards in barley—to the Manhattan Malting Company, Samuel Flower, M. M. Ferguson and J. W. Dawes, all of the Gallatin Valley. There were two awards on timothy—one to Thomas Blake and one to Mrs. E. Dukes, and one on blue joint to Bailey and O'Donnell, Billings.

"I feel quite content with the success attained," continued Sutherlin. "We got more awards than the Dakotas and a number of other states."

Statewide agricultural organizations were forming rapidly in this period. The Montana Stockgrowers had been a powerful force since its beginnings in 1879 and formal state organization in 1885. The Montana Wool Growers Association was organized in 1883. The Montana Horticultural Society was formed in 1897 and the State Board of Horticulture was created by the Legislature in February 1899. A broader membership organization, the Farmers Alliance, had organized its first chapter in 1889 and expanded to a small state organization in 1892. The Montana Agricultural Association formed in Helena in March 1902. A result was the creation by the Montana Legislature in 1893 of the State Bureau of Agriculture, Labor and Industry. Its two chief jobs in agriculture were minor, however, consisting of making annual statistical reports and promoting an active immigration policy for homesteaders and agricultural settlers.

Astonishingly large and successful community and county agricultural fairs were first held in the 1860s. Not until 1903, however, did the Legislature establish the Montana State Fair, held annually in Helena "to disseminate knowledge concerning, and to encourage the growth and prosperity of all agricultural, stock grazing, horticulture, mining, mechanical, artistic and industrial pursuits in the State of Montana." The board of directors consisted of a member chosen from each county in the state.

The Experiment Station

The highly practical farm people of Montana were not excited over the opening of the Agricultural College in 1893. Enrollment for the four-year academic program in agriculture was low. But a brisk demand soon appeared for short courses of a practical nature. There was also a continuing demand for assistance on the part of individuals and communities.

The Experiment Station activities, therefore, fitted the immediate needs of rural Montana. The first staff consisted of S. M. Emery, director and hor-

*Specific reference to sources has been largely omitted, but enough information has been given concerning date, type of report, person reporting, etc., to identify the source in the Archives in the Montana State University Library where the original manuscripts and notes are also on file.

ticulturist; Luther Foster, agriculturist and botanist; Dr. Frank Traphagen, chemist; and Dr. W. L. Williams, veterinarian. Director Emery had a distinguished background of professional and public service in Minnesota. He had been engaged in private horticultural work centering in Great Falls for several years; he knew the state well and sensed many of its important problems. This was equally true of Dr. Traphagen who had established a reputation as a master teacher and a tireless experimental chemist at the College of Montana in Deer Lodge. Professor Foster, who came from South Dakota, organized and directed the first short term of the College in 1893. Veterinarians were in short supply, and after a long search, Dr. Williams was employed and arrived from Michigan in the fall of 1893. In 1894, A. M. Ryon, who served as president of the College in 1893-1894 and who had also been at the College of Montana, began work in the important field of irrigation engineering.

This small staff, all of whom also taught college courses, set out with great vigor to conduct scientific research on the many untouched problems confronting agriculture in Montana. From the beginning they were both pleased and harrassed by requests for information and assistance, and for appearances at community institutes, schools and fairs. They welcomed these contacts as opportunities to discover what farmers and ranchers felt their major problems were, and often found some exciting experimental work taking place on the farms.

Early Farmers' Institutes

To meet the widespread interest at the Experiment Station, the staff turned immediately to the Farmers' Institute method with which they were familiar and which enabled them to work with larger numbers in a systematic way. Sixteen institutes were held during the winter of 1893-1894. A sharp financial panic depressed the land, and no funds were available for travel, lodging or remuneration of speakers. The institutes were held, therefore, in the nearby western valleys. This was economical, and it was also in these valleys that the major diversified farming was taking place.

In the Gallatin Valley, the institutes were held in Bozeman, Springhill and Manhattan. In the adjacent Jefferson Valley, meetings were hosted in Willow Creek, South Boulder, Whitehall and Waterloo, and in nearby Sheridan and Twin Bridges. In the Madison Valley, institutes were organized in Ennis and Meadow Creek. In the Bitterroot, Emery worked enthusiastically on horticultural problems in Hamilton, Stevensville and Missoula. Meetings were also held at Townsend and Great Falls.

This promising beginning slowed down as the staff turned major attention to designing and assisting in the construction of economical buildings, preparing experimental plots, making tests and meeting increasing daily requests. This left little time to prepare new materials for the institutes. In 1894-1895 only three institutes were held—in Columbia Falls, Kalispell and Big Timber. No institutes were held in 1896-1897, but 11 meetings the following winter reflected continued interest. Political and personality conflicts, not immediately related to the Experiment Station work, caused the resignations of A. M. Ryon, Luther Foster and S.M. Emery by 1900. No institutes were held in the years 1897-1899 but eight were conducted with manifest interest in 1899-1900. By this time the agricultural community had acquired a firm confidence in the capable and cooperative staff at the Station.

Early Publications

To assist in answering the requests which came from institute participants as well as those who could not attend, the Experiment Station immediately began to publish bulletins. They did not wait for the results of personal research to point up local problems, but used basic research with which they were familiar. Montana Experiment Station *Bulletin No. 1*, issued in March 1894, titled "Organization—Announcements," alerted the people of the state to the available services at the Station. Smut was causing a great deal of trouble at the time, and Luther Foster quickly drew together a short *Bulletin No. 2*, "Smuts of Wheat, Oats and Barley," issued in May 1894.

Foster found that the large number of farmers who had come from the Midwest retained an interest in raising hogs. Since diversification was widely favored at the time, Foster issued *Bulletin No. 3*, "Pig Feeding," urging the use of the small grains which Montana produced so abundantly. Williams soon discovered that glanders in horses was causing heavy losses, and he issued *Pamphlet No. 4*, "Glanders" in June 1894. These bulletins had the effect of giving people confidence that the Station specialists were aware of their problems and were willing to give assistance rather than merely to pursue a cloistered career of research.

A Farmers' Institute Board and Program

The Farmers' Institutes made sufficient impact upon the state to cause the Legislative Assembly to enact a statute signed March 14, 1901, creating a Board of Administration for Farmers' Institutes. The sum of $2,000 was appropriated for its use. The board was to consist of the presidents of the Montana Wool Growers' Association, the Montana Livestock Association (Stockgrowers) and the Montana Horticultural Society, together with the governor of Montana and the director of the Montana Experiment Station as *ex officio* members. The Montana Agricultural Association and the Montana Dairymen's Association, which were being organized, were also to have membership. In the 1903 Legislature, the Montana Registered Cattle Breeders' Association was also added.

The board met on June 10, 1901, and elected A. L. Stone of Missoula, president of the Horticultural Society, and Samuel Fortier, who had become director of the Experiment Station, secretary. Since $1,000 of the small annual appropriation had been designated for a printed report, Fortier agreed to serve without compensation as secretary and managing director of the institutes.

A revitalized series of institutes began in November 1901, with 17 in all held during the winter. The first report, published in November 1902, contained edited summaries of major papers. The one-and two-day meetings addressed technical problems of production during the day, and at the evening dinner meetings considered general educational problems, agricultural resources and the improvement and beautification of farm homes and farm yards. Women were urged to attend, but came in small numbers in the early years. Women speakers were sought but were difficult to obtain.

The first report listed 16 subjects which had been presented and several speakers who had led the discussions. Experiment Station and other staff were usually present to assist, but local citizens were used extensively. The subjects covered included:

1. Alfalfa—I.D. O'Donnell of Billings was by far the most popular speaker on this subject and is usually considered the originating and outstanding alfalfa grower in Montana. W. W. Wylie of Bozeman, also an alfalfa enthusiast, spoke frequently to groups.
2. Stock and Stock Feeding—Robert S. Shaw, Station agriculturist, was the chief speaker. He always tried to have local speakers present the results of their experiments with which he had often assisted.
3. Irrigation—Samuel Fortier came to the Station with a national reputation and was in steady demand as an authority.
4. Grains and Grasses—Robert S. Shaw also took the lead in this subject. A. J. Noyes of Dillon often assisted.
5. Soil Fertility—Dr. Frank W. Traphagen, Station chemist, saw alkali as a chief villain in the Montana soils. He spoke so frequently on this subject that he was soon dubbed affectionately "Alkali Ike." William Flannery of Bozeman frequently discussed "Summer Fallowing."
6. Dairying—R.N. Sutherlin of White Sulphur Springs, copublisher with his brothers of the *Rocky Mountain Husbandman*, gave vigorous support to the institutes and often talked on phases of dairying. Mrs. E. Broox Martin of Bozeman was a stickler for cleanliness, and occasionally spoke on "Choice Butter Making."
7. Root Crops—Professor Shaw recognized the importance of pota-

toes and boosted their production. Dr. Traphagen very early saw the possibilities of sugar beets and actively promoted experimentation.

8. Insect Pests—Dr. Robert A. Cooley, Station entomologist, usually directed this subject. He used local men, women and youth extensively to assist with striking exhibits of pests and their destructive work.
9. Horticulture—This was one of the most popular subjects because of the desire of the Midwestern settlers for the fruits they hoped to raise. Director Emery had stressed this field and J. W. Blankinship, Station botanist, carried on. Mrs. J. C. Fergus of Whitehall was interested in shade trees and flowers and Olney Taylor of Park City created interest in fruit culture in eastern Montana.
10. Food Adulteration—This was less frequently discussed, usually by Dr. Traphagen.
11. Poultry—This was a field of steady interest, but not many felt capable of discussing the scientific aspects of breeds,housing, feeding and disease. H. C. Gardiner of the Station spoke most frequently.
12. Bees—Bees were considered a highly desirable spare time enterprise, but there were few authorities. Ralph Benton of the Station and M. Kirchner of Miles City spoke of winter protection, undesirable flowers and predators.
13. Education—Among a wide range of speakers, College President James Reid spoke on "Industrial Education," J. T. Smith of Livingston spoke on "Farmers' Reading," and E. O. Clark of Big Timber discussed "Business Methods."
14. Domestic Science—Mrs. S. M. Emery, who had helped organize some of Montana's first women's clubs, spoke frequently on this subject. Mrs. Frank Traphagen spoke occasionally on "The Dignity of Housework," as did Maud Wisner of Bozeman on "The Ethics of Domestic Science."
15. Agricultural Resources of Montana; and
16. County Statistics—These were popular subjects for evening discussions and were usually handled by well-informed people from the area in which the particular institute was being held.

Records of attendance were highly incomplete until the organization of the Farmers' Institute board following 1901. A steady increase reached a peak in 1910-1911 with the Northern Pacific "Better Farming Special." The institutes proved to be a ready-made method of working toward larger production in World War I, and were particularly useful for the emergency war meetings. The annual reports showed attendance as summarized in Table I.

TABLE I

Farmers' Institute Attendance, 1901-1918

Year	No. of Meetings	No. of Sessions	Attendance
1901-02	17	--	-----
1902-03	33	43	1,543
1903-04	39	72	3,439
1904-05	47	100	6,946
1905-06	63	133	7,890
1906-07	61	130	7,541
1907-08	71	154	12,113
1908-09	--	119	10,961
1909-10	76	137	17,091
1910-11	101	175	30,344
1911-12	139	292	20,977
1912-13	162	272	18,067
1913-14	112	207	20,934
1914-15	144	185	18,934
1915-16	161	206	21,334
1916-17	109	160	17,971
1917-18	103	187	10,649
1917-18— Emergency War Meetings	340	340	87,889

Finances always lagged behind needs, but were supplemented by local support and generous voluntary assistance. A summary of state appropriations is as follows:

Year	Annual Amounts
1901-1903	$2,000
1903-1907	$4,000
1907-1909	$7,500
1909-1911	$8,000
1911-1918	$10,000

Organization and Action

The Experiment Station personnel, whose members soon found themselves immersed in Farmers' Institute and Extension work almost beyond the time available, constantly looked for organizations which would be helpful. Fortunately, these began to emerge with increasing rapidity early in the century. The membership of the Farmers' Institute board in 1901 was previously noted. Added in 1902 to the board, the Montana Agricultural Association, its president, W. M. Woldridge of Hinsdale, and its numerous functional committees proved to be of great assistance. The institute's tenth biennial report for 1905-1906 noted a striking increase in national farm organizations with which state groups were affiliating. These included the Farmers' Protective Union, the American Society of

Equity, the Farmers' Alliance and the Montana Farmers' Union. The report commented that the Farmers' Union formed in Cascade County in 1905 was the "latest and most active."

The Farmers' Institute law stated that an annual institute should be held in each county. Funds did not reach that far in spite of exceeding the budget almost every year. Silver Bow and Granite counties were usually bypassed since their interests were largely in mining.

Director Emery took the position early that "half a dozen progressive farmers" at a meeting might be worth more than a large crowd. Nevertheless, since state funds and personnel were being used, and the railroads were now giving assistance, as well as the communities, he insisted that the community should support the meetings; not only the farming population, but the businessmen as well. When Frederick B. Linfield became director, he took the same position.

Frederick B. Linfield, who had come to the College in 1902 to replace Professor of agriculture Robert Shaw, was one of the most important figures in the progress of agriculture in Montana throughout an entire generation. He was one of several staff members who had been trained in Canada at the Ontario Agricultural College in Guelph. He had served on the agricultural

F.B. Linfield, dean of agriculture, 1913—1937.

staff in Utah before he came to Montana. When Samuel Fortier resigned as director of the Experiment Station in 1903, Linfield took this office and also that of director of the Farmers' Institutes. His gruff voice did not carry well, and he seldom spoke to large groups, but in small meetings and personal conversations, his genial presence, sincere quest for additional information, and cogent reasoning made him extremely effective. In his extensive correspondence and professional articles he was persuasive, indeed. His superior administrative ability enabled him to take the additional office of dean of agriculture in 1913, which he retained until he retired in 1937. In semi-retirement, he continued to give direction to Station publications until 1942.

When large numbers came to the Farmers' Institute meetings, they became an excellent political forum. The institute officers tried desperately, and in the main successfully, to keep political action out of the meetings. During the 1901-1902 sessions, however, a couple of issues took political form. R. N. Sutherlin of the *Husbandman* was a dairy enthusiast, and at the Missoula Institute he introduced a resolution, adopted with only one dissenting vote, that a petition signed by the institute members be forwarded to our United States senators urging them to support a bill to tax colored oleomargarine, since it "competes with good butter in a fraudulent fashion." This action took on additional overtones when the Wool Growers protested, their association having recently formally opposed this tax.

Another petition was circulated at this meeting protesting any bill which would provide "for the leasing of the public domain, in whatever form." Many stockmen felt that this would be generally detrimental and would drive the small stockman out of business, but stockmen were to be found on both sides of this issue.

Director Linfield, in his low key diplomatic manner, urged in the 1904-1905 institutes that such questions be studied rather than be made the subject of precipitous political action. This attitude, he felt, should also be taken toward the tensions created by the rapidly increasing number of farmer-homesteaders who were also annoying the stockmen.

The increasing number of questions which were channeled into the College and the Experiment Station staff led the thoughtful director to present a paper in 1906-1907, titled "Some Needs in Agricultural Education," which contains the first local mention of a county agricultural agent or "expert."

> Farming is becoming more and more a science that requires a broad and thorough training to be fully comprehended. Through force of circumstances but few farmers can give the time and energy necessary to fully master the science of agriculture. To have available an agricultural expert, with whom he could advise as to the best agricultural practices and management, would add 10 percent to 20 percent to the income of the farm.
>
> It has appeared to me that it is in this direction that we must work for the development of the agricultural extension work. The local agricultural ex-

> pert must have a broad training backed up by thorough practical experience. He would, moreover, have to make a thorough study of the local agricultural situation so as to be able to advise intelligently. He should be in close touch with the Experiment Station and the Agricultural College. In fact, he should be an outpost for bringing home to the people the work of these institutions, urging all who can to take full advantage of the work these institutions are doing.
>
> How this agricultural expert may be provided for the district is a problem for future solution, but a few suggestions are submitted. At the start he may be engaged by the county high school, giving half of his time to the high school and half to the farming community. It would require an experienced high-priced man, but that would be no disadvantage for the high school. Again, this man may be engaged by the county as a part of their general education system to instruct in the public schools of the county and also be available for general advice to farmers. There is also the possibility of the state taking up the matter under the Agricultural College Extension work by dividing the state into districts made up of several counties. During portions of the year these men could be available for general Farmers' Institute work over the state. Again the farmers, in any given community might combine together and engage an agricultural expert to advise them in their business. Nearly all large business enterprises have at present experts in various lines continually on their payroll. These men devote their time to special study of various phases of their business, that they may advise and direct the various superintendents of the divisions of the enterprise.

With this paper Linfield brought Montana into the mainstream of national thinking as it was being shaped by Kenyon Butterfield in the Land Grant College Association.

At another meeting in the same year, Director Linfield continued to probe into basic needs. Here he stressed efficient farm management and the use of the best known farm practices. Management had been given almost no attention and Linfield set the stage for what was to become Montana's greatest contribution to the Extension movement in the years under discussion:

> Probably the direction in which the lack of efficiency costs us most dearly is with ourselves. First, we have no clear idea of the problems that so frequently confront us, or how to grapple with and solve them, or we fail to handle our time, our labor and our capital with proper business tact and management...In the present stage of the development of scientific agriculture, nothing will return larger dividends to the farmers than a generous investment in scientific agricultural facts and principles. My observation in this western irrigated country is that in no other place is there a greater difference between the results obtained by the good and the poor farmer. In no other place is ignorance so costly, or the reward of intelligent effort more pronounced. In no other place does the study and application of correct methods of farm practice pay a larger dividend.

The published report of the Farmers' Institutes, in a printing of over 5,000 copies, immediately became a useful device for agricultural education throughout the state. When certain subjects began to attract active interest, these were given more space. I. D. O'Donnell soon became "Alfalfa Bill" because he stirred up so much interest in this crop. It seemed desirable, therefore, to ask him to expand his discussion into a special 50-page section in the third annual report for 1904-1905.

This pattern met with approval. The fourth annual report developed "The Farm Home," and the fifth annual report that of "Sugar Beets." When a full-time director for the institutes was named, the annual report was changed to consider only one subject. The sixth annual report for 1907-1908 was devoted to "Dry Land Farming" and designated as *Montana Farmers' Bulletin No. 1.* A first special *Bulletin No. 2,* "Better Roads," was also published, since this was an urgent issue at the time. The seventh annual report of 1908-1909 was "Poultry." Following 1909, the lack of time and funds made it necessary to restrict the annual reports to a very brief summary and a financial statement.

The increasing burden of the voluntary and gratis work of the staff was given frequent consideration, but nothing was done about it. Many carried teaching, administrative, research and extension duties. Civil engineer E. Tappan Tannatt, whose enthusiasm led him to write the bulletin "Better Roads," was also the irrigation specialist. Both subjects were popular and he was in great demand as a speaker. On December 8, 1905, he protested to Director Linfield that his teaching required six hours per day in the classroom. Then he had extensive work in connection with irrigation, highways,and other civil engineering work, and in addition he traveled on institute work. When he had notice of a change of schedule which would prevent him from reaching home before 3:30 a.m. the day before Christmas, after traveling all night with three changes of trains, and would cause him to leave home at mid-day on Christmas and travel all night to meet his next appointment, Tannatt said, "I feel that there are limits which the state has no right to exceed."

Linfield had complained in a similar vein a year before to new institute board member Woldridge concerning his own overload. "During the past year," he wrote, "I have had to put in two shifts a day much of the time, and it is only the Montana climate that has kept me going."

Director Fred S. Cooley

The 1907 Legislative Assembly yielded to Linfield's prodding and increased the institute budget from $4,000 to $7,500. Linfield immediately proposed to the board that a full-time director be appointed to assume the duties then carried by a number of the staff as a voluntary service. The board agreed and Fred S. Cooley became superintendent of the Farmers' Institutes on September 1, 1907. Cooley had graduated from the Massachusetts Agricultural College at Amherst in 1888 and was assistant professor of agriculture when asked to come to Montana. He was a capable and dedicated man, judicious in the choice of his staff, wise in adjusting his program to the people's needs and desires, and particularly adept in work-

F.S. Cooley, first director of Extension, 1914—1924.

ing with young people. By 1914 when the Smith-Lever Act provided for an Extension Service, Cooley, who had long before developed a full-fledged Extension program, was the logical director. He remained as director until his retirement in 1924. One of his last services was to write a thoughtful 24-page "History of the Montana Extension Service" emphasizing the period from 1914 to 1924.

Railroad Special Trains

The western railroads were, of course, vitally interested in the continued increase in the numbers of farmer-homesteaders. They had assisted Farmers' Institute personnel through low rates or free passes, as well as aiding students to attend regional and state fairs. In 1908 the first of the special agricultural trains was run. It was sponsored by the Northern Pacific Railway and used the dairy industry as its subject. The enterprise was a major effort for Station dairyman W. J. Elliott. Almost the entire staff joined in preparing the exhibits, and several accompanied the train to assist

Northern Pacific special train featuring dairy cattle in 1908.

with the lectures and demonstrations. In 1909 the railroads demonstrated their good will by providing lecturers for institutes.

In 1910 the Northern Pacific ran the "Better Farming Special." This was billed as "a Farmers' Institute on wheels," and, with 15 staff members on board, it attracted some 7,292 visitors at 44 stops in a 12-day run in June. A similar theme attracted a record number of people, 28,900, in 1912. In this same year both the Milwaukee and the Great Northern ran successful trains, and the Milwaukee followed with a highly successful train in 1913. The last of the trains was fitted up by the Chicago, Burlington and Quincy in 1915. Hardin was its only stop in Montana, however.

Dry Farm Experiments

The railroads also gave major support in supplying experimental farms or tracts for the study of dry farming. Director Linfield recalled that when he came to Montana in 1902 such a study had been planned:

> So arrangements were made in the spring of 1903 to secure a lease on some 20 acres of land on a dry farm near Wayne Siding some 25 miles southwest of Great Falls in Cascade County. Dry farm cropping was started that spring and during the summer a dam was constructed across a coulee to hold back run-off water for irrigating a garden and other crops near the farm buildings. Director Fortier planned to extend the dry land field studies over

the state but my contention was to first build a foundation for agricultural research and education at the College before outside studies were undertaken.

Linfield succeeded Fortier in 1903, and he remarked that "the dry farming problem continued to make demands." The Northern Pacific Railway initiated a plan for cooperation with the U.S. Department of Agriculture and the Montana Experiment Station at a meeting in Bismarck in January 1905. The Northern Pacific guaranteed $2,000 per year, the federal government offered $1,000 the first year, and Linfield was asked to try to secure $2,000 from the Montana Legislature then in session. He secured only $1,000, but a beginning was made on a program of leasing a number of dry farms in various parts of the state for purposes of research and demonstration. Over a seven- to 10-year period, private agencies contributed about $50,000 of which the Northern Pacific gave $25,500, the Great Northern $8,000, and the Milwaukee $5,000. The cost to the state was about $100,000. By 1910 the Farmers' Institute board observed "a general apathy of the people in the vicinity of the demonstration dry farms," and they were rapidly abandoned after 1915. By that time the permanent branch stations had been established where, said Linfield, experiments could "now give place to better methods."

The dry farming movement owes a great deal to the energetic Hardy Webster Campbell, originator of the "Campbell Farming System." A native of Vermont, he undertook dry farming in North Dakota, then settled near Lincoln, Nebraska. He had an inventive mind and as a result of his experiments he became a proponent of the "dust mulch" method. He designed tillage implements which were adapted to arid conditions and were widely used.

Campbell was persuasive and by January 31, 1906, a large number of agriculturists came together in Denver to form the Campbell System Farming Association. This led in turn to a series of famous Dry Farming Congresses.

The Dry Farming Congresses

The first Dry Farming Congress was called in Denver on January 31, 1907. It took on the nature of a fair since exhibits of produce grown under dry farming conditions were displayed together with the new implements and methods of soil culture. Montana participated in this congress in its second meeting in Salt Lake City in 1902, and also in the third congress in Cheyenne in February 1909.

The fourth Dry Farming Congress and International Dry Farming Exposition met in Billings, Montana, in October 1909. Delegates from several foreign countries were present and neighboring Canadians came in numbers.

The famous conflict of opinion concerning the type of publicity which should be given to the agricultural conditions in the arid west took place at this meeting. Both James J. Hill, who had retired to the position of chairman of the board of the Great Northern, and his son, Louis H. Hill, president of the railway, spoke urging increased efforts to improve agricultural production in the plains region, and to overcome the impression that this was a difficult area in which to farm. Louis Hill urged that the name "Dry Farming Congress" was negative in its approach and should be changed. Said Mr. Hill:

> It is desirable to start with a good impression, and do not give the people whom you expect to come here an impression that they are going to a dry country. . . Those of us who are acquainted with the country for 20 years or longer know that there is no reason on earth why this land cannot be cultivated with just as good an opportunity and as good a chance to harvest a crop with as great a certainty as in any of the states east of the Mississippi River. . . Montana is neither arid nor dry.

Governor Edwin L. Norris of Montana, president of the congress, in his address of welcome also took exception to the term dry and warned that Louis Hill had informed him that unless a name were adopted which "will more correctly describe cultivation with moisture stored in the soil," his railway would probably withdraw support from the congress. Paris Gibson described with enthusiasm his experience in raising wheat near Great Falls where, over a period of 20 years, he had averaged 22 bushels per acre without irrigation, in spite of several years of drought. If he had used the methods and implements now proposed by Hardy Campbell, he did not doubt that the production would have exceeded 33 bushels per acre.Louis Hill again spoke at length and with conviction concerning the productiveness of the State of Montana, quoting figures from the federal Department of Agriculture to show that Montana's production per acre of rye, oats, wheat, barley, flax and especially potatoes, was well ahead of the national average, as well as of the Midwest.

Director Linfield, in a short speech, patiently explained that the term dry farming did not necessarily describe the country, but had originated to designate the methods used to farm lands which could not be reached by irrigation methods. Since the congress was presently concerned with this type of land and was using this name, he felt that it should remain.

Then Linfield continued to the point of direct disagreement with Mr. Hill. "There are," said Linfield, "really three types of farming being carried on in Montana: First, many who have come from the East and Midwest are farming as they did there, as if this were a humid area; second, irrigation is being developed in the valleys; and third, there are those on the plains lands and benches above the irrigation ditch who recognize that 12 to 15 inches of rain is normal and are planning accordingly. The point for us to remember," Linfield emphasized, "is that farming in this country is different from that in the East. . . not less scientific. . . but a different kind of farming above the ditch. . . The point for us to consider is, how are we going to leave that impression more forcibly upon those people who are coming here?" Professor Thomas Shaw of the University of Minnesota, who had participated widely in Montana Farmers' Institutes, supported Linfield's point of view.

The name of the congress was not changed, and apparently the Hills bore no malice. Within a month after the congress, Shaw was engaged by the Great Northern to participate in all of the Farmers' Institutes in Montana during the winter of 1909-1910 season. Montana continued to take an active role in the organization, which in 1918 merged with other associations to become the International Irrigation Congress.

While discussions were taking place concerning the worth of arid lands and the best method of cultivating them, Montana products were given first place in major expositions over the nation. First prizes were received for hulled barley, hulless barley, best 100 pounds of wheat and of oats, for potatoes, for the best and widest use of alfalfa as food for man and beast, for the largest and best exhibit of products from any state, and to Judith Basin County for the most beautiful, elaborate and attractive booth at the show. In 1912 a number of first prizes were won for quality stock.

A Department of Agricultural Extension

In this active climate of agricultural progress, the demands for extension activities continued to increase. Superintendent Cooley annually urged that a separate staff be allocated for his institute work, and Director Linfield, whose Experiment Station staff was regularly raided to carry on the programs, agreed.

The problem was national in scope and as early as December 1909 President Hamilton reported to the State Board of Education that the November meeting of the American Association of Agricultural Colleges had made two recommendations of interest. The first was that each state establish a formal Extension Department to include all agricultural instruction given off campus. This Extension work would be coordinated with the Farmers' Institutes and have the same superintendent.

A second recommendation was that Congress should be asked to make an annual appropriation of $10,000 to each state to assist with Extension work. The president outlined to the board the extensive work being carried on with which he was thoroughly familiar since he, with the governor of Montana and the director of the Experiment Station, comprised the Farmers' Institute board, and he frequently took part in institute programs.

Hamilton continued to press the need for expanded Extension staff and services in frequent reports to the state board. December 1, 1911, he reported that the Smith-Lever Act, designed to give federal assistance to states for Extension work had passed the House of Representatives, and would probably pass the Senate within the month. The Farm Management Division of the USDA also had a large appropriation for Extension work in farm demonstrations which it proposed to make available to states.

However, a requirement for participation was that the College should have an Extension Division and sufficient funds, which President Hamilton estimated at $5,000 each year, to support the superintendent. Hamilton described the importance of the work in Extension as "scarcely second in value to either the College or Station."

After a number of years of steady educational work, the 1913 Montana Legislative Assembly took double action to promote Extension work. A Department of Agricultural Extension was approved for the College, and county commissioners were permitted to use general funds of $100 per month toward the expenses of a county agricultural instructor, upon petition of 51 percent of the farm voters of the county.

Director Linfield was greatly pleased with this legislation, which he had so frequently urged over almost a decade. A long-time friend and confidant, John P. Fabrick, often remarked that the director's steady plea was, "We must learn more about Montana." Until climate, soils, water use, insect prevalence and other factors had been investigated, Linfield urged a cautious approach to new ventures.

Linfield was also, no doubt, one of the first men in the nation to visualize the worth of a resident local agricultural "expert" as he called him, and to urge that such a resident expert, advisor, teacher, and demonstrator replace the sporadic institute approach.

F. S. Cooley summarized Linfield's contribution in his "History of the Montana Extension Service" in 1924. Cooley, of course, was also a pioneer in Extension work, and his contribution was substantial, too. Cooley commented:

> Perhaps the greatest single contribution to the agricultural development of Montana has been made by Director F. B. Linfield of the Experiment Station. His sagacity, enterprise, and sound judgment have stamped themselves indelibly, not only on the work of the College and Experiment Station with which he has been connected for over 20 years, but on the agriculture of the state. Professor Linfield organized the agricultural faculty of the College and the research staff of the Experiment Station. He developed the Farmers' Institutes. He established demonstration farms in cooperation with the Northern Pacific Railway to determine the agricultural possibilities of the state. He enlisted the support of the United States Department of Agriculture in branch experiment stations in different agricultural regions of the state to study the best crops and practices. When others were carrying on a frenzied campaign of land selling and colonization, he was working indefatigably to discover how the settlers might be enabled to succeed and prosper in their new ventures. No more important work has been wrought in Montana than that performed in the interest of its agriculture by Director F. B. Linfield, one of the ablest research executives in America.

With the establishment of an Extension Department and the participation of the counties in 1913, Montana was in a position to take full advantage of national assistance when it became a reality the following year. The 20-year period, 1893-1913, of "voluntary" work on the part of all of the Experiment

Station staff members and many others, had been arduous but profitable. In the beginning days, both ranchers and scientists had been here only a few years. A close relationship grew up which has continued to this day. Substantial beginnings were made very early in most of the areas for which the Extension Service has received richly merited praise: improved farm and ranch practices with plants and animals, more efficient management, enlarged leadership opportunities for farm youth, improved health, and marked advances in the quality of living in rural homes.

Montana, in short, developed a relatively mature Extension program as a grass roots movement. This was not done in isolation, because its leaders and many of its able citizens were participants in the activities taking place over the nation. Montana was ready to move into the cooperative federal-state-county extension movement without pausing for a change in step.

II. Organizing the Montana Extension Service, 1913-1917

The establishment of a Department of Extension for the state in 1913 came as a result of increasing interest in the Farmers' Institutes and allied Extension work, stimulated by 10 good crop years beginning in 1906, broken only by a drought in 1910.

The institutes grew in popularity when the practice of bringing in out-of-state speakers was initiated in the 1908-1909 season. "In one old institute town," observed Director Cooley, "where meetings had heretofore been unsatisfactory, the attendance was so large this year that at least 50 people were unable to get into the hall." Visiting speakers were not always easy to obtain. H. L. Dean, horticulturist from Michigan, came for a two-month period in response to an urgent plea occasioned when a Wisconsin specialist cancelled out after he saw the schedule. Cooley was not unkindly disposed toward the drop-out, recalling in his annual report of March 22, 1909, that:

> During one week, seven days, our daily average was 120 miles by rail, 15 miles by team, and one institute session each day. On one occasion we got off the train to go to a hotel at 9 o'clock in the evening, found the hotel quarantined, the only hotel in town; got back on the train and went 110 miles out of our way to get a bed.

Dean, undismayed by the rigors of work in Montana, proved to be so versatile and capable that he was induced to become secretary of the Horticultural Society and inspector of orchards. Dr. A. A. Brigham of the South Dakota Agricultural College at Brookings was also an assisting lecturer in 1908-1909.

In the 1909-1910 season, following the Dry Farming Congress in Billings, Professor Thomas Shaw of the University of Minnesota was provided by the Great Northern Railway for the entire season, and Hardy Campbell of Lincoln, Nebraska, together with Montana's experimentally-minded dentist, Dr. W. X. Suddeth of Broadview, were made available by the Milwaukee Railroad. Dr. Suddeth was particularly interested in dry land grain and sheep ranching. Shaw returned in 1910-1911 and 1911-1912 and was assisted by a capable group of specialists.

Farmers' Week at the College

Another innovation, the first of the Farmers' Week meetings at Montana State College, occurred January 22-29, 1913. This was initiated in connection with the second annual Country Life Convention; a first meeting had been held in Great Falls in 1912 as an offshoot of the White House Conference. These meetings, designed not to help the farmer raise better crops, but to call his attention to opportunities for better business and better living on the farm, incorporated a wide range of subject matter such as health, roads, schools, recreation, and the country church, as well as crops and crop production.

The eight-day meeting attracted some 200 out-of-town registrants and a notable list of speakers, many of whom were used for institutes over the state. Speakers included O. C. Gregg, former head of the Minnesota Farmers' Institutes; Perry G. Holden, agricultural philosopher from Iowa State; L. R. Alderson, state superintendent of schools in Oregon; Oliver Wilson of Iowa, master of the National Grange; President J. H. Worst of the North Dakota Agricultural College; A. E. Chamberlain, former director of Farmers' Institutes in Minnesota; and Joseph E. Wing, well-known agricultural writer from Ohio. Women's work was stressed by Mrs. Alice Peloubet Norton, specialist in dietetics and health from the University of Chicago, and Miss Mary G. Deem, head of the Department of School of Arts of the Valley City Normal School.

The Second Annual Farmers' Week at the College, held January 21-29, 1914, was the highlight of the institute season which saw 112 institutes with 207 sessions and 20,962 in attendance. Farmers' Week attracted the annual meetings of the state Grange, the Montana Seed Growers, the state Veterinary Association and the state Dairymen's Association and a total of 300 registrants. Sessions were built around daily themes such as Rural Organization Day, Farm Homes Day, recreation, health, rural schools, and on Sunday, Country Church Day was observed. Visiting lecturers included old favorites such as Professor Thomas Shaw; O. C. Gregg; T. A. Hoverstad, agriculturist for the Soo Railroad; and new personnel such as Frank W. Waugh, head of the Department of Horticulture of Massachusetts Agricultural College; Dr. Jesse Macy of Grinnell College; and Mrs. Bertha Dahl Laws of Minnesota. Governor Sam Stewart and State Superintendent of Schools H. A. Davee headed an impressive list of Montana speakers including College staff members.

Director Cooley was interested in emphasizing women's work, and in the early institutes he drew heavily upon the capable staff in home economics, hoping to increase the attendance of women. In January-February 1911, Miss Mina A. Stoner, former head of the Department of Domestic Science at the University of Wyoming, joined the institute circuit. She addressed 27 high schools, 26 ladies' meetings, and conducted a short course in the Victor High School. Mrs. Una B. Herrick came to the College in January 1911 to take charge of a new women's dormitory. Her natural interest and a background in public service enabled her to take a lead in institute programs. She, Mrs. Laws and Miss Lucille Brewer of Montana Sate assisted in the 1914 Farmers' Week. Superintendent Cooley hoped to stimulate the opening of many domestic science courses in the high schools through holding short courses, but lack of funds prevented this expansion.

In the requests of President Hamilton and Superintendent Cooley to the State Board of Education for enlarged support, Cooley stressed his desire for a separate department of Extension to carry on the varied types of work he planned to pursue, such as:

1. Community short courses or movable schools
2. Better farming trains
3. Women's institutes and short courses
4. Correspondence courses
5. Permanent educational exhibits
6. County teachers' institutes
7. School corn clubs
8. High school lectures
9. High school short courses
10. Preparation of publicity releases

Short Courses

Short courses and "movable schools" up to one week in length, devoted to a special subject, were popular for several years. These were most successful in the rapidly developing fruit area of western Montana where the Montana Horticultural Society assumed major reponsibility. Director Emery had initiated meetings dealing with the multiple problems and skills involved in handling trees and fruit. Two such schools were held in the winter of 1909-1910, one in Hamilton and another in nearby Stevensville. Another two were held in 1911-1912 in Hamilton and Kalispell, where William Heikle of Pullman, Washington, came to give instruction on the packing and marketing of fruit. In the autumn of 1912, a series of shorter meetings were held in the western valleys during the apple season, August 26-October 1, on the grading and packing of apples. Kalispell, through its adjacent compact population and the facilities which the town and the high school furnished, was the best center for short courses for a decade or more.

Cooley soon found that the one-week short course was successful for some subjects and in some communities and not for others. He remarked, "Farmers live too far from larger towns to enable them to go in for a week at a time, and facilities for holding schools in the country are poor." Securing staff for a number of the week-long meetings was also difficult. Such schools for women were successful over a period of years, however. Nevertheless, a number of highly successful short courses were held each year until they were interrupted by World War I. The attendance reached 475 in Kalispell in 1915, and a record number of 11 meetings were held in 1916-1917, with 350 attending in the small town of Belt.

Farmers' Picnics

The vagaries of Montana's weather and the availability of county agents to handle local arrangements caused a reassessment in 1911 of the time of year to hold farming demonstrations or one-day institutes. This timing was particularly important in the rapidly expanding settlements in the eastern counties. Here a single community covered an area larger than Connecticut or Delaware, and the period between the end of planting time, say June 1, and hay harvesting in early July was often a good time for a meeting. This in turn led to the farmers' summer picnics which had a tremendous vogue the country over in the relatively uncomplicated era before World War I.

The record is highly incomplete, and many picnics of 150 to 350 held in

Scene from a Farmers' Picnic, 1913.

various localities meant more to community development than those which drew larger numbers. Some incredibly large picnics were hosted in a number of Montana communities, as the following partial listing shows:

Place	**Year**						
	1910	**1911**	**1912**	**1914**	**1915**	**1916**	**1917**
Moccasin Station	2,500	3,000	4,000	3,500	3,500	4,000	3,000
Roundup	100	600	1,500	2,000	1,500		1,500
Stevensville		1,500	1,500				
Huntley Project	500	3,000		350	800	3,500	

Families or neighborhood groups shared picnic lunches, usually supplemented by a 4-H concession stand. Talks were given and demonstrations were held, often several going on at the same time; new machinery exhibits were sometimes available and if time could be spared from the serious business of the day, recreational sports contests were arranged.

Subject Matter Specialists

In 1911, the Legislature approved repeated requests for money by appropriating limited funds to the Farmers' Institute board to employ the first full-time subject matter specialist. Dr. W. J. Hartman, a veterinarian and general purpose livestock man, was employed to assist with the innumerable requests which had made many inroads into the time of the livestock men in the Experiment Station. Hartman found time to encourage the many farm youth whom he found engaged in stock raising projects. Since farm youth were also a continuing interest with Superintendent Cooley, Hartman soon discovered that he was commissioned to give major assistance to starting boys' clubs, and he had a busy time at fairs judging their stock entries. Hartman was another staff member trained at the Canadian Ontario College of Agriculture at Guelph. He also had an additional veterinarian degree from the Chicago Veterinary College. He remained Extension specialist until he took over the position of emergency agent in Gallatin County during World War I. He resigned on July 1, 1918, to operate his own farm in the Gallatin Valley.

Another special interest which broadened in 1911 was flax growing. This resulted in the employment of M. L. Wilson, a Fallon rancher, to discuss flax culture in the 1911-1912 institutes. Increased interest then led to the contribution of $2,500 annually by the American Flax Association for promoting flax growing in Montana. The sum of $500 was earmarked for

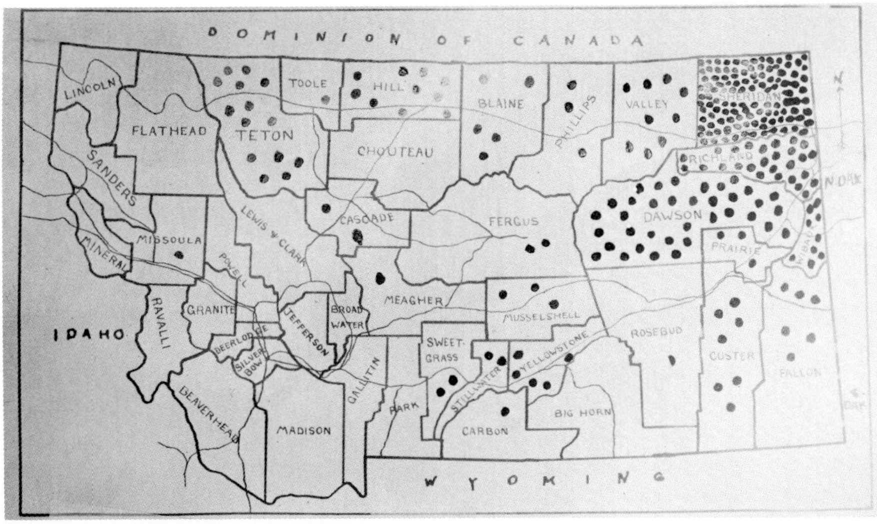

Distribution of flax production in 1914. Each dot represents 10,000 bushels.

distribution to sprayers for treating flax seed for wilt and canker. Some $1,800 might be applied to the salary of a field man who, among other things, would promote flax growing. This sum was applied to M. L. Wilson's salary and he was attached to the Experiment Station as an institute lecturer and supervisor of demonstration farms.

M. L. Wilson

Since Milburn Lincoln Wilson was to play such a dominant part in the pioneering development of rural Montana, indeed of rural America, an introduction is desirable here at the risk of later duplication. Wilson was born on a farm near Atlantic, Iowa, October 23, 1885. Following graduation from Iowa State College in agronomy, and work in Farmers' Institutes with nationally known Perry Holden, he came to Montana in 1907 to homestead near Fallon. Here he began dry farming, and soon married a charming and able young lady of the community, Miss Ida Morse. He soon encountered many problems in his dry farming and since he was one of the very few in the entire Montana farming population with a college degree, he became a focus for questions from the community. His experience in Iowa made him feel free to take these questions to the Farmers' Institutes and to the Experiment Station. The leaders of these agencies found him to be tremendously stimulating and willing to undertake imaginative and thorough experimentation; this in turn led to his joining the institute staff in 1911.

Wilson had a naturally philosophic mind and shared his father's fondness for Abraham Lincoln, hence his middle name. Wilson's collection was one of the best private holdings of Lincolniana in the nation. It probably also led to his deep interest in government, history and people. When he returned to graduate school at Chicago, Cornell, and Wisconsin, where he obtained a master's degree, his work was largely in economics and philosophy. At Wisconsin his instructors included Richard T. Ely, who was refining the emerging subject of economics, and Henry C. Taylor, often credited with originating the field of agricultural economics. Through his philosophical approach to the practical problems of agriculture which he found in Montana, Wilson also made a major contribution to the development of agricultural economics. He was often accused of being an "impractical dreamer," but he selected an unusually capable group of practical men to work with him, he secured financial assistance for his projects, he was adept in using political means in attaining desired goals, and he carried out major studies of an immensely practical and important nature. He was always an original thinker and a genial leader of men. He made national and international contributions as he moved through several positions at Montana State College, noted below. After providing important guidelines for presidential candidate Franklin Roosevelt's agricultural policy statements, he served in the Department of Agriculture with Henry Wallace beginning in 1933 and eventually became undersecretary. He became director of the Federal Extension Service in 1940, and later went intermittently on extended excursions for the Ford Foundation to Russia, India and other nations as consultant and advisor.

First County Agents

The creation by the Montana Legislature in 1913 of an Extension Department in the Division of Agriculture at Montana State has been noted. The accompanying act allowing the county commissioners to use $100 per month from their general fund toward employing a county agriculturist led Superintendent Cooley to act at once.

On April 15, 1913, he employed two county agents. M. L. Wilson, still working with the Flax Association grant, was assigned to Custer and Dawson counties, and Carl H. Peterson to Fergus County. Fergus was the first county to appropriate the $100 per month for this purpose in response to the petition of 51 percent of its rural voters. Wilson began his work in Miles City on May 1, and because of his previous work of a somewhat similar nature since 1911, has usually been considered the first county agent in Montana. Carl Peterson, manager of the Experiment Station farm on the Montana State College campus, set out for Fergus County on May 7th.

Custer and Dawson counties were a natural choice for Wilson since he previously lived in the area. In his 1924 *History of the Montana Extension Service*, Cooley recalled conditions at the time:

> Mr. Wilson was assigned as county agriculturist to Custer and Dawson counties with headquarters at Miles City and Glendive. . . The territory included in those two counties now comprises the counties of Custer, Dawson, Wibaux, Richland, McCone, Garfield, Prairie, Fallon, Carter, and Powder River. It was roughly 200 miles square and contained 40,000 square miles, or 25 million acres of land. Most of it was, or had recently been, open range without fences. There were no graded roads, only trails across the prairies. The Glendive-Jordan Trail was famous for having been marked with a plow a quarter of a century before. The wheel tracks, with a faint overgrown furrow between, marked a trail out of the ordinary, a comfort to the traveller, because it gave confidence of reaching one's destination. Streams were forded at low water. In high water crossing was dangerous. When the homesteaders and settlers came, just before and after Mr. Wilson's taking up county agent work, the open range began to be fenced. Trails leading as straight as possible to their destination were cut by strands of barbed wire. The new highways following section lines led over buttes, across sloughs and coulees, and were hard to travel. Nevertheless, Mr. Wilson served his people in this vast area so faithfully and well that county agent work continued in both Custer and Dawson counties and five county Extension agents are now employed in that territory.

As Wilson started his journey to the new job, he carried what has come to be referred as Director Cooley's famous "letter of May 1," containing instructions to the new job. A rather long letter contained the following pungent paragraph:

> There may be no occasion for office hours at Fallon or for the maintenance of any formal agriculturist's office at that place. Probably for the present your office will be chiefly under your hat. Make arrangements for about four cooperators from among the best and most interested farmers in each of about 10 or 12 townships. Study comparative cost and crop yields between engine and horse tillage. Promote boys' corn contests. In an unobtrusive way use your influence in favor of agriculture in the rural schools. Keep a list of farmers' wants and things for sale. Observe degree of success of alfalfa, corn, peas. Be on the lookout for good seed. Promote use of good sires. Visit cooperators each 14 to 20 days. Cultivate farmers' organization for educational and social benefits. Have a few most useful bulletins with you for distribution on your travels. Make out daily reports.

Wilson must have felt that he was making rapid progress when on May 20 he was authorized to purchase a desk at Miles City and have it delivered to his office with the Chamber of Commerce. Considering the size of his territory, his instructions and his natural inclinations, his office continued to be largely "under his hat."

The *Miles City Star*, on May 1, contained the story of the arrival of M. L. Wilson, and his discussion with the Chamber of Commerce:

> Mr. Wilson, of the Bozeman Agricultural College, was introduced by Chairman A. J. Freeman as the agronomist who was assigned to Rosebud, Custer and Dawson counties to acquaint the farmers with better farming methods. Mr. Wilson talked for 20 minutes and made clear his opinion that combined farming and stock raising, under the new conditions, had become absolutely necessary and would be the eventual solution of the problems which confront the old settlers, as well as the "honyocker" at present in

eastern Montana, as to what is the most profitable use that can be made out of Montana land.

The solution, according to Mr. Wilson's ideas, lies in the growing of corn, construction of silos, and the feeding of dairy cows or small bunches of cattle. . . Mr. Wilson said he would like to have a committee of the chamber appointed to confer with him relative to the formation of a better farming club, and the chair appointed Messrs. Butler, Auld, Collins, Daniels and McLean.

In explaining the purpose of Extension work, Wilson often expressed it as he did in Miles City, as the effort to promote "the most profitable use that can be made of Montana land."

The *Star* continued to report Wilson's planning activities. He met with the Better Farming Committee at 4:30 in the afternoon to draw up plans for a county agricultural program. He particularly urged diversification. Grain feeding should be encouraged and thus marketed through livestock. The pit silo was a favorite at the time, and he described it as being economical in construction and highly efficient in use. He urged diversification to the point of suggesting that the name of the committee be changed to the Diversified Farming Committee, and that the community groups to be formed should be called Diversified Farming Clubs.

With his Iowa background, Wilson did not need direction from Cooley to urge the growing of corn. Emphasis was being given to corn growing at the time, prompted by the annual gift, begun in 1911, of James J. Hill and the Great Northern Railroad, consisting of $1,000 for prizes to boys throughout the state for producing a superior product. These prizes pointed up a situation which was to be of great importance; namely, that parents' interests follow closely the enthusiasms of their children, and that youth could be depended upon to follow directions more closely than their elders. The Hill project had centered upon individuals; Wilson proposed the broader group approach—that a Boys' Corn Club receive the prize rather than the individual.

Peterson in Fergus County

Carl Peterson in Fergus County found the Chamber of Commerce in Lewistown as helpful as Wilson had found the chamber in Miles City. Indeed, the Lewistown chamber had guaranteed the $100 per month until the petition, which the chamber had initiated, was accepted by the county commissioners. The news media simplified the procedure to the point that the chamber rather than the state became the sponsoring agency. The *Fergus County Democrat* of April 29 announced: "The Chamber of Commerce takes the most important step ever made for the benefit of farmers. Farm expert employed."

An issue in mid-May noted:

Among the Hobson visitors on Tuesday was Carl Peterson who has been

> engaged by the Lewistown Chamber of Commerce to look after the better farming movement recently inaugurated in Fergus County.

In spite of the best intentions of the Chambers of Commerce, this association in the press and public mind continued to the extent that county agents were expresssly forbidden to accept space in chamber offices in spite of the convenience and economy. The occasional friction which arose between town and country in some communities completely destroyed the usefulness of the county agent when the rural people felt he was too closely allied with the "Main Street" promotion agency.

In the Judith Basin, Peterson found conditions in sharp contrast to those in Custer and Dawson counties. Large groups were easily brought together to view demonstration plots on farms near Lewistown, Moore, Roy and Hobson, and massive farmers' picnics soon assembled each summer. Peterson's experience on the Station farm was translated into plans for soil study and tillage experiments. He concentrated on alfalfa rather than corn.

In a May 1914 report made to the College Executive Board, Director F. B. Linfield summarized the year's accomplishments of the county agents:

> Our experience the past year has demonstrated clearly the value of the county agriculturist who is alive to his possibilities. M. L. Wilson in the eastern part of the state is starting a revolution in farm methods, through the encouragement given to growing corn and alfalfa, the fattening of hogs and the beginnings of dairying. For instance, at Wibaux, from last year's experience the implement dealers stocked up with two corn cultivators for the coming season's trade. Two weeks ago they had sold 17 and had orders for more. Many dealers in other towns report a similar experience. What corn and hogs will do in the eastern part of Montana was particularly illustrated a few days ago. A sheepherder who had worked for us on the dry land farm north of Harlem moved to Rosebud County some five or six years ago and took a homestead of 160 acres close to our Forsyth substation. He had absorbed some good farm practices. Last week he offered Professor Atkinson $3,500 cash for 160 acres of land adjoining his homestead. Professor Atkinson said he had a fine home and good barn and outbuildings—as prosperous looking a place as any person could wish to see, and corn and hogs did it.
>
> In Fergus County the agriculturist has pushed the seeding of alfalfa and this year over 3000 acres will be seeded in various parts of the Judith Basin.
>
> We appear to have been fortunate in the two men first selected as county men and I trust this good fortune will continue with us as the man is nearly everything in this work.

Taylor and Blanchard

Two additional county agents were appointed under the direction of the Farmers' Institute and Extension Department before the Smith-Lever Act was passed. The vast distances of southeastern Montana proved to be impossible to cover with only one person to meet the urgent requests of the vigorous and rapidly increasing population. On April 1, 1914, John (Jack) C. Taylor was named agriculturist for Dawson County, and M. L. Wilson withdrew to Custer County. The Custer County of the time still contained the area of the present Custer, Powder River, Fallon and Carter counties. Taylor, who was to become the second, and long-time director of Extension in Montana, was a native of the state and the first agent to be trained at Montana State College, where he graduated in 1912. He had been raised on a large wheat ranch in the Bear's Paw Mountains in northern Montana, the son of state Senator and Mrs. L. B. Taylor. His early experience served him well as did his close association with Wilson's methods and philosophy.

In response to an urgent request from Cascade County, R. A. Blanchard was named agriculturist for that county beginning May 1, 1914. He quickly won support over the entire county and instituted a progressive program.

The Smith-Lever Act, May 8, 1914

The passage of the Smith-Lever Act establishing the Extension Service was the culmination of a decade of effort on the part of the Association of Agricultural Colleges and Experiment Stations. The association was organized in 1905, and by 1906 a committee on rural education was formed under the chairmanship of Kenyon Butterfield, president of Massachusetts Agricultural College, and included major leaders of agricultural education. All were well aware of Seaman A. Knapp's recent innovations in Louisiana. The committee recommended:

> That each college in the association organize as soon as possible a department of extension teaching in agriculture, coordinate with other departments of the agricultural work, with a competent director in charge, and if possible with a corps of men at his disposal.

President Woodrow Wilson made a ceremony of signing the bill on May 8, 1914, commenting, "Next to the Federal Reserve Act, this is our greatest contribution to the national welfare."

The act appropriated $10,000 annually to each state without requirement for matching funds. Additional appropriations would be made each year to each state on the basis of its rural population over a 10-year period, if the state, its counties and various agencies would match the additional sums. The annual increase varied from about $2400 to $3000 for Montana, until at the end of the period the total would amount to $40,000 annually. A key paragraph in the act outlined the procedure and purpose:

> That cooperative agricultural extension work shall consist of the giving of instruction and practical demonstrations in agriculture and home economics to persons not attending or resident in said colleges in the several communities, and imparting to such persons information on said subjects through field demonstrations, publications and otherwise; and this work shall be carried on in such manner as may be mutually agreed upon by the secretary of agriculture and the state agricultural college or colleges receiving the benefits of this act.

Cooperative Extension Work

The act was distinctive in at least two respects. First, it provided the most massive program of adult education the world has ever seen. Second, it brought greater cooperation between various levels of government and people's organizations than had ever existed before. Federal, state, county, farm and civic groups, and the people on the land were to work together. The bill had been before Congress for several years and represented the best thinking of agricultural leaders, all of whom recognized that rural people did not take kindly to directions from on high. The provision for cooperative funds as well as effort, at the bedrock of county government, had been an integral part of the plan from the beginning. The movement toward acceptance by Congress had been slow enough to give the states time to develop the pattern of support which the act required. The act was administered expertly in many ways. The provision for a gradual expansion to be matched by the states prevented the often strenuous and wasteful attempt to meet big goals too quickly.

Since the Experiment Station, federal and state, had encouraged Extension work, the unity of the two services was maintained at the federal level to a high degree. In Montana, an easy transition was made when F. S. Cooley, superintendent of the Farmers' Institute and head of the Department of Agricultural Extension, was made director of the new Extension Service. His report of February 28, 1914, written with the new act in mind, showed how closely the state program already approached that outlined by the Smith-Lever Act:

> There is a close relationship between the Farmers' Institute as it has developed here and the Extension Service of the agricultural college. In a cooperative way the Farmers' Institute has undertaken to hold farmers' short courses, to prepare Extension circulars and matter for the press, to run better farming trains, to promote a redirection of rural education, to discuss educational matters with school superintendents and teachers, and to arrange for boys' and girls' industrial contests. This idea of cooperation has led to the appointment of the superintendent of Farmers' Institutes as director of the Extension Service in the College. The appointment was ratified by the Farmers' Institute board. Its effect is to merge all these Extension activities under one management, thus avoiding duplication.
>
> The integration of Farmers' Institutes with general Extension Services does not in any way reduce the importance or efficiency of the former, but rather engages the efforts of the entire Extension staff for institute work during the season. Instead of a superintendent and a few casual, temporary lecturers, there will be a large, effective permanent organization that will be expected to make farmers' meetings an important interest. By using county agricultural agents and state specialists, the personal relationship with farmers will be increased and the staff will be working at closer range. It is the plan to divide the state and give personal supervision of institutes to county agents, who will have the support of the central office and its staff. It is believed that this plan will result in more meetings, larger attendance, more permanent and effective farmers' organizations, and better adaptation of instruction in the needs of the community.

President James Hamilton of Montana State College signed the Memorandum of Understanding in Denver on August 3, 1914. This was the formal agreement in which the state affirmed compliance with the provisions of the Smith-Lever Act. Director Cooley summarized the essential features of the act as:

1. The appropriation of a large sum of money to states for agricultural extension work.
2. A cumulative principle whereby this appropriation is increased yearly for 10 years, thus providing for steady growth and avoiding the waste of public money from the sudden expansion without adequate planning.
3. A partnership with the states giving the latter an equal share in the responsibility for the work. All appropriations under this act in excess of $10,000 to each state must be offset by an equal appropriation by the state for the same purpose.
4. Division of these federal funds is according to the agricultural population. Cities and towns of over 2500 inhabitants are not included in computing the share of each state.
5. Emphasis on employment of local resident agents (agriculture, home economics, or club) as the medium for extension work.
6. Creation of a federal administrative and supervisory office to direct and regulate the work of the states, approve the projects, and audit the expenses and examine the accounts.

Director Cooley reported to President Hamilton in the autumn of 1914 on the progress which had been made in organizing the new service and emphasized again: "As I called to your attention in a previous report, this is the most important agricultural education work that has ever been undertaken."

Building an Extension Organization

Financing the new service was readily identified as a major problem. Director Linfield and Director Cooley worked closely to use every source possible. Already appropriated federal funds were accepted and channeled into Extension projects: $6,600 for each of two years to initiate studies in farm management; $1,500 each year to help provide a dairy specialist, and allocations to three reclamation projects in the state. Mention has been made that the early demonstration farms had largely served their purpose by 1914, and Director Linfield reported to the State Board of Education that he was holding $10,000 for their disposal for its best use in new or old programs.

Director Cooley proceeded to expand his state organization. In addition to the county agriculturists then serving, M. L. Wilson in Custer County, Carl Peterson in Fergus, J. C. Taylor in Dawson, and R. A. Blanchard in Cascade, George Cook was assigned to Gallatin County in September, and Charles Hanson began work in Missoula County in November. Plans for a rapid expansion in county agent work made it seem desirable to bring M. L. Wilson into the state office September 1, as county agent leader.

The state staff consisted of Director Cooley and Dr. W. J. Hartman, veterinarian and livestock specialist and spare time organizer of boys' and girls' clubs. Through Hartman's capable work, assisted by the county agents, arrangements had been made to employ Miss Augusta Evans as boys' and girls' club specialist. Miss Evans, a graduate of Smith College in 1911, had three years experience in rural extension and rural school improvement work. She received $1,200 per year from the federal Farm Management fund, and her travel costs were paid from the Farmers' Institute fund.

Miss Kathrine Jensen was secured as home economics specialist at a salary of $1,800 paid largely from the Smith-Lever funds. Miss Jensen was a graduate of North Dakota State College, with a master's degree from the University of Illinois. Howard C. Cannon was named as the first farm efficiency specialist, and Frank M. Hillman was engaged in October as the dairy specialist.

Since the important dairy work of the state took place largely in the Flathead and Bitterroot valleys, Hillman was located at Kalispell with offices in the highly cooperative Flathead County High School. In 1916 he was transferred to the State College and his activities were broadened to cover the entire state. He emphasized a three-point program: cow testing associations, the first one was a Hamilton-Stevensville unit; silo demonstrations; and a "good bulls" campaign.

Providing Public Information

The attitude in the state concerning the expanding influence of the Experiment Station and the new Extension Service is reflected in the state press. Local papers, and there were many of them, were naturally receptive to the county agents' news releases concerning agricultural meetings, farmers' picnics and club activities. The *Montana Farmer*, which began publication in Great Falls, July 1, 1913, used quantities of state and county office releases; its editor, Harvey P. Griffin, was a former staff member at Montana State. He recalls that the rather ample magazine sold for $1.00 per year. He had little assistance in writing articles and he welcomed the staff articles which he also regarded as important for the Montana farmers.

The first article of note appeared September 27, 1913, an editorial on "The Agricultural College and the Farmer," which was highly favorable to the contribution which the College could make to Montana agriculture. October 4, M. L. Wilson contributed an article on "Building Pit Silos on Montana Farms," one of his favorite subjects at the time. In the same issue was a news item, "The State College at the State Fair." The October 11 issue contained three items relating to Extension work. A photograph of Dr. R. A. Cooley appeared with a short sketch of his work as state entomologist. Another short note outlined the plan for Farmers' Institutes, and a third

Trench silo, Pete Hoogland farm near Malta, Mont. First filled in 1923.

summarized the boys' and girls' contests at the State Fair. By the end of the year, 15 articles written by research and extension men appeared, discussing the entire gamut of issues from J. C. Taylor's "Growing Alfalfa in the Milk River Valley," to F. S. Cooley's "The Study of Agriculture in Elementary Schools," to Alfred Atkinson's "Great Interest in White Clover," and O. B. Whipple's "The Protection of Strawberries."

The Sutherlin brothers of the *Rocky Mountain Husbandman,* strong and strongly opinionated supporters of Montana agriculture since 1875, gave almost no early recognition to the expanding Extension service, although they had been deeply involved in the Farmers' Institutes. As the impact of the European war deepened their policy changed, and by mid-1916 they were making full use of the College releases.

Expanding Extension Work

The program of the Farmers' Institutes continued in full force and expanded to use the new Extension personnel. Director Cooley reported on September 1, 1914, that during the past six months, since March 1, some 76 institutes had been held, attended by 9,489 people. Speakers had also been sent to five farmers' picnics during the summer, attended by 10,150 people. Whenever possible the county agriculturist managed the program which Cooley said, "made possible a larger number of meetings and a closer relationship to local problems and needs." In his report of February 28, 1915, for the past year, Cooley called attention to the additional meetings which the agriculturists had held in their own counties: "About 20 or 25 farmers' meetings have been held in each of the four counties having county agriculturists in addition to those receiving support from the institute funds.'' Superintendent Cooley had indicated in his September 1914 report that he planned to largely abandon out-of-state speakers in favor of his county agriculturists and specialists. However, his February 28 report noted that for the large number of 144 institutes, attracting 18,934 people, he had again used additional speakers, including Professor Thomas Shaw, who was by now almost a native.

The brief but carefully recorded daily and weekly reports of M. L. Wilson as county agent leader noted many activities. Those for the week ending February 20, 1915, were perhaps a little above average, but not much. On Monday he was with Director Cooley in Helena testifying to the need for the proposed budget before the legislative Appropriations Committee in the House. On Tuesday he was in Moore assisting agent Peterson with an institute at which 125 attended. Wednesday, he was in Lewistown where the county Equity Society had a meeting and also made up a major share of the 500 who attended the institute. Thursday, Wilson and Peterson were out in the Basin again at Roy, where 150 people attended a Better Farming Meeting. Wilson talked on ''Corn and the Agent's Work,'' and assisted in forming a cooperative seed corn purchasing association. Friday, the two men met with the Progressive Farmers' Club and planned for corn and silo demonstrations. They closed out the week with a meeting at Winnett, where with 180 farmers they organized another farmers' club, and arranged for the cooperative purchase of seed corn.

On Monday, March 1, Wilson was on home ground in Fallon for an institute, and assisted the Executive Committee of the Farmers' Club in planning a farmers' elevator. Tuesday, he was in Terry conferring with the chairman of the county commissioners on the prospects of employing an agent. Wednesday, he appeared before the board of the newly formed Wibaux County with a petition containing the names of 57 percent of the farmers asking for an agent. On Friday he spoke at a Sidney short course on corn and silos. Saturday, he explained the county agent work to the Richland County board. ''Think they will act favorably,'' he said. However, neither Wibaux nor Richland was to have an agent for several years.

On March 13, he noted: ''Took all day to drive from Dennis to Wibaux. Roads very bad.'' Saturday, March 20, he was in Chicago where he arranged with the Flax Development Committee of the American Paint Manufacturing Association to continue their support to the extent of $2,300, of which $1,650 would be used to continue the employment of a county agent who would also assist flax development.

On April 19 he spoke to the annual Montana Stockgrowers Association meeting in Miles City on feed production and the work of the county agent. He admitted, ''In all of my public talks, no matter what subject is assigned me, I explain the county agent's work.'' On the 22nd he was in Glendive where he purchased a Ford five-passenger for $580 for Piper, the new county agent, and taught him how to drive the machine. Wilson noted happily that Hill County wired they had made an appropriation for a county agent and asked that a man be placed as soon as possible. Lewis and Clark County also reported an appropriation to begin May 1. During the week of May 8, Sweetgrass and Teton counties' boards had also asked for the assignment of agents.

Wednesday, June 15, found Wilson attending and speaking at the Watkins picnic in the present McCone County area. "Cold day. 125 present," he noted. Thursday, he moved on to the Whitney Creek picnic, "one of the oldest and best in the state. 300 present." Friday, he continued the circuit to the picnic of the Clear Creek Farmers' Club, organized by agent Jack Taylor the year before. "400 present. Very successful." Saturday, he made an unusual notation, "Resting up." The summer picnics extended into August and on the 7th he was at the Roundup farmers' picnic held about 20 miles north of town where approximately 2,000 people were present. The Commercial Club provided $2,500 in prizes. Said Wilson, "It was one of those crowds which do not come to be talked to, so I made no public address. There is strong sentiment all over the county for an agent."

Burrs Under the Saddle

In spite of the crusade which both Cooley and Wilson carried on to secure the expansion of the county agent program, the movement proceeded slowly. In 1915 three new counties employed agents: F. B. Carpenter in Lewis and Clark, John M. Davis in Teton, and Paul P. Banker in Hill. In 1916 only two counties added agents, E. W. Anderson in Sheridan and C. A. Bush in Flathead.

While 12 counties had accepted the program from 1913 to 1916, only eight agents were in active service in 1916. Wilson lamented, "A number of agents have not worked out." He interviewed candidates widely but felt that agents performed best in the country in which they had been raised, and that experience in actual farming and ranching was a must. He urged a program of training at Montana State for county agents, and for high school agricultural teachers. County commissioners were often hard to please even when a candidate had reasonable qualifications. Numerous difficulties arose to plague the agent.

October 9, 1915, found Wilson in Missoula, for instance, trying to adjust a difficulty caused by the agent assuming the detailed management of the county fair, which had suffered an $8,000 deficit. Whatever the causes, the agent was blamed for "extravagance and poor management," and the commissioners were in a mood to drop the entire Extension program. That decision was avoided, but a firm policy began to emerge; namely, that the county agents should give every assistance in planning local projects, but should not under any circumstances accept the confining task of actually managing them.

The aggressive action of the Chamber of Commerce in Lewistown, particularly, has been noted. In some cases the club, through the secretary, claimed the full-time service of the agent for its exclusive use. Cooley, in his *History of the Montana Extension Service*, recalled some instances with indignation, but concluded that "Such misunderstandings have been few and rare, and. . .generally, their spirit has been fine and their support exceedingly helpful. Occasionally their zeal has been a handicap."

In some counties the Extension Office urged the county commissioners to make a place for the agent in the courthouse, whereupon, also in some cases, the commissioners seized upon the agent as their chief errand boy, and threatened the entire program if he did not render prompt service when requested by them or their chief political supporters. Cooley had some unpleasant memories:

> Some county commissioners are opposed to Extension work. Most of them are honest and try to carry out the wishes of the majority of voters and taxpayers as they size up the county sentiment. Some are against the work on account of the expense to the county. A few are just stubborn and ignorant and unprogressive. Quite a number are bossed and do what the boss tells them. When a prominent stockman or banker or merchant in a county compels the commissioners to vote against the county agent there will be no such agent employed. That is just what has happened in a number of instances. It has prevented the inauguration of county Extension work. It has thrown the work out where already established. The uncertainty and instability of county support is a serious obstacle to county agent work.

The farm organizations occasionally provided another hurdle in the acceptance of the county agent. Montana appears to have been more fortunate than most. "All, with but one exception, supported the movement," said Wilson. This one was the relatively short-lived National or Farmers Society of Equity, not to be confused with the largest organization in the state, the American Society of Equity, which was highly cooperative. The more radical society took the position that the county agent was a lackey of big business, and that the effort of the Service to increase production was aimed at accumulating surpluses and decreasing prices. A resolution, at a St. Paul meeting, December 9, 1915, voiced this protest:

> RESOLVED: That the propaganda miscalled Better Farming, originally financed by Big Business but whose County Agents are now maintained at the expense of the farmers, is not to the best interests of the farmers of the Northwest. The problems which are pressing hard for solution are those of marketing, transportation, and distribution. Better prices for our produce and not more produce at less prices should be the aim of every farmer. We therefore oppose the movement for better farming believing it is against the best interest of the agricultural classes.

This National Society of Equity was second largest in the state at the time, with 2,700 members, compared with 6,000 for the American Society, and was a power to be reckoned with.

The Farmers' Union, which had only about 800 members at the time, but with an effective concentration, also raised a frequent voice of protest and as an organization was suspicious of the county agent. Wilson related in his April 1, 1916, report his experience of winning friends in this group:

> During this month the state leader was invited to speak before the county convention of the Farmers' Union of Toole County at Sunburst and discuss the county agent matter. Considerable interest was shown and after the state leader's talk a vote of the 50 delegates present was taken, it being announced by the chairman that all of the locals had been urged to instruct their delegates on whether or not they wanted a county agent. The vote showed 42 instructed against an agent and eight for. After the vote was announced, one delegate arose and said, "I voted as an instructed delegate. We had a different idea at our local than the explanation which was made today. I move that we vote as uninstructed delegates, and, if the motion carries, that the delegates here present explain the real work of an agent to their locals and circulate a petition to the county commissioners to grant us an agent."

The second vote was 37 for an agent, and 10 against an agent, and three not voting. A favorable discussion followed, said Wilson, and plans were made to circulate a petition.

Even at retirement from a long and successful career, when a person may be unduly inclined toward mellowness, F. S. Cooley in his *History* recalled other difficulties which his beloved Extension Service had faced:

> There were serious obstacles to the organization of the Montana counties for county agent work. The first of these was local apathy and even opposition on the part of farmers. Stockmen as a rule are against county agent work, and it is well-nigh impossible to secure favorable consideration in some of the leading livestock counties. Farmers are by no means unanimously for the work. Their opposition has prevented the appointment of county agents in many counties.
>
> Farmers' objections are based on (a) their natural conservatism, (b) their belief that they cannot learn anything more, especially from a young man college educated, (c) the expense and cost of taxes, (d) misinformation, some of it from organizations to which they belong, (e) personal dislike for the agent or those who favor his employment. Farmers are slower than businessmen to take up the county agent idea. Often they have been against the agent until after they became acquainted with him and his work. Later they have become staunch supporters.

Natural conditions also often interposed obstacles:

> Another preventive of County Extension is the agricultural and financial poverty of counties. There are not enough successful farmers to justify the expense in some counties. Other counties are heavily in debt and have too little revenue to afford it. The marginal counties will fluctuate up and down with farming and business prosperity. Some Montana counties will possibly never be able to justify extension work. In a few counties where Extension work can be well afforded and ought to be made use of, it is prevented by the stupidity and ignorance of the county commissioners or their advisors.

Cooley and Wilson were sure that expansion should not be too rapid.

They estimated that it cost $1,750 to equip a new agent the first year with office furniture, photographic equipment, surveying and other tools of the trade as it was then practiced. The usual beginning salary of about $1,600 at that time had to be added, making a total cost of $3,350. If the agent was successful, a full-time secretary became highly desirable, and other costs of operating the office did not decline. Since the increase in the Smith-Lever funds each years was not more than $3,000, this when matched by the state or county did not cover the cost of adding two agents per year. When counting his eight agents in 1916, Wilson hoped to add four in each of the next two years through increased state and county support.

And so it was that as the year 1916 closed, the Montana Extension Service had, in one year and a half of slow stable growth, obtained eight county agents and a state staff of eight, all dedicated to the vigorous expansion of the Smith-Lever purpose: "To aid in the diffusing among the people of the United States useful and practical information on subjects relating to agriculture and home economics and to encourage the application of the same."

III. Rapid Maturity: World War I, 1917-1920

The Extension Service was shocked out its early period of slow, steady growth with the country's declaration of war on April 5, 1917. Following the outbreak of war in Europe in August 1914, numbers of official notices had expressed the desirability of increased production, and predicted higher prices. A prominent theme of President Wilson's successful 1916 presidential campaign, however, had been "He kept us out of War." American agriculture was not prepared to be plunged directly into a major war. M. L. Wilson's April 1917 report outlined in a restrained manner the way in which Extension was thrust suddenly into the middle of the enormous national mobilization effort:

> Shortly after war was declared, the governor appointed a State Defense Council, and this council in turn appointed three members of a County Defense Council in each county. At the request of Chancellor Elliott, Director Cooley and myself appeared before the State Defense Council and a plan was discussed for placing a temporary county agent in every county for the five months, April, May, June, July and August, as well as the organization of this work, which contemplated that the College should pay the salaries of such agents out of college funds, and that the boards of county commissioners would be asked to furnish automobile transportation and $100.00 a month for five months; that the state was to be divided into districts and a member of the faculty be placed as a supervisor in each district.

Much of April was spent in scrounging for emergency county agents whose job was to be, according to Cooley's Memorandum of Purpose sent to the State Defense Council:

> To help arouse an appreciation of the present national crisis; locate and distribute seed and information regarding programs for crops; work for the increase of crop production; encourage the increase of meat and wool production; make food and crop surveys; assist in finding and placing farm laborers; cooperate with bankers and farmers in financing farm operations; promote the general food preparedness plan; cooperate with other county agents in the defense program...

Only two men were employed from out of state in spite of extensive searching. Smith-Hughes instructors in the high schools were pressed into service; a junior and three seniors in agriculture at Montana State College were employed. Only one person was employed who did not have some college course work. To complete the quota in record time, Extension and Experiment Station specialists were drafted. Dr. Howard Welch, veterinarian and long-time staff member, recalled the historic occasion:

> In April 1917, Extension Director F. S. Cooley called a meeting of the agricultural staff, and made the bald statement that we, all and each of us, would henceforward drop all our activities and become county agents, proceeding to read a list of our names, and the county to which we were assigned. There was a stunned silence, then someone weakly asked as to when this thing would be initiated—to which Cooley replied, "tomorrow morning." I can't remember who was sent to which county, except that I drew Madison. We had very foggy instructions, and still foggier ideas of how to carry them out. We were to "organize" our counties into defense committees, etc., and many a bewildered rancher or businessman found himself the chairman of a war committee without a program.

On Saturday, April 28, Cooley and Wilson spent the morning with the five leaders who would supervise the work in the districts into which the state had been divided. These men were Alfred Atkinson, C. N. Arnett, O. B. Whipple, J. C. Taylor and W. L. Beers. They were to begin their tour on Monday. A fairly definite program of work was outlined, whereby these men would meet with county agents, most of them new, and advise with the County Defense Council and the commissioners. Plans were laid for emergency work, primarily increased production and raising of the best seed possible. Widespread publicity was encouraged through posters and other means to acquaint the people with emergency needs, methods and personnel.

The war emergency, without doubt, advanced the cause of Extension by years. There was immediate and universal recognition that this agency was ready-made for the urgent task at hand, that of reaching every farmer and rancher in the nation. Woven into his June 30, 1917, report was rejoicing by Wilson that his new agents had rendered valiant service in the short time they had been at work, because, he said, "Nearly all the old farmers were inclined to regard the work with suspicion."

Organization for War

Since most of the emergency agents did not reach their posts until early May, instant action was of the essence if anything was to be accomplished in 1917. County agents' offices immediately became seething centers of action as meetings were arranged to urge expanded production and then find seed, labor, machinery and credit. The advantages of the statewide organization were immediately apparent as surplus seed and labor were quickly located and made available. The statistics of accomplishment from Wilson's August 1917 report reveal only a small part of the drama involved:

8,628 farmers had been assisted in securing seed.

250,041 bushels of seed grain, corn and potatoes had been secured.
409,611 additional acres had been planted as a result of the war campaign.
3,565 farms had been assisted in securing additional labor.
1,033 farmers had been assisted in securing purchase credit for seed and machinery.

The "Madison Avenue" publicity techniques of recent years may well have had their origins in the encompassing war information effort directed by the Extension Service in the summer of 1917, surely the greatest of its kind in Montana. Following the rapid-fire action in locating county agents and stimulating crop production, the Department of War became conscious that the Extension Service in the Department of Agriculture had the promotional facilities it needed. An enormous nationwide program was instituted to inform the people of the issues of the war, of the need for increased production and the preservation of food, and to secure the full cooperation of every citizen in the war effort through maintaining a high morale, increased health and well-being, participation in Liberty Bond purchases and armed services support. Wilson summarized this effort in his June 30 report:

> The chief work (in June) was in looking after the organization of the 400 farmers' National Defense Council meetings, more popularly termed farmers' patriotic picnics. These meetings were instigated by the State Defense Council. They were planned to come off in the latter half of June and the fore part of July. The method of formulating the organization of this undertaking was perfected in May, when the county agent leader held district conferences with all of the county agents. The dates and places where the picnics were to be held were arranged for by the county agent. The county agent then appointed a committee or leader within each community to care for the details of the picnics. With this organization it was possible that the ideas of the State Defense Council would easily be carried out clear on down to the individual communities. Furnishing a corps of speakers was quite a task. Twenty different groups of speakers were selected, with from three to four in each group.
>
> The program was arranged as follows: The subject of the patriotic speaker was "The Meaning of the Great War," which aroused interest in our nation's great undertaking. This speech was usually given by some distinguished man with ability along this line. Food economy talks and canning demonstrations were given by a competent woman from the College or the community. The agricultural talk was presented by a speaker from the College, or occasionally by some prominent farmer, who presented the facts about the food shortage, his subject being "The Role of the American Farmer in the War." The county agent usually acted as chairman, and gave a timely talk about the problems of his county.

A total of 337 meetings were held in 37 counties during June and July, with an attendance of 85,678, which represented about 90 percent of the state's farm population. These meetings served not only to publicize the war, but also the Extension Service, which was completely new in many of the counties. Through the organization and conduct of the meetings, the county agent and his people became better acquainted, and the people of the community came to know each other. The work of the home demonstration agents at these meetings was particularly well received. Their suggestions relieved the austerity of the meatless, wheatless and sweetless war meals, and their demonstrations of the new pressure cooker revolutionized the preservation of food for the farm family.

By the war's end the Extension Service had expanded from its eight state staff members and eight county agents in early 1917 to a total of 85 members consisting of:

1 extension director	2 dairy specialists
4 county agent leaders	1 marketing specialist
26 county agents	4 rodent control specialists
2 state home demonstration leaders	2 entomologists
	1 agronomist
15 home demonstration agents	9 clerks
3 state club leaders	16 county club leaders

Finances also expanded rapidly as the state and the counties made increasing use of the new agency, and as the federal government made war emergency funds available. The total budget for the year closing June 30, 1918, was $207,812.03. Of this, the state made available:

State Extension Service	$38,500.00
Farmers' Institutes	10,000.00
County appropriations	26,000.00
	$74,500.00

The federal government contributed:

Smith-Lever funds	$17,866.69
Department of Agriculture for special projects	15,750.00
War Emergency funds—approximately	99,695.34
	$133,312.03

The combined efforts of the Experiment Station and the Extension Service bent to the task of gearing the state to the larger agricultural program to assist with war needs. Major war emergency activities for 1917 included:

1. Increased acreage campaigns
2. Seed distribution
3. Organization of Farm Bureaus to cooperate on war programs
4. Labor distribution
5. Administration of county and state seed loans

Regular long-time projects included:

1. Crop improvement
2. Livestock improvement
3. Rodent and insect control
4. Boys' and Girls' Club work
5. Federal Loan Associations
6. Extension Schools and institutes

With the planting season only a few days away when the county agents reached their posts, their first efforts were to secure seed and labor. They immediately encountered the need for credit in a large majority of cases. Many of the new homesteaders were already deeply in debt, and few farmers had sufficient reserves for the large additional expansion which was being urged. There were assurances that the federal government would make loan funds available, which it did. In the meantime, local chambers of commerce and banks gave full cooperation. Fred Bennion, emergency agent for Yellowstone County, reported as early as May 1917 that through the efforts of the County Defense Council, "The people of Billings raised a guarantee fund of $25,000 for grain and seed loans to homesteaders and people whose notes were not bankable."

Drought in 1917

By midsummer the encouraging early progress in agricultural circles to meet war needs ground to a shattering halt when the most severe drought in a decade descended upon the land, deepening steadily as the summer lengthened. Not since 1910 had there been a threat of drought, and 1916 had been particularly lush. The thousands of homesteaders who had come into the state since 1914 were convinced that Jim Hill was right in acclaiming Montana the equal of the Midwest. They were utterly unprepared psychologically or financially for the realities of an arid region. By August Wilson commented tersely that county agents "have assisted in the formation of 75 National Farm Loan Associations." He noted in July:

> In many instances farmers staked everything they had on this year's crop, and with a failure apparent, banks and merchants have tightened on credit so that a good many farmers find it difficult to secure necessary supplies.

The numerous farm loan associations proceeded to develop a strong program for 1918, which would surely be a better year. Wilson reported in December 1917 concerning Richland County as a typical example:

> A survey indicates that three-fourths of the cultivated area in a number of districts will not be put in unless seed is supplied to farmers on some terms more favorable than the usual 10-12 percent.
>
> In all, nearly $1,000,000 has been applied for in this county. . . Owing to the slowness of the Federal Farm appraisers to reach Richland County there will be no end of hardships.

The situation worsened steadily, and Governor Stewart called a special session of the State Legislature, held February 14-25, 1918. The first request was for laws to relieve the distress of agriculture. A lien law was enacted which gave prior claim for loans which were made for seed and planting purposes. Counties were authorized to hold elections to secure permission to bond in order to make loans for seed and planting expenses. The

Legislature appropriated $500,000 for assistance to agriculture, in addition to $11,956 for the Extension Service and $8,805 to the Farmers' Institutes. Since the State Council for Defense had been organized hurriedly at the insistence of the federal government, it was given legal status as a war coordinating agency, status which would terminate three months after the formal close of the war.

When the drought continued in 1918, state and county funds were extended to the limit, and in late summer President Wilson supplemented these to the amount of $5,000,000 from the war emergency funds. County agents were spending as much as 50 percent of their time in negotiating loans for their farmers. In the late summer of 1918, some $300,919 from the emergency funds was issued to 1,480 farmers to seed 101,621 acres of winter wheat and rye. In the spring a federal office was established in Great Falls and 6,725 farmers received $2,150,000 for 470,000 acres of spring wheat seeding. A total of 2,000 drought permits were issued which secured freight rates at one-half price for shipment of feed grains and forage, and costs were lowered by $200,000.

Late spring 1919 was even dryer than 1917 or 1918, and Governor Stewart called another special session, held July 29-August 11. Since the war was over and federal funds were no longer available and state funds were low, the responsibility was again placed largely upon the counties. Bonding elections held soon after the special session approved $4,500,000 for relief purposes. A grain grading and warehousing act firmed up standards and enlarged facilities provided by previous laws. The governor urged a program of road building in which employment would relieve prevailing distress, and a State Highway Commission was established. The usefulness of the irrigation projects which were in operation was apparent; a Montana Irrigation Commission was established and the Irrigation District law was expanded. The severe curtailment of available forage caused a widespread movement to sell livestock. A clearing office was established in Billings in cooperation with the U.S. Bureau of Animal Industry and Markets to secure the best prices possible.

One redeeming feature of these years was that high prices prevailed. Stockmen had an early advantage in this situation, but when three years of drought depleted the ranges and a late spring in 1920 forced continued feeding of hay at $50 per ton, both stockmen and grain growers were at the last ditch. Drought conditions lifted just enough in 1920 to provide encouragement, but not enough to give real relief. And then prices fell disastrously. Director Cooley described the situation in his *History*:

> The markets had gone down. Great war stocks of wool became available for manufacture and prices declined. Sheep dropped to a small fraction of what they had been worth a few months previous. Beef markets were glutted and slow. There was no market for cattle. Hides could hardly be given away. Horses were a drug. Whole flocks and herds would not bring enough to pay the winter's feed bill. . . to have slaughtered livestock, horses, cattle and sheep wholesale in the fall of 1919, and accepted a total loss would

have been less disastrous financially than to have bought feed to bring them through the winter.

I met a stockman who could have drawn his check for $100,000 in the fall of 1919, a year later he was cooking in a restaurant without a dollar left. Another stockman, whose assets were in six figures in 1919, made affidavit in the spring of 1920 that he could not sow his crops without a $75.00 seed loan. Nearly all of the stockmen were in distress. Many of them hopelessly insolvent. Millions of dollars of investments were swept away. Hopes that things would get better were not realized. They kept getting worse.

Those who say that farming is hazardous in Montana under dry land conditions, who think it folly to cultivate land wthout irrigation, who think this land is suitable only for grazing, should remember that stockmen suffered even greater losses during the droughts of 1917, 1918 and 1919 than did the farmers. During the 40 or 50 years previous, the success of stock growing had been fully demonstrated. Montana was a wonderful grazing state. Cattle and horses and sheep reached their best development. Fortunes were made, cattle barons flourished. They had firmly established themselves. Nevertheless, ruin overtook the livestock industry and even the reserves accumulated during years of plenty could not save them. Losses were overwhelming. Dry farmers lost less only because they had less to lose.

As a result of the general economic situation of the depression in agriculture, of a maladjustment of freight rates and of three consecutive years of drought, Montana conditions were much accentuated. Deflation was well-nigh complete. The leading farm industries of livestock and grain growing were hardest hit. Farm indebtedness was in excess of any hopes to pay. Taxes were delinquent.

Thousands of families abandoned their farms, leaving teams and implements and what they could not take with them. Banks and loan companies and merchants suffered with the farmers. Failures in business were numerous. County debts were heavy and county funds were lost through bank failures.

This was a grim situation and, as county commissioners surveyed a bankrupt condition, Extension Service leaders feared ruthless curtailment of county agent support. They were greatly pleased when 23 of the 26 counties retained their agents. Cooley cited one county which had suffered a $100,000 loss in a bank failure, whereupon funds for the county agent were lopped off. Before the agent left, the commissioners questioned their judgment. The petitions of county farmers caused them to reconsider, and he was retained. Extension leaders were also encouraged by the retention of eight home demonstration agents. Only one Boys' and Girls's Club leader was retained, but the high regard for the program resulted in its incorporation into high school agricultural programs along with volunteer leaders to keep it in full force.

The new Extension Service had met the emergency conditions of the war and the drought in a superb manner, and most of its efforts had been restricted to these situations. However, rather early, the Service had worked with community groups to develop local interests and resources. Cooley's 1917 report indicated a few of these "normal" activities:

In Flathead County the agent worked with an Idaho Falls seed company to contract with 37 farmers for 422 acres of seed peas. He estimated that this would add $50,000 a year to the income of the county. In Custer County, agent C. M. Yerrington served as secretary and general purpose man for the cooperative Yellowstone Valley Seed Growers' Association which sold 76,000 pounds of graded alfalfa seed in Montana and in 16 other states. An estimated $3,500 was saved to the growers through cooperative management. Yellowstone County organized a Bean Marketing Organization which stimulated the growth of beans to a point where 50,000 bushels were handled. In Madison County 250 carloads of potatoes were sold by the Potato Growers' Association. In Ravalli County six carloads of apples were marketed to other cooperative associations in Montana. In Jefferson County a cooperative dairy and cheese factory was built at Elk Park. In Musselshell County a badly needed rural telephone system was established.

The Farm Bureau

As has been noted, a basic feature of the Extension Service was the closest cooperation possible with the people on the land. Through contact with farm organizations, individual cooperating farmers and better farming committees, the county agent tried to keep in touch.

The idea of a formally organized group of farmers in each major community, with representation on a county-wide committee, took form in Broome County, New York, in 1911, soon after the first county agent in the northern states began his work. By 1917 New York had 39 counties organized, representing 40,000 farmers, and the movement was named the Farm Bureau. This movement took two forms. One was the formation of a new community and county pattern. The other was the use by the county agent

Bulletin board in local grain elevator in Phillips County, Mont.

of the community and county branches of existing farm societies, such as Equity or the Grange, or if several of these organizations were working in a single community, a federated committee was formed.

Almost immediately the commercial and political action policies of the older farm associations rose to plague the Farmers' Institutes and the county agents' programs which were established and funded by the state. In the official sessions of the Montana Farmers' Institute state meetings, conflicting resolutions and petitions were circulated and debated hotly. Since the institutes were promoted by a state board and were intended to be purely educational in nature, this action was highly divisive and disturbing. The Sutherlin brothers of the *Rocky Mountain Husbandman* have been mentioned. Their support of the institute and their equally strong support for the dairy industry and its opposition to oleomargarine was only one illustration of the issues which arose. Since this condition prevailed over the country, the States Relations Service of the USDA, in general charge of implementing the Smith-Lever Act, recommended in 1914-1915 that a new organization, the Farm Bureau, be built up in each county to support Extension Service activities.

Wilson took a cautious approach. He felt that the movement was experimental and that the loose representative committee which he had organized in 1913 was probably sufficient. On May 15, 1916, his weekly report noted that G. E. Piper, new agent in Dawson County, wished to organize "what he calls a Dawson County Farm Service Bureau. I am afraid his plan tends toward commercialism," said Wilson, "but a demand is coming for some organization. We must soon formulate a state policy toward this." Piper proceeded to develop the committee, and it appeared to function smoothly.

Two specialists in Farm Bureau organization came from the Washington, D.C. Extension Office to Montana for the August 1916 conferences, and then worked with county agents for a time. Lewis and Clark County pioneered in effective organization. First, a representative planning committee set up five districts, and named a member from each to make up a county executive committee which in turn named a committee in each of the major communities, resulting in 17 districts replacing the original five. A Farm Bureau organization then formed in each of these districts, coordinated by a county committee, to help carry out the work of Extension. Dawson County followed this pattern, as did Cascade and Flathead counties. Wilson felt: "Farm Bureaus should not be pushed to the extent that the older farm organizations feel they are competing." Cooley was emphatic:

> The Farm Bureau does not seek to crowd out the Society of Equity, Farmers' Union, Grange and other farmers' organizations. Its function is rather to federate existing agencies for rural progress and to connect these with state agricultural institutions, the U.S. Department of Agriculture and the general government policies.

Cooley admitted that the bureau units would have been helpful in the war effort in the summer of 1917, but there had been no time to organize them. By August other representatives from the Washington office came to promote the idea. At the January 1918 annual conference, plans were made for a campaign which began in February. During the summer, 19 county organizations were completed, making 23 in all; one in each county where a county agent was employed. Cooley remarked:

> A working organization of over 10,000 progressive farmers in 23 counties now assists the agents in formulating plans and directing Extension activities that are carried out under the direction of the Farm Bureau. In other words, the farmers are directing their own work with the agent acting as the manager for putting the plans into operation for the betterment of agriculture within the counties.

Bess M. Rowe, state home demonstration leader, in her report for 1919, commented that the Montana Farm Bureau now included women as well as men, and that:

> This resulted in one big thing being accomplished—it has changed the point of view of our Farm Bureau women as to the function of the home demonstration agent in the county. . . The women have come to feel that the work is *their* work—that we are in the state for the sole purpose of helping them carry out their plans for home and community betterment and of helping them to see and to solve their problems.

A list of memberships showed a range from 627 members in Valley County to 125 in Hill County. A monthly county Farm Bureau Bulletin with its buy and sell exchange list, community experience exchange, and notice of meetings served effectively in unifying the counties for action programs. Since the war and drought years provided plenty of opportunity for such activities, a state legislative statute was approved February 13, 1919, providing for the incorporation of the Farm Bureaus, authorizing commerical

A group of Farm Bureau women attending a conference in Bozeman, Mont., 1919.

business operations, and outlining as its purpose:

> A corporation to be known as the County Farm Bureau may be organized in any county to develop and to carry out a county program of work in cooperation with the Board of County Commissioners of said county, the Montana State College of Agriculture and Mechanic Arts of the University of Montana, and the United States Department of Agriculture, for the advance of agriculture and home economics, the promotion of better understanding between the citizenship of town and country, and the development of a wholesome community life.

The place which the bureau took in the county depended upon local leadership, of course. Director Cooley came to give it high priority. In his report for 1917-1918, he discussed the bureau at length and commented: "Situations have been met by the Farm Bureau that could not possibly have been handled by the county agent alone. A valuable feeling of proprietorship of county agent work has been created among farmers." He made the sweeping statement which is repeated in essence in the reports of the period: "All the county agent's activities must now meet with the approval of the Farm Bureau."

Wilson's 1919 report was somewhat more restrained. The state law was in operation and Farm Bureau offices were housed in courthouses in 18 of the 23 counties, all that had room for such an office. Said Wilson:

> With its agents and workers financed from public funds, the Farm Bureau must remain educational and non-partisan, avoiding controversial questions and business activities. The opinion prevails, however, that the active assistance of state and federal Extension agents is essential or desirable for the further satisfactory growth and development of the Farm Bureaus. The farmers feel that it is necessary for them to organize for mutual welfare. On the whole, the farmer has not had the training nor has he a sufficiently wide perspective to enable him to organize on a correct and substantial basis. The individual farmer feels that he cannot cope with the conditions in which he finds himself nor with the corporate interests that have developed for handling the products of the farm. There is danger that farmers will organize on a selfish basis as some existing organizations have done.

Wilson felt that instead of the farm bureaus setting up business enterprises in their coprorate name, their members as individuals should take part in "subsidiary marketing organizations," in which they had the greatest interest.

In February 1919, a three-day conference of Farm Bureau presidents from 21 counties, all of the bureaus which were active at the time, met in Bozeman to organize a State Federation. Careful preparation had been made for the meeting, a constitution was adopted and a program of work was planned for the year. Plans for the year were greatly facilitated by Extension Service *Bulletin No. 30*, issued January 1, 1919, titled "A Program of Work for the Montana Farm Bureaus." The state leaders, at least, envisioned the Farm Bureau as made up of action committees, each assisting in some specific phase of Extension work. Cooley explained the functioning of these committees by referring to the Committee for Farmers' Institutes:

> By referring the selection of places for Farmers' Institutes to the Farm Bureau committees on Farmers' Institutes, by securing their advice and assistance with respect to the subjects desired, speakers wanted, and dates preferred, a stronger local support can be worked up for the farmers' meetings.

This bulletin listed with appropriate discussion 23 subjects for Extension work, any or all of which might have action committees organized within the community and county Farm Bureau structure:

1. Farm Bureau finance
2. Gopher and prairie dog extermination
3. Agronomy
4. Home economics
5. Social life
6. Boys' and Girls' Club work
7. Fairs
8. Marketing
9. Farm management
10. Farmers' meetings and Extension schools
11. Projects and correlation of work
12. Agricultural history
13. Office organization and methods
14. Agricultural engineering
15. Horticulture
16. Insect pests
17. Publicity
18. Bee keeping
19. Botany
20. Crop reports
21. Dairy work
22. Livestock
23. Relation of county to national Farm Loan Associations

Wilson's 1920 annual report devoted 20 pages to the Farm Bureau. "The

Extension Service booth at the Midland Empire Fair, 1918.

Farm Bureau is now firmly grounded in the older counties. It is looked upon as an Extension institution for doing Extension work," he said. Farm Bureau units were operating in 366 communities, 294 community meetings of a combined social and business nature had been held with 24,253 attending. Twenty-two counties held 139 summer picnics attended by 29,636. Fourteen tours had been made to demonstration plots, silo demonstrations, and branch Experiment Stations. Fifteen counties held or cooperated in fairs, and eight counties joined in hosting five week-long Junior Camps for 12- to 16-year-old youth, which were considered highly productive. Membership during the year increased, however, by only 104 to a total of 10,601.

The Farm Bureau entered a distinctly new phase on February 12, 1919, when representatives from 12 states met in Ithaca, New York, to discuss the possibility of a national federation. The basic future policy was the chief subject of discussion. Many favored primarily an educational approach; others urged a program for nationwide cooperative marketing and political action to secure improved business and economic conditions. Specific plans were made for a forthcoming national meeting in Chicago. Here the delegates from the East, South and West favored the educational approach, but the Midwest strongly urged an action program. With an attendance of some 500, Illinois alone had 220 delegates and the actionists prevailed. Director Cooley attended from Montana. A temporary organization was formed, a proposed constitution was sent to the states, and a meeting for a permanent organization was called for Chicago on March 3, 1920. A plan for a strong national structure was developed and offices were set up in Chicago and Washington, D.C.

The president of the Farm Bureau, chosen for his middle-of-the-road position, had a full appreciation of the importance of Extension support and on January 1, 1921, he sent a bulletin to all county agents. It read in part:

> The county agent is the keystone of the federation. . . The American Farm Bureau Federation is exactly what the individual county Farm Bureaus make it. And the county Farm Bureau, I have found again and again, is just what the county agent makes it. . . The county agent is the strong right arm of the American Farm Bureau Federation. . . Ask him to continue to help the American Farm Bureau Federation so that the Federation may help him and his people.

As the Farm Bureau Federation began lobbying activities, the Congress instituted an inquiry. In the hearings, other farm organizations testified that the bureau was in competition with them and doing the same things. The bureau was accused of working to limit production in order to increase price, and some congressmen equated this with bolshevism.

Threatened with curtailment of funds for Extension work, the Department of Agriculture formulated a "Memorandum of Agreement" in May 1921, which stated that Extension personnel would aid the farm population in every way possible, but would "not themselves organize Farm Bureaus

County agent's office in Richland County, Mont., 1919.

or similar organizations, conduct membership campaigns, solicit memberships, receive dues, handle Farm Bureau funds, edit and manage the Farm Bureau publications, manage or take part in other Farm Bureau activities which are outside their duties as Extension agents.''

Since the county agent had been doing almost all of these things, the Farm Bureau organizations declined in number and membership for a time. The relationship had been so close and so helpful that Extension Service personnel continued to work with the bureau organizations as a matter of habit for several years, although other organizations protested. J. C. Taylor, in his first report in 1924 and for another year or two, referred to the assistance given by the Farm Bureau and the desirability of its remaining strong, but always asserted the desire of Extension to work with all of the farm groups. Former Director Cooley, in his 1924 *History*, acknowledged, ''The Farm Bureau is peculiarly the child of Extension work. It has been a most valuable auxiliary to the Extension Service.'' He also insisted that Extension had wished to cooperate with all farm organizations, and in the situation where several groups were competing on none too friendly terms, he maintained, ''the greatest unresolved problem in agriculture as I see it is the organization and leadership of farmers.''

Close cooperation between Extension and the bureau rather naturally continued, but special care was taken to remove any formal or financial connection. Occasional continued criticism called attention to the statute of 1919, which still remained in force. In 1949 Director Robert B. Tootell and the president of the Montana Farm Bureau, supported by the Farmers' Union and the Grange, formulated an act which the Legislature passed removing the statute.

Mignon Quaw's "Long Trail"

State and community leaders were impressed with the progress made in community cooperation which came out of Farmers' Institutes and other meetings where social aspects were emphasized. Sparseness of population, poor roads, long work days, and few schools and churches in many areas prevented the rapidly growing population from getting acquainted even within communities. A result was the establishment of a specialist in recreation. Miss Mignon Quaw took this position, along with that of assistant state leader in home economics serving from 1918 through 1921. She was particularly well suited for this work through personal characteristics, membership in a large Gallatin Valley family noted for its leadership in social functions, and her ability to create musical and dramatic productions. Miss Quaw developed suitable programs of recreation for many types of farm gatherings from country schools to state conventions.

As the movement to form Farm Bureau units began in the spring of 1918, there was a need for some device to bring people together and to entertain and instruct them. County Agent F. M. Hillman suggested a Farm Bureau dramatization as such a device. Miss Quaw volunteered to write the play, and during the first 15 days of October she wrote "The Long Trail." The cast of seven was composed largely of state specialists who could combine their work with the play. "The Long Trail" was performed 33 times in the state, with a total attendance of 14,692. The community Farm Bureaus were in charge of arrangements, and often this was the first occasion which brought all of the members, men and women, together in a cooperative effort. The community bureau paid $15.00 for the cost of shipping equipment, and this sum actually netted a profit for the production group. A small admission charge was made which provided a welcome sum for the local organization.

The play made an impact through its emphasis upon the importance of agriculture and of the community organization. Three acts separated time intervals of 25 years each: 1868, 1893 and 1918. The theme and Act I recalls that of the many trails marked out by man in his march toward civilization, agriculture is the oldest and most widely traveled of all. Men of vision have left blazes along the trail: the wheel, the alphabet, the printing press, the harvester, the land grant college, and the Homestead Act. The nation which has been first to recognize the importance of agriculture has made the greatest and fastest progress.

Act II emphasized education. Doctors, lawyers, engineers, ministers and teachers have progressed in proportion to available education—why not farmers? The Agricultural College was established in 1893. A dearth of knowledge in scientific agriculture was apparent, so the Montana Experiment Station was established in the same year. In 1895 the rural free delivery system was initiated.

Act III focused upon the war days and the urgent need for more knowledge, more production, more cooperation. The Extension movement has been under way in various forms for 25 years. In the educational effort all taxpayers support the schools. Not everyone can go to college, but the farmer is particularly fortunate in that Extension, assisted greatly by the postal service, brings the college resources to him. A great chain of agricultual education has been forged—the United States Department of Agriculture, the agricultural colleges, the Experiment Station, the Extension Service.

The entire chain is ineffective unless its most important link is strong, the people themselves. The Farm Bureau is now the link which is enabling the people to organize in their communities in a more effective way than ever before in history, and to take full advantage of the facilities offered by these other links. The Farm Bureau means progress. Civilization's trail of agriculture is a long trail, but an extremely useful and productive one.

In a gentle but effective and insistent manner, Miss Quaw urged the close cooperation of all farmers in improving farm conditions and their strong support for educational and social programs in the states.

Mixing gopher poison, Lewis and Clark County, Mont., 1919.

Rodent Control

Early Extension workers found that often issues which were not seemingly important obtained more support than some of the more basic projects. Rodent control was one of these peripheral enterprises which attained spectacular proportions. It came to have particular appeal as a war measure for increasing production.

Carl Peterson apparently had a major part in starting the rodent control project. While supervisor of the Montana State College farm he found that it was infested with gophers, and the Experiment Station personnel assisted in working out a successful and economical method of extermination by using poisoned grain. When he became county agent in Fergus County in mid-May 1913, he found gophers and other rodents to be a real menace. The mature animal was estimated to destroy a bushel of wheat per year. Gophers and prairie dogs were also accused of consuming up to 50 percent of the grass in the vicinity of their villages which were usually located in the most fertile areas.

Adults could not be induced to engage in extermination campaigns in sufficient numbers, however, so Peterson hit upon the idea of making this a project for the boys' and girls' clubs. Then he secured the cooperation of the local commercial clubs for awards and prizes. By May 15, 1915, the small community of Judith Gap was overrun on its first "Gopher Day" when the youth of the area brought in 32,000 gopher tails for a compensation of one cent per tail and spent the afternoon competing in athletic contests for suitable awards. Many other communities also organized in the war on the rodents.

The need for a massive cooperative effort became apparent immediately since the migrating nature of many rodents and their high reproductive rate made it necessary to achieve a 75 percent kill in a large area to be effective for more than a year or two. The Legislature enacted an extermination law in 1917 authorizing the county commissioners to appoint exterminators upon the petition of 10 resident farmers. Commissioners might levy one-half mill to pay for poison, salaries and other expenses, and they might make assessments against railroad and non-resident land holders whose lands could best be administered in this way, and also against other lands where owners were non-cooperative. This law was strengthened in 1919. Rodents comprised a sizeable family, and in the law were described as striped gophers, flicker tail gophers, pocket gophers, Columbian ground squirrels and prairie dogs.

In 1917, 12 counties cooperated in mixing and distributing baits to cover 950,000 acres. R. A. Cooley, state entomologist, was in charge of the 1918 campaign. He estimated that 27,600,000 gophers were destroyed, which at a conservative estimate of 10 cents damage per gopher totaled $2,760,000 in return for the rather substantial cost of $82,000 spent on the campaign. By 1919, 23 counties and 18,759 farmers incorporating 6,409,489 acres showed an estimated savings of $1,887,866. Another campaign was waged in 1920 when "Gopher Week" was declared in April, resulting in savings estimated at $2,000,000. Even so, not everyone was happy. Cattlemen with extensive range did not feel the pinch of the grass consumed by the rodents and were opposed to being assessed for extermination. Absentee owners who had purchased land for speculative purposes also objected.

The Indians picked up the campaign with considerable enthusiasm, particularly the Blackfeet. A campaign had been initiated in 1921 to encourage every family to plant at least five acres in garden, corn and wheat. During the week of the Fourth of July tribal festival, an infestation of gophers destroyed a large percentage of those crops. One full-blood chief noted the threat, took his family to the celebration, but returned home immediately and waged war on the invaders. His fine crop made such an impression that the Indians voted to postpone their next annual celebration until the crops were in less danger. As a result, in 1922 they took more corn to the Heart Butte mill for grinding than the community could consume. A Hedgesville farmer summarized what appeared to be majority opinion of both Indians and whites concerning the annual campaign values: "It is the nearest thing to a miracle ever seen in Montana. . . the cost of the gopher campaign is as nothing in comparison to the saving of grass and grain, to say nothing of the lesson we have learned in organized work for the common good."

The coming of the rains later in the 1920s lowered the pressures, and a quiet campaign for control rather than extermination was carried on. However, the rodent population slowly built up again until it appeared to explode in the drought years of the early 1930s. In 1934 funds from the Montana Relief Commission were made available and another extermination campaign destroyed an estimated record 9,000,000 pests. Extension Service *Bulletin No. 83*, "The Prairie Dog in Montana," issued in March 1927, went through repeated printings over several years. The persistent plague in the 1930s, however, was that of grasshoppers. The experience with poisons, mixing stations, distribution methods and cooperative field teams worked out in the rodent program was invaluable in the face of this even greater danger.

Good Roads and the Model T

Lack of transportation and communication facilities posed one of the major problems in carrying on the work of the Extension Service in its formative period. This was given early attention in a forceful way by the Farmers' Institute board when, in 1907, it issued *Bulletin No. 2*, "Better Roads." This was written by Professor of civil engineering E. Tappan Tannatt of Montana State College who was critical of the slow progress. He called attention to Montana's particular difficulties of vast area, small population, low taxable valuation and temperature extremes. He urged that a special study be made of these temperature factors, as well as soil variation, rather than an attempt made to build roads quickly by the same methods used in the Midwest.

Tannatt was also concerned about the attitude of conflicting groups toward road building. The commercial groups in the towns and cities were the strongest boosters for good roads into their towns. Farmers in the fertile valleys supported a good roads movement as did the mining interests, although both groups were troubled by the cost factor. Forestry and lumbering groups were opposed, probably because of costs of building and fencing and restrictions on the use of heavy loads. Tannatt maintained that the stock growers on their large ranges were the chief offenders in slowing down the good roads movement. He acknowledged that they were merely operating in their own best interests, as were the merchants and farmers:

> In the eastern and central portions of the state we find vast stock interests represented. Those who are engaged in such pursuits desire an open country, and have little need for, nor do they desire, good wagon roads. Good roads would attract the settler and would result in the settling of the country, the fencing of the water rights and the contraction of the ranges. Good roads would be followed by laws requiring their protection and care, and these roads and laws would prove largely a source of annoyance to the stockmen.

Tannatt noted that Montana had only 22,419 miles of improved roads, less than half that of either North or South Dakota, and only 65 miles of surfaced roads. Indeed, Montana was at the bottom of the list of five northwest states in the road category. The average cost of improved roads at $1,146 per mile would be immediately compensated for by an average increased valuation of $9.00 per acre for land in the service area of the road. The better roads subject was often formally discussed in the Farmers' Institute meetings, and Tannatt reported that county commissioners in every county were overwhelmed by requests for new and better roads. He lamented that little was known of the principles of location or building of roads, and he called for the creation of a State Highway Commission to establish standards and coordinate the structure.

The coming of World War I, which brought demands for assistance the new Extension Service could render, also caused the increased use of the low costs Model T Ford. The automobile, of course, also added immediate urgency to the better roads movement. The light, simply constructed and easily maintained Model T was invaluable in enabling Extension people to contact farmers and ranchers at home, and also in assembling them in frequent meetings. Wilson's weekly reports during his first year as county agent leader reflected the trend. On April 20, 1915, when he was in Glendive to meet the incoming county agent, G. E. Piper, he reported:

> Mr. Piper arrived on the noon train. We purchased a Ford five-passenger for $580. I gave my personal note for same, but expect to make arrangements with Mr. Piper to own the car. We desire to start the policy of the agents owning their cars and pay them a mileage which will pay their expenses and pay for a Ford car in two years.

County agent getting performance record on a large tractor in Phillips County, Mont., 1919.

Wilson remained an additional day and taught Piper how to operate his car, and introduced him to the Glendive people.

In mid-May Wilson was in Missoula with the new agent. "Mr. Hansen has had his auto but a short time; therefore he is not as yet thoroughly acquainted with the country, its problems and needs," he said. In June Wilson noted that Paul Carpenter, the new agent in Lewis and Clark County "owns his own auto and is to receive actual expenses for the operation of the same, $5 per month for depreciation and $2.50 for storage." In July he was in Havre with another new county agent. "Met Mr. Banker. Helped him buy a Ford Model T car," Wilson commented. The next day he was with Mr. Banker who was learning to drive his car.

The use of the new car reflected the conditions of the time. In July 1916, Wilson was in Glendive and reported, "Attempted to go to Sidney in auto, but was compelled to abandon trip. Bridges out." E. R. Jackman, college junior, serving as emergency agent in Blaine County in 1917 and later prominent in the Oregon Service, commented:

> Transportation with a Ford without benefit of starter, and lights ran only with the engine. Roads were terrible and when one stopped to proceed more gingerly the lights went out. Therefore, there was no recourse except to go fast and keep the lights going. Most of us lived.

It is now difficult to believe that railroads ran many passenger trains everyday; hence transportation to county seats was relatively easy for the specialists, but once there the automobile and the county agent became doubly important. In spite of slow trains and poor roads, the mileage logged by the early Extension people was incredible. A few selected months in 1915 indicate Wilson's travel pattern for that year:

Month	**Miles by Railroad**	**By Automobile**
January	1,400	90
February	2,165	200
March	1,300	60
April	2,081	50
May	4,389	780
August	2,442	908
Year's Total	**25,496**	**4,350**

The automobile made the favorite Extension method of teaching, the demonstration, feasible. Wilson described an imposing trip out of Havre, July 31, 1916, with Agent Paul Banker:

> Auto tour inspecting farms. Viewed 10 farms, 30 demonstrations, and rode 120 miles. . . about 15 autos participated. Would have been fully 75 had it been held a month earlier. Plans were made twice before for this trip, but rain prevented.

Postwar Adjustment

Director Cooley, in his report for the year ending June 30, 1919, commented, "Extension work during the past year experienced unusual vicissitudes." First, the war worked up to a peak and everyone in agriculture made every effort to provide greater production. Then, rather rapidly, the end of the war came. The efforts of the Extension Service to help guide the adjustment to peace conditions were severely handicapped by the nationwide outbreak of influenza which prevented meetings for several months. The continued need for large production required large plantings, but the drought period from 1917 to 1920 prevented achieving this."

With the close of the war and the reduction of federal funds, Extension activities came suddenly to depend upon state and county support. As Director Cooley laid careful plans to secure the necessary support for continued substantial activity from the 1919 Legislature, he met one of the frequent opposition movements. "The Extension Service has been the object of numerous and bitter attacks by a group of people, chiefly politicians. It was the avowed purpose of these people to destroy the Extension Service, place the county agent work under the direction of chambers of commerce, and prevent the passage of adequate appropriations for Extension work," he said. A legislative investigation was held in which nothing concrete came to light, and the final result, to Cooley's delight, was increased confidence in the integrity of the Extension Service.

Cooley characterized the third period in the year as one of reorganization. He reported with great satisfaction that every county agent active at the end of the war was retained, and only two home demonstration agents were dropped, and this came about because of the acute financial conditions within the counties. A new state statute allowed the county commissioners to appropriate, at their judgment, from general funds or special levies for Extension work, a marked improvement over the previous statute of 1913. This, together with the law authorizing the incorporation of farm bureaus to assist in Extension work, gave strong local support to the movement.

Throughout the period 1914 into the early 1920s, the Extension Service developed quickly under continuously strained conditions. In this early period of organization, war and drought, it had gathered a staff and initiated a program which commanded extensive support even in the face of organized opposition. And during this time the Experiment Station and the Extension Service, working together, had been given sufficient opportunity to build solidly upon sound, experimentally based information. Cooley summarized this movement:

> Among the projects that have occupied a permanent place in the Extension program have been the organization and proper functioning of the county Farm Bureau; the control of destructive rodents, such as ground squirrels, gophers, etc.; Farmers' Institutes and Extension schools; grain and crop demonstrations; forage plants, among which are sunflowers, corn, alfalfa, and sweet clover; the dairy projects, including silos, cow testing associations, bull circles, and general feeding methods; livestock, including livestock standardization, improved sires, and livestock marketing associations; cooperative marketing of potatoes and livestock; cooperative purchase of seed, forage, and feeds; farm accounts and farm management; irrigation; boys' and girls' club work; home labor savers; school lunches; and home problems of food, clothing and health.

This background gave the Extension Service a magnificent array of projects which occupied a permanent place in its programs as it set out on what its leaders hoped would be a period of normalcy.

IV. M. L. Wilson and National Agricultural Policy

A particularly capable group of men have contributed to the growth of Montana agriculture in the closely related fields of research and extension. Four of these men who worked in the formative period demand special attention because of the quality of their leadership. F. B. Linfield was the catalyst who quickly recognized the worth of new ideas and superior ability, and channeled the work of many people toward successful accomplishment. M. L. Wilson was a genius in collecting and refining new ideas and was caught up in the national trends of the day, to which he made a major contribution. J. C. Taylor, third county agent and director of Extension, 1924-1945, consolidated the innovative work of the pioneer period and contributed substantially toward a mature organization. Cooperating with all these men and making their efforts effective through his administrative skill was Alfred Atkinson, president of Montana State College from 1919 until 1937.

The entire period of their constructive work was one of steadily erupting new ideas, of rapidly broadening national influence, of unbelievable progress, and also of catastrophic drought, depression, war and despair. The closeness and almost complete harmony with which these men worked for so long is distinctive in itself. Utterly unlike in many respects, all were conservative, all were highly sensitive to public reaction, all were dedicated and determined. That they were good friends as well as close associates also speaks well for them. The state and the nation will be eternally in their debt.

Wilson Formulates Ideas

The chief innovator, the yeast of the mix, was the thoughtful, restless, friendly Milburn Lincoln Wilson. As has been noted, Wilson was born near Atlantic, Iowa, in 1885. Following graduation from Iowa State College in 1907 with a degree in agronomy, he was financed by his father in the purchase of a small farm near Randolph, Nebraska. Here, with additional leased lands, he enjoyed good crop years and obtained experience as a tenant-farmer. In 1908 Wilson heard enthusiastic reports about the new large

M.L. Wilson, Montana's first county agent.

steam tractor farms near Beach, North Dakota. Wilson came to investigate and found, of course, a vast choice of land which could be homesteaded, with quantities of low cost lease and purchase land also available.

At Iowa State Wilson had known Alfred Atkinson, who came there from Canada to earn a second degree, also in agronomy. Atkinson then came to Montana State College in 1904. By 1907 he was in charge of the agronomy work in the College. The Experiment Station needed help, and he offered a position to Wilson when he graduated. This, no doubt, led to Wilson's coming to the College in his 1908 visit to the northern plains. He was again offered a position; again he declined.

Following a good crop year in 1909, Wilson sold the Nebraska farm at a profit and came to Montana. He homesteaded near Fallon, purchased adjacent railroad lands and leased additional land to make up a 1000-acre unit. Half of this he hoped to irrigate eventually by pumping water from the nearby Yellowstone River. Wilson recounts in his Columbia University *Oral History* narrative how he entered into a partnership with A. C. Gifford, rancher, banker and real estate promoter of the Fallon area. This gave Wilson the opportunity to satisfy his intense desire to experiment with machine farming. His first major purchase was a 110-horsepower Case steam tractor. He and Gifford used this in the spring of 1910, together with LaCrosse disc plows. Wilson wrote Atkinson, "We are pulling 10 feet of plows followed by a disc harrow weighted, then an Emerson clod crusher which is followed by a single disc drill and drag." He had many questions to ask about methods and the typical extensive Wilson correspondence followed concerning experiments, success and failure.[1]

The year 1910 was dry in the Fallon area. Wilson and Gifford had planted much of their land to flax, but the crop was so poor it was not worth harvesting. Wilson found himself financially embarrassed, and in the fall he accepted the third offer from Montana State College. He was given charge of the extensive series of corn judging contests for youth arranged by the Farmers' Institute. These meetings were the culmination of the first year of what was to be an annual project in corn growing by boys and girls with prizes provided by a gift from James J. Hill. From this project Wilson acquired an enduring enthusiasm for corn in Montana, which no doubt came from his early life in Iowa.

In the spring of 1911, Wilson took a full-time position with Montana State as a field agent in dryland farming for the Agronomy Department. He supervised 20 temporary field stations which the Experiment Station had set up over the state. He was allowed time to supervise his own farm and he eventually proved up on his homestead in 1914. During the winter he took a major part in conducting Farmers' Institutes. This work was ideal for putting him into close contact with the problems and the people of the entire state. Wilson came in that opportune time when research and inquiry dominated, and this exactly suited his temperament. He joined enthusiastically in the quest with Linfield—"We must know more about Montana."

Just at this time the Agronomy Department was trying to promote summer fallow. During the summer of 1908 an experimental plot near Forsyth, which had been summer-tilled in 1907, yielded three or four times as much as adjacent traditionally planted lands. However, since the spring of 1911 promised ample moisture, Wilson found that most of the farmers with their Midwestern backgrounds felt that fallowing was a waste, and therefore almost a sin. He could find only four farmers who were practicing summer fallow. Wilson thereupon modified the fallow idea by experimenting with cultivated row crops which would make lighter demands upon moisture. Corn, he felt, would be ideal, or perhaps flax or alfalfa planted in rows.

His first experience with flax had been unfortunate, but he retained confidence in it as a good cash crop to alternate with corn, and he promoted it for a number of years. Alfalfa also had a great potential, he felt, as a row crop to alternate with wheat. He was perplexed that so little had been done in identifying suitable varieties and growing seed when promising varieties appeared.

He entered into this experiment with zest because not until World War I did he feel that wheat was the most suitable crop for Montana dry lands. He believed that the uncertain and relatively low yield on dry lands, low prices, and high freight rates would prevent it from being profitable. In the spring of 1913 he observed that wheat acreages were increasing throughout

A fair display promoting summer fallow.

the world with prices remaining about the same, but he feared they would fall. He estimated that the cost of raising a bushel of wheat in Montana was 56 cents. The average price was about 63 cents. The profit of seven cents would not support the operation.

Corn could be converted into meat in the feedlot, he pointed out, at more profit than raising wheat. For a number of years he searched for seed from many countries of the world, looking for varieties which would withstand drought, wind, hail and other weather vagaries of the plains. He helped develop several varieties which were widely used until the hybrid types appeared.

His interest in machine farming led him to be constantly alert for new developments. The expensive steam tractors could be justified only if they could be equipped with suitable large scale auxiliary machinery. In 1911 he found a duck foot cultivator in Great Falls which he felt would be useful in summer fallow. By 1914, he was conducting demonstrations with this plow. Wilson was completely ambivalent in his use of both Extension and Experiment Station facilities. As county agent and county agent leader, his work was full-time Extension, but he kept any number of Experiment Station personnel busy testing his ideas.

Wilson immediately showed an ability to interest others in his projects and to secure their active support. He had a large part in introducing the age of grant-in-aid since he secured a large part of his own salary in his first few years from flaxseed oil companies. During his first year of service he proposed to the Northern Pacific Railway that it assist in farm demonstrations for producing alfalfa seed in the cultivated row technique. He secured quick rapport with farm groups and with individuals in their homes or in the field. He often made preliminary quick contact with many farm people preparatory to having his field men work with them on projects. Their first question was, "Where is M. L.? When will he be coming this way?" Pleasing in appearance, mild in manner, unhurried, completely sincere, often rambling in speech illustrated with the regular circular motion of one or both arms, his audience or the individual farmer was drawn easily and completely into his orbit of thinking.

The College began closing its demonstration farms in 1912. These were established largely on short term lease to experiment with a single idea. Wilson was enthusiastic about the diversified ranching practices in his early years and wanted to work with individual ranchers and farmers on this approach. Note has been made of his appointment as the first agricultural county agent in May 1913 in anticipation of the Smith-Lever Act. Here he quickly initiated numerous ideas, many of which were to be of enduring significance. He insisted that the farmers themselves make final decisions, that stock ranchers as well as crop farmers be given consideration, that townspeople and businessmen also be aware of farm problems.

Wilson soon proved his flexibility in the face of rapid developments. In 1913 he was not yet aware of the full implications of open range stock operations. He proposed smaller herds under fence, cared for by the forage and grain raised under diversified activities. This made sense to the incoming Midwestern homesteaders who were also interested in launching into the cattle business, but with limited resources. When they quickly got a glimpse of the seemingly unlimited possibilities of the open range stock operations, they abandoned the diversified activities until the severe drought periods forced a reconsideration.

Wilson's faith in diversification was given another hard blow by the World War I demands for wheat at highly inflated prices. His logical reasoning that meat rather than wheat was best for Montana had to be revised within a year. He commented, "Now, because of abnormal world conditions, wheat is one dollar per bushel and hogs five cents live weight at the county shipping point. It is readily seen that this condition in a new state, with farmers eager for cash crops, has given a stunning blow to a well conceived plan of county agent activity in developing farm diversity."[2]

Wilson sensed when he came to Montana that merely solving one problem, then another, without relating them did not produce the answers everyone was searching for. He had been trained in agronomy and he was fascinated with its possibilities in the new land. He wrote to Atkinson in 1911, however, saying he felt that the Agronomy Department was doing excellent work in advancing crop production. But, he continued, "What we want now is the business side of it," and suggested statistical studies in cooperation with the cost of production work in the United States Department of Agriculture, with which he was also keeping contact.[3]

In informal remarks at the symposium in his honor at Montana State University in July 1966, near the end of his eventful career, he recalled this early search. Listing the segments of research such as fallow and mechanized power, he said, "All these were elements more or less known individually but unknown as to their combined potential." And, he continued, "The 'X' factor was management and when that key was found, Montana forged ahead as never before and faster than any other sector of American agriculture."

As county agent leader, he continued to emphasize the importance of the business side of agricultural operations. At the close of World War I he was given frequent leave during the years 1919-1922 to study and teach at major American universities where the new field of agricultural economics was emerging. His ability to combine a theoretical approach with practical work in the field rapidly gave him an important role in this new movement.

The 1922 Triangle Farm Success Survey

Wilson's insatiable curiousity led him to channel many questions to the United States Department of Agriculture. Here he soon came into contact with Dr. W. J. Spillman, chief of the Office of Farm Management. Spillman was as innovative as Wilson, and just at this time was urging that in the new Extension organization, county agents make a study of the successful farms and secure enough actual detail to attempt to chart the major causes for success. Positive application might then be made which would assist farmers who, in the same neighborhood, were less successful. This idea particularly attracted Wilson because he had worked out plans to make just this kind of study in southeastern Montana when he was made county agent leader.

Wilson's enthusiasm for the new agricultural economics enabled him to persuade the administration of the College, by April 1, 1921, to create the position of Extension economist, with himself in charge. In consultation with President Atkinson and Directors Linfield and Cooley, Wilson proposed to work out an adjustment which would assist Montana farmers "to farm so as to stay," as he later phrased it. He made a trip to Washington's wheat empire, to eastern Oregon, and to Alberta, Canada, where he obtained new ideas and found encouragement for some of his older ones.

Considerable attention just then was focused upon the large "Triangle" area extending over all or parts of eight counties in northern Montana. From the apex at Great Falls, this fertile high plains section extended north and west to Shelby and eastward to the Highwoods and the Bear's Paw mountains southeast of Havre. This vast expanse, half the size of Wilson's home state of Iowa, had, in the 1912-1919 period, attracted hordes of Jim Hill's homesteaders to its gently rolling, prairie-like northern plains. During the lush years through 1916 the settlers had prospered.

But, by 1921 the drought of several years had intensified and frequent "black blizzards" shifted the fertile topsoil day by day; discouraged bankrupt "honyachers" were leaving by the hundreds. Journalist Garet Garret reported what he termed a national scandal in a series of lurid articles for national magazines, notably the *Saturday Evening Post*, which he titled "That Pain in our Northwest."[5]

> You may drive through the Triangle now for 40 or 50 miles along the new bonded Roosevelt Highway, and seem to see nothing but abandoned towns, abandoned farms; abandoned banks along the road with pieces of white paper pasted on their door panes, and steam tractors sinking in the fields like lost locomotives.

However, sprinkled rather generously over the area, seemingly without distinct advantage of location or proximity to water, were ranches where the operators were holding their own, even if they were not making striking advances.

The Northern Montana Branch Experiment Station had been established south of Havre in 1915 and the superintendent was discovering some of the basic characterisitics of the region. William Reed, a farmer on Big Flat near Turner, had also organized a Summer Fallow Club in 1921, which was verifying Extension's urging that new methods could make a difference. J. W. Freeman, a member of the State Board of Education from Great Falls, who was deeply interested in dry farming, secured a grant from the state board for a farm success survey to be made in the Triangle by the Extension Service with M. L. Wilson in charge.

The survey began on April 27, 1922. A list of 200 farmers who had managed to stay on was compiled by banks, insurance companies, businessmen and county agents. This was reduced to 130 who were still either free of debt or had an indebtedness which was normal in terms of expansion and easily within their ability to repay. An extensive questionnaire was prepared which was taken to each of the cooperating farmers by a small trained staff to obtain the greatest amount of uniformity.

The qualifications and experiences of the farmers and their families came under scrutiny, as did types of crops, number of acres and yields, marketing, farm buildings, farm power, implements, types of tillage, diversification of crops and livestock, shelter belts, irrigation and other pertinent information. Interpretation was made against the background of the work of the Extension Service and the Experiment Station. This included several findings by the Northern Station, most important of which, perhaps, was that the soils of the great basins in the northern plains retained high degrees of fertility since they had not been leached by excessive rainfall. An intensive study of weather conditions over a 40-year period showed that the current dry spell of some six years in length was not abnormal. Then M. A. Bell's famous tree ring study made in 1936 traced periods of damaging drought back as far as 1784. The influence of Extension agronomist A. J. "Ole" Ogaard's continuing work on summer fallow, strip farming and mulching experiments was also incorporated.

Only 26 farmers appeared to be successful from a strict viewpoint. These were the farmers who, after paying all expenses and allowing for depreciation on buildings and machinery, made 6 percent interest on their investment and $500 or more additional for their labor.[6] Insofar as was possible, a study was also made of the farmers who had failed to operate profitably. These findings supported the conclusions made from the study of the successful farms.

The findings of the survey were summarized as Extension Service *Bulletin No. 66*, "Dry Farming in the North-Central Montana Triangle," in 1923. Five principles were formulated from the study to guide the future development of the area:

1. Build reserves of soil moisture, feed and cash during good years for use in less favorable years.
2. Wheat was recommended as the most suitable cash crop for the area, but to provide a certain margin of safety for the dry years, diversified farming should be adopted with farm gardens, milk cows, chickens, turkeys and pigs to provide supplemental income.
3. Grazing on a large scale should be adopted when tillable lands were unavailable in sufficient quantity or were marginal in productivity.
4. Since the yields per acre were likely to be low, the largest units of tillable land possible should be acquired. Use of horses or tractors should be geared to produce the greatest amount of power possible at the lowest possible cost. Tillage methods such as summer fallow should be adopted, not only because this conserved moisture, but also because it broadened the labor pattern and allowed one farmer to cover a larger acreage efficiently.
5. Moisture conservation should be practiced in every way possible, not only through tillage practices but also through irrigation from dams storing flood and waste water. Stock water dams, where feasible, should also be constructed.

This Triangle study became a landmark, incorporating as it did many of the new methods of production, and the idea that management might be the key, the "X," factor to a successful operation.

A Department of Agricultural Economics, 1925

The importance of the field of economics and management was recognized in 1925 when the Purnell Act made federal funds available for the establishment of a Department of Agricultural Economics in each of the land grant colleges. M. L. Wilson, naturally, was made head of this department which incorporated the functions of college teaching, the Experiment Station and Extension work in this field. This gave Wilson the staff, and through an excellent capacity to delegate responsibility he obtained the time to work with the ideas which were being generated with astonishing rapidity in the state and nation.

The attempts of the economists to discover how larger profits could be made on the wheat crops in the Triangle area were hampered by the distressingly low price of agricultural products. To help examine and ease this situation, national in scope, the Bureau of Agricultural Economics, which had been established in 1922, instituted an annual Outlook Report. This was based on a study of national and international supply and demand. As the report was issued each year, a meeting of state leaders was called in Washington to assist these men in interpreting the report in their states. Each state then issued a report relating it to local conditions. The first for Montana was prepared in February 1925, written by M. L. Wilson and H. E. Selby. Wilson felt the Outlook Report method of informing the agricultural community was highly important. In Montana a similar report was often issued monthly until 1941 when the War Board program and production goals took its place. It was revived in 1946 as a circular titled "Looking Ahead with Montana Farmers" and was continued until 1959 when changing conditions made it unnecessary.

The implication of the Outlook Reports was a voluntary reduction or increase in production to meet changing market demands. Wilson had this in mind when in 1927 he was called upon to write Extension *Bulletin No. 84*, "An Agricultural Program for Montana." This came out of a two-year study and *Bulletin No. 81*, "Basic Facts About Montana's Agriculture," issued in December 1926, which was studied intensively over the state for several months. Better years had come again and production had gone up strikingly. Farm prices continued to be low, however, and the specter of recurring drought was always present. *Bulletin 84* suggested that four lines of action appeared to offer promise in working out a viable agricultural program for Montana: 1. national legislation; 2. cooperative marketing; 3. increased efficiency; 4. adjustments in production.

The *Bulletin* stressed the fourth suggestion since it was almost wholly and immediately within the control of the rancher. The question of diversification arose again, and the recommendation was made that on dry lands emphasis should be placed on wheat. Just four years before, the Triangle study had recommended diversification wherever possible. The open-mindedness of the agricultural leaders in Montana later enabled them to keep abreast of national policy and make an impact upon it. A complete departure from diversification was not made, however,and livestock were recommended wherever feasible, but a high priority was given to wheat.

The *Bulletin* also reviewed methods and urged that one-third of the wheat acreage be held in summer fallow each year, one-third be planted on summer fallow, and the remaining one-third planted on clean stubble where a crop had been grown on summer fallow the preceding year. "Stubble mulch" was receiving favorable attention at the time. The Triangle study had demonstrated that one of Wilson's objections to wheat, that of high labor costs, could be substantially reduced by the summer fallow methods which spread the labor time of preparing seed beds over an entire year. The total time required was also greatly reduced by the new dry land implements and the new sources of power such as the big hitch for horses and the new tractors.

The Fairway Farms

The discussions leading to the formulation of "An Agricultural Program for Montana" in 1927 had their background in another experiment. This was the Fairway Farms project which began in 1924 and continued for a decade or more. The major emphasis of both the Experiment Station and the Extension Service had been, up to the close of World War I, upon increased production. The tremendous expansion during the war had caused large purchases of land and machinery involving heavy indebtedness. When the demand and the price fell, and drought came, widespread bankruptcy followed. Then the Montana wheat farmers, for instance, farmed more intensively and produced in 1923 a total of 52,486,000 bushels in comparison with 40,360,000 bushels in 1920, and the prices dropped further.

The lack of flexible credit facilities had long troubled farm leaders. The number of rural bankruptcies in Montana in the early twenties became so large that banks and insurance companies also experienced financial disaster. One of every two mortgaged farms in the state was foreclosed upon between 1921-1925, and more than one-third of Montana's banks closed between 1920-1924. Even where credit was available, the problem of advancing from a farm laborer to a tenant and then to farm ownership was extremely difficult. Farm leaders feared that America was developing a farm peasantry as a permanent part of our system. Farm leaders were looking for many answers.

The idea of the Fairway Farms originated with Dr. H. C. Taylor about 1919 when he was professor of agricultural economics at the University of Wisconsin. Seeking to apply the principles which he taught, and using his own funds, he established in the midst of many unsuccessful farm operations near Madison, Wisconsin, an efficient unit where the amount of land, the types of crops, machinery and methods were successfully woven into a profitable dairy farm. The operator who first rented soon became a tenant-purchaser and then owner in rapid succession.

Taylor then conceived the idea of applying this plan on a large scale. He dreamed of a $1,000,000 corporation which could continuously develop suitable units for a single-family operation, incorporate good management practices, and enable the tenant through flexible and long-term credit facilities to advance to ownership in a "fair" way. Just at this point Taylor was called to Washington to help reorganize the Bureau of Agricultural Economics. In 1923 he was secretary of a group which presented his plan to the Laura Spellman Rockefeller Memorial Foundation. Its director, Dr. Beardsley Ruml, was interested. Ruml had grown up in the Iowa corn belt and had been dean of social sciences at the University of Chicago, another cradle of the new rural economics and rural sociology.

Before Taylor had finished his formal proposal to the Foundation, he came to the northern plains in October 1923 to investigate the severe drought conditions. In Montana his host was Chester C. Davis, the state commissioner of agriculture. Davis was thoroughly familiar with Wilson's Triangle study and he pointed out the successful farms and briefed Taylor on the findings recently published in *Bulletin No. 66.* Taylor immediately saw that model experimental farms could be developed here which would serve as models for the Great Plains as his dairy farm near Madison had served for the Midwest. Wilson was asked to come to Helena for an exchange on how management might assist the farms and farmers of northern Montana. The three men were guests of Governor Joseph Dixon in the executive mansion, and during a long evening they discussed many plans for improving agriculture in Montana.

Wilson was easily sold on Taylor's idea, and upon his recommendation the Rockefeller Foundation offered a grant, which Wilson preferred to have as a loan, of $100,000 for the project. Additional loans totaling $50,000 were made. On March 24, 1924, in Bozeman, Articles of Incorporation were drawn up for a Fairway Farms Corporation. To give the corporation a statewide character, J. L. Humphrey, a Great Falls investment banker with considerable unhappy experience in financing dry land farms, was chosen president. Chester Davis was to be vice president. M. L. Wilson was to be secretary of the corporation and manager of the project. These men, together with Augustus Vaux, a successful farmer and stockfeeder from Sidney who was named a director, comprised the executive committee. Charles Vandenhook, president of the Commercial National Bank in Bozeman, became a director and treasurer, and Walter Hartman of Bozeman was named attorney for the company. To give the corporation the national character which the plan envisioned and which the foundation desired, three nationally known economists were named to the board: Henry C. Taylor, Dr. Richard T. Ely of the University of Wisconsin, and Dean L. C. Marshall of the University of Chicago. Wilson had worked closely with all of these men.

Organizational work continued throughout 1924. The assistance of county agents, banks and insurance and loan companies was sought in locating desirable plots over the state which would provide variety in size, soil, tillage characteristics, availability of water and, in some cases, adequate grazing areas. Tracts with suitable farm buildings proved to be particularly hard to find. Farm buildings were moved, repaired or constructed, and wells were drilled to provide the greatest convenience and efficiency possible. A procedure of purchasing smaller home units and leasing adjacent tillage or grazing lands appeared most economical and practical. After a year of search, nine tracts were obtained.

Finding suitable tenants was equally difficult. To make the experiment valid the officers attempted to secure families of "average" capability. However, they also felt that this should include experience and responsibility. A great deal of attention was given to the conditions of the contract with the tenant, and the time table designating when various provisons would take effect. Contracts varied depending upon the size and value of the farm tract and the estimate of its productivity. The Federal Farm Loan system, set up in 1916, required that the tenant have at least a 50 percent equity in his operation before it advanced loans. The corporation wished to experiment with lower requirements, but it did require that the tenant have capital in some form. This could be in cash, machinery or livestock, and was considered an initial payment in the transaction. Then a tenant-purchase contract suitable to both parties was drawn up. Many of the provisions of these contracts were later adopted wholesale by the Farm Security and Farm Home Administrations.

The project suffered a severe blow when its continuity of management was broken by Wilson's departure for Washington on August 1, 1924, before the choice of farms and tenants was completed. At Henry C. Taylor's urgent request, Wilson went to head up and consolidate the work of the Office of Farm Management and Cost Accounting in the Bureau of Agricultural Economics. Sherman Johnson served as acting head of the Department of Agricultural Economics at the College in Wilson's absence, and gave over-all direction to the Fairway project. Virgil D. Gilman, farm management demonstrator, assumed much of the detailed management. J. L. Humphrey, president of the corporation,also gave a great deal of his time to the enterprise.

Wilson remained in Washington until July 1926, but he returned several times to advise with the farms, and he kept closely in touch through his usual extensive correspondence. In a report of progress for the first two years, he summarized, in 1926, the purpose of the Fairway Farms:[7]

> The Fairway Farms Project is an experiment in farm tenancy and farm ownership. Its objective is to facilitate the climbing of the "agricultural ladder"... It was organized for the purpose of carrying out the first of what, it is hoped, may prove to be a series of experiments to find and demonstrate better practices in adjusting the relations of farmers to the land... It is quite as much a demonstration of the proper size of farm, of methods of selecting and testing men as tenants and prospective owners, of the efficient method of equipping farms, and of the best type of farming suited to each of the regions in which the farms are located, as it is a demonstration in fair methods of adjusting the relations of landowners with tenants who wish to become landowners. Through all these problems which the corporation has attempted to solve, the farmer's as well as the nation's, runs this idea of better standards and methods in the farming business whereby capable men may more easily climb the "agricultural ladder" to ownership.

Wilson was absent again in the crucial summer of 1929 when he went to Russia to advise on their large scale wheat production projects. By 1930 he became engrossed with national policy into which he was drawn almost full time. In 1932 he was active in national agricultural planning, and with the formation of the Roosevelt administration in March 1933, he went to the Department of Agriculture. Elmer A. Starch, who had come to the Department of Agricultural Economics in 1926, soon became Wilson's chief assistant with the Fairway Farms. He also became head of the department in 1933 and assumed general charge of the project.

Wilson was historically minded and was intrigued with the Indian lore of the region. He had hoped that each of the nine farms could be given an Indian name from an adjacent tribe with the appropriate meaning of "Fair Way." Many of the suggested names were unpronounceable and the idea was used in only a few instances.

Only one farm was located in western Montana—the Woanlo farm of 360 acres located just south of Heron in Sanders County near the Idaho border.[8] The test here was to see if the large cutover forest areas could be adapted to grazing and dairy purposes. The soils proved to be poor and the work was strenuous, slow and expensive. The well recommended tenant-purchaser was handicapped by illness and became over-extended by his expansive tendencies. By 1943, however, he completed the purchase of a major part of the farm, and his operation answered many questions.

All of the other farms were in the northern and eastern parts of the state because experiments with arid conditions had highest priority. The Cloverleaf farm in the Triangle area, 10 miles southeast of Rudyard, was of great importance. A home unit of 1,440 acres was acquired at a cost of $9,385, and additional land was leased to make a four-section tract. Swen Twedt, an intelligent, experienced farmer, was given a tenant-purchaser contract. He possessed goods valued at $1,400. Wheat yields averaged 10.08 bushels per acre from 1925 to 1930, and by 1929, Twedt had accumulated an equity of $7,000. Then the drought deepened and by 1936 Twedt was indebted to the corporation and to banks and merchants in the amount of $17,645. The total value of the Cloverleaf holdings at the time was estimated at $11,662.

The circumstance of a drought which hit so hard that even a capable and established farmer such as Twedt was unable to escape complete financial disaster made a crash impact upon the thinking of Wilson and his associates, many of whom were in national policy-forming positions by this time. An evaluation group said of Twedt and the Cloverleaf in 1937:[9]

> Everybody in the party agreed that Mr. Twedt was doing a good job of farming the land. With yields which he had during the first four years, he would have paid out in a few years and owned the farm, but the conditions during the past six years have run him into debt far beyond the value of the land. . . This is the one farm now in the possession of the Fairway Farms Corporation that can be said to be operating in accordance with the original plan. Swen Twedt, the operator, a splendid farmer, is so overwhelmed with debt that there is no possibility of his paying out unless some debt adjustment is made, and not then, unless better years are ahead.

This is one of several stories from the Fairway Farms, however, which had a happy ending. Adjustment was made in 1943 with a new purchase contract for $11,500. Better years came and Twedt did become the owner of the Cloverleaf. His heroic effort has been a legend in the Triangle since that time.

The Comanche farm in the "Wheat Basin" some 25 miles north of Billings fell upon hard times in the drought, but the purchaser completed his contract by securing employment with the Resettlement Administration for several years. A Richland farm, seven miles northeast of Sidney, defeated two tenants and tested a third severely until the rains came and he was able to complete his contract, expand his holdings, and share in the proceeds of two oil wells drilled on the farm. A Rosebud farm in the Yellowstone Valley near Forsyth was designed to test corn-hog operations, but the lands proved to be unsuitable and the project was soon abandoned. The Wyota farm, two miles south of the town of the same name, was all under irrigation. A hay-dairy enterprise suffered many adversities before the tenant completed the purchase at a considerably reduced price. The Taiyan farm near Fraser, and not far from Wyota, experimented with alfalfa and suffered the same difficulties of drought, cold and several tenants. The Yanktonai farm near Whitetail proved to be so distant and so unpromising that a suitable tenant was not found and it did not get beyond the planning stage.

The Lone Warrior farm on the Fort Peck Reservation near Brockton provided more experimental data than any of the other farms. Two and one-half sections were purchased and an additional section and one-half were leased. The Campbell Montana Farming Corporation had raised wheat on these sections during the period 1917-1922 as a war project, and the quality of the soils was pretty well-known. Delay occurred in securing title and leases from the Indian Service, and in the meantime the corporation had encountered some of the tenant problems previously noted. No house was available and wells were not reliable. The corporation then decided to place this farm under management from the Experiment Station, using employees who would follow orders closely and keep detailed records of the experiments undertaken. This farm went directly to the new gasoline-powered tractors. Machinery companies made gifts of new implements which assisted the experimentation greatly. Elmer Starch gave detailed direction to the Lone Warrior in its early years. He commented:[10]

> One of the Fairway Farm units was converted into a specialist job analysis management technique laboratory. It was decided that the operations of the farm would be evaluated job by job. For instance, a plowing job done with each of two, three, four or six bottoms would be measured in terms of time and cost per acre for each . . . About 3000 job analyses were run during the years 1928-1936, and other verification tests were run in 1931-1932.

Starch's carefully supervised operation resulted in Experiment Station *Bulletin No. 278,* "Farm Organization as Affected by Mechanization," published in May 1933. This was a landmark study which showed the farmer in some detail what he might expect to achieve in the transition to the new machines.

On all of the Fairway Farms a number of the newer ideas which had emerged in the Triangle farm success survey and elsewhere were used. Strip farming was given exhaustive trials. Various types of mulches were used on these strips to assist in the main objective, which was to conserve moisture. The former favorite, the clean dust mulch, proved to be defective since it blew easily. The clod and stubble mulches gave the appearance of careless farming, but resisted blowing. This type fitted particularly well into the new shallow plowing which was found to be desirable, and for which new cultivators were quickly invented. The costs were at least $2.00 per acre less than by the older methods.

Summer fallow, especially when produced with the new wide swath machines, served a second purpose—that of enabling the single-family operator to use all of his time to full advantage. During what were otherwise slack summer days he was busily engaged in preparing the seed bed. If yields were to be low in arid regions, then more acres must be planted, and this was a way to do it. This summer operation assisted in still a third way by having the ground ready for either fall or spring planting at the most advantageous time. M. A. Bell's tests at the North Montana Branch Station had been conclusive in the value of the timeliness of planting and harvesting. Bell demonstrated that earlier planting by as much as two weeks—about May 5 rather than May 20—produced as much as seven and one-half more bushels of wheat per acre. He also showed that an earlier date for the beginning of summer fallow plowing—in May rather than in July—resulted in an increased production of up to four bushels per acre. The element of timeliness alone could make the difference between a comfortable profit or a crippling failure.

Wilson's continuous search for greater sources of power led him quickly to the various hitch arrangements used in Oregon and Alberta, which he developed in his big team hitch. By this device, from six to eight to 16 and 20 horses might be hitched to a single machine. This had been used in the Triangle and the early Fairway Farms were equipped with many big teams. Almost immediately the new medium-sized internal combustion gasoline tractors became available and a rapid but expensive shift was made. Starch, in his *Bulletin No. 278*, comments that the new lighter tractors appeared in 1924, and by 1930 there were some 19,000 on 14,000 farms. The new lighter combines appeared in 1925 and 2,195 were in use by 1929. The need for horses declined further by 1929 when 14,615 trucks were in use in the state. This was a major revolution, and the changeover was costly. In the end the advantage of motorized power and transportation on a wheat ranch was substantial, since maintaining a large number of horses was expensive and time consuming.

In a reconsideration of the Triangle survey, Wilson made a considerable shift from his former position of emphasis upon diversification. He became persuaded that usually when the major crop failed, the diversified crops also failed. By field trials he double-checked experiments, again using corn, flax and alfalfa, all in row crop plantings, as an alternative to summer fallow on the Fairway Farms. As a result, specialization became accepted practice, especially in relation to wheat. Wilson and his associates continued to have an interest in partial irrigation in order to provide a base for

Big team hitch in action pulling a combined harvester and thresher. (With diagram insert)

dairying and limited grazing, and this practice was later tested extensively.

The Fairway Farms were not successful financially. Indeed, no farms in arid areas were profitable in the terrible years of the experiment. But the project yielded valuable information on dry farming techniques and farm management just in time for further research on a national basis. The Rockefeller Foundation turned over its equity in the Fairway Farms to the Farm Foundation in Chicago, largely because Henry C. Taylor, who had initiated the project, was now its executive director. The Cloverleaf and the Lone Warrior continued to be preferred farms. They were sold in the early 1940s for substantially more than the purchase price.

A major result of the experiment was M. L. Wilson's firm conviction that so long as major regions of the nation were subject to catastrophic forces, their problems were of national concern.

Unfortunately, none of the principals have written in detail their experiences and reflections on the Fairway Farms. Since many of these men continued in positions of leadership in agriculture, the best record is in their thought and action. Wilson wrote in November 1929:[11]

> The past five years have been probably the most dynamic in the history of northwest farming. There has taken place a complete and almost revolutionary shift from the kind of farming which has been carried on here for 30 years over to motorized mechanized farm organization. This change has introduced all kinds of problems. . . We are of the opinion that the Montana State College and Fairway Farms Corporation are the only agricultural

agencies in the United States that have given any considerable attention or research to the problem of farm reorganization which meets the changes that are taking place through the introduction of large scale low cost machinery.

Elmer Starch commented in 1969:[12]

> Fairway Farms is thought of as just a string of farms where some modifications in the "usual" were being adopted. Actually they were a laboratory where management principles were recast in terms of the anticipated potential of mechanization, specialization, new forms of financing, etc. They were not just farms; they were the "wind tunnels" in which new theories were put through the proving stage.

Russell Lord, in his *Wallaces of Iowa,* has yet another interpretation:[13] "These farms were not only experiments in rehabilitation. They were experimental plantings in applied economics. They were, as nearly as I can determine, our first definite American experiment in regional land use."

The farm management experiments of the day also incorporated the earliest sustained attempts to focus upon human labor as a basic factor of real value in agricultural cost accounting. In 1924 labor had been estimated at 15 percent of the cost of producing wheat. By 1930 this was reduced to 7 percent and in some cases as low as three percent. Wheat was being produced on the Lone Warrior with only *three minutes* of human labor for each bushel. By the older "peasant" methods of hand and horse labor, at least *three hours* were required. Five years earlier on the same farm, 10 to 12 hours would have been required, and the average over Montana then was estimated at seven. By rather detailed calculations the labor costs for producing wheat had been cut by two-thirds and the total cost by one-half. Starch jubilantly recalled that although "through all time the cardinal principle of farm life was many crops and long hours," this no longer prevailed.

An increase in the size of the farm unit was a result of the fuller utilization of labor. The determination of the proper size was a major objective of the Fairway Farms. A first recommendation was that an 800-acre unit was a minimum for a profitable single-family farm, and all of these acres should have a high productivity. Such a unit could support machinery up to a value of $4,000. The Federal Farm Land Bank accepted the 800-acre unit farm with a three-plow tractor and good management as a base for their loans. *Bulletin No. 278* enlarged this recommendation and held that a 1,600-acre unit, 800 in fallow and 800 in crops each year, was more than twice as good as the single 800-acre unit, and that a 3,000-acre unit was still better.

The operation of an acreage of this size with the accompanying machinery immediately raised the question of credit, and Wilson, with considerable regret, recognized another aspect of the changing times:[14]

> We, therefore, see but little place in Montana for the wheat farmer who doesn't have sufficient capital to undertake such an enterprise and as far as our wheatlands are concerned we are inclined to believe that our problem is more a recombination of farms and consolidation of individually owned

tracts of land, so as to make the land fit the complex of machinery.

Tangible results of the Fairway experiments were made available as rapidly as possible. Two special trains, one in 1929 and another in 1930—"Low Cost Specials"—carried as many samples of the new machinery as could be wedged in. Through pictures, lectures, and special bulletins and demonstrations the new methods were described. The annual fairs were widely used. *Bulletin No. 278* attempted to include enough detail to enable ranchers everywhere to determine:

> How large a land unit was necessary to support a family; what type of equipment should be purchased and how much; then with this equipment, how large an area can the family labor operate; what is the best combination of land, labor and equipment; what low cost methods can be incorporated to advantage; and finally, how can each job be analyzed for efficiency.

The wide interest which prevailed in the efforts to improve the agricultural situation is often illustrated by the story of the 1929 Field Day on one of the farms when plans were made for 200 visitors. When 2,500 came, Starch recalled, "We were embarrassed when the well went dry at 1:00 o'clock."

The decade of the "troubled twenties" which brought the realization that prolonged periods of drought were also normal, and stimulated the various imaginative attempts to meet these periods, did place the Great Plains on a sounder basis than ever before. For many years farmers had thought of 160 acres of land as providing the ultimate in security, all and more than a single family could manage with the methods of the day. But on the low

Steam engine pulling an early combine, 1919.

yield plains this was not enough. For many years farmers had incurred indebtedness with a firm commitment to meet annual payments. On the plains this could result in disaster for lender and borrower.

The grasp of the multiple problems of the Great Plains and a demonstration of improvements by the Montana planners was the essence of their genius: low production per acre, therefore greater acreages; increased acreage, therefore greater sources of power through the big hitch with horses, or with the new machines; summer fallow to preserve moisture and to utilize labor to the fullest; shallow plowing and rough mulches—fast, economical and the best practice; flexible and long-term credit; the combination of irrigation, dry farming and grazing as an adaptation of diversification; the emphasis upon the age-old admonition that the laborer, including the owner and his family, is worthy of his hire—any one of these could produce a major change. All of the improvements focused in a single operation with capable management produced a revolution. But if, in spite of the best efforts, conditions such as drought or low prices, disaster still stalked an entire region, then the appropriate governmental unit should render assistance.

Four Germinal Ideas

The Montana Experiment Station and Extension Service men were well pleased with the advancements which had taken place in production. They were still concerned about the narrow margin of financial security which agriculture in the plains retained, and therefore the constant threat of tragedy in human terms. The search continued for methods of adapting single-family farming to the Great Plains with security, profit and largely independent action. When the striking technical advances still did not insure this, Wilson began to look increasingly toward cooperative effort, and, with reluctance, toward implementing government assistance if and when necessary. Four developments of significance were taking place in Montana in this period.

One was a land classification study based upon physical characteristics. This was begun by soils specialist Leonard Gieseker in 1921, and continued by him for 30 years. Gieseker established four grades of farm land and five grades of grazing land. Where this classification scale had been completed before the drought years, the farm management people found it to be extremely useful. It established a scientific base for determining the best use of the land, its sale value for mortgage negotiation and taxable valuation.

A second movement was the organization of the Mizpah-Pumpkin Creek Grazing District southeast of Miles City. Many private owners had developed traditional units during the moist years spread over 108,000 acres. Interspersed were tracts held by the state, the Forest Service, the Northern Pacific Railway, and by the county through earlier tax delinquency—all tightly fenced. Drought and low prices rendered the small private holdings unusable. The cooperative grazing district association formed in 1928 placed the entire area into a single management unit and restored productive conditions. This experiment provided the background for a State Grazing District Law in 1933, which in turn stimulated the enactment of the federal Taylor Grazing Act of 1934.

A third inquiry was related to taxation and tax delinquency, closely associated with the numerous mortgages and their foreclosures during the dry years. In Teton County, the county agent, Robert Clarkson, and the county assessor, Otto Wagnold, gave special attention to the fact that certain lands kept showing up regularly on the delinquent tax rolls, with resultant loss to the owner and to the county. They studied the Gieseker land classification charts and held a series of community meetings which eventually covered the county. A careful study of the farms and ranches, field by field, was made by the owner and his neighbors. The accuracy of the classification was checked, known and estimated yields considered, and possibilities of improving the land and its use were weighed. The current taxable valuation was often adjusted to a large degree. This led to a revision of the base and method of establishing the taxable evaluation and levying taxes, which was adopted by the State Board of Equalization. This study provided Wilson with additional ideas concerning land use adjustment, farm security measures, soil conservation practices, and other innovations which were being considered.

A fourth idea was a do-it-yourself tax assessment on the part of land owners and operators. This resulted from the extreme need for economy in county government and the difficulty of securing a large number of field assessors who could reach a satisfactory agreement with landowners and farm operators. At about the same time, county assessors, James Adams in Sweetgrass County and C. A. Bishop in Stillwater County, came up with the idea that the taxpayer should prepare his own assessment list. Safeguards provided included a possible spot check by the assessor's office, open books in the office for a neighbor's inquiry, and possible publication of the voluntary assessment in the local press. Violent disagreement as to the feasibility of the plan arose, but it was adopted. Everyone was surprised when the completed reports showed an increase in the taxable valuation of the county and rather close accuracy when spot checks were made. The method was adopted by the State Board of Equalization for use throughout the state. This experiment gave Wilson confidence that the American farmer would rise to the challenge of full and honest cooperation in tabulating production allotments.

Formulating National Policy: Domestic Allotment, Subsistence Homesteads, Resettlement

Agricultural instability continued in the state, and when a farm bloc originated in Congress in 1921 and the McNary-Haugen Plan emerged, the

Montana agricultural leaders became increasingly involved. The Montana men came to favor the Peek-Johnson plan which called for some type of export corporation which would use the tariff approach to assist agriculture. In the state this soon took the form of the Montana Agricultural Export Corporation League. Much of the publicity work was done by M. L. Wilson, Chester C. Davis, Lester Cole, editor of the *Montana Farmer,* and W. L. Stockton, president of the Montana Farm Bureau Federation. The Montana league was important in this movement in the Pacific Northwest and gained national recognition.

Wilson's service in the U.S. Department of Agriculture, 1924-1926, was highly important since agriculture continued to be in trouble. President Coolidge steadily vetoed the several McNary-Haugen bills, but a number of creative minds were at work in the department. Wilson added his ingenuity to that of Henry C. Wallace, former secretary of agriculture, his son, Henry A. Wallace, whom he had known at Iowa State, and also Henry J. Spillman and Howard Tolley. He also continued to work closely with Dr. Henry C. Taylor, partially responsible for the Fairway Farms. Russell Lord, in his *Wallaces of Iowa,* draws attention to the early development of New Deal ideas and of Wilson's part in them:[15]

> Many measures which have come to be thought of as New Deal measures were taking form in the United States Department of Agriculture back around 1924. The devices which Wilson took a lead in developing were: tenant rehabilitation, experiments in land planning, balanced harvests by domestic allotment (later the Triple A) method, and subsistence homestead projects of rural-urban design.

This story of Wilson's influence in national agricultural policy is only indirectly related to that of the Montana Extension Service; yet, because it is a seldom-told story, it is included since his ideas sprang directly from his experience in the service. In addition to the Triangle Study and the Fairway Farms experiment, Wilson was greatly impressed in 1925 and 1927 by the cooperation of the Montana farmers in formulating "An Agricultural Program for Montana," in which some 1,200 farm people participated and recommended adjustment in production, in the business organization of individual farms and in the general practices of definite communities. His feeling for democratic action is expressed in his comment:[16]

> When such programs are developed in each community, and when each community can be better informed as to its advantages, its competitors, and its place in the whole scheme of agricultural production, then all farmers will be able to make sounder judgments as to what and how much to produce, and how the farm as a whole may best serve the needs of society and will be a successful business unit returning an income which will enable the farm family to get its share of the good things of life.

When the crash of 1929 deepened into another period of lowering agricultural prices, Wilson, Tolley and others looked toward legislation more dynamic than that being requested by the Hoover administration. Wilson's visit to Russia in 1929 caused him to reject his former export plan.

An interesting arrangement for feeding two cutter bars instead of one feeding into a single thresher unit.

He felt that Russia's gigantic plans for agricultural production and increasing tariff controls which he sensed in Western Europe would nullify large and continuous exports from the United States.

The important movement toward domestic allotment began in 1930 when Wilson and Tolley supplied many ideas for the Christgau Bill. Victor Christgau was a congressman from southern Minnesota with a farm background and training as an agricultural economist with John D. Black at the University of Minnesota. Black had gone on to Harvard and had become one of the inner circle of the "new idea" men. In 1929 he published *Agricultural Reform in the United States* in which he outlined a version of the domestic allotment plan that he credited to Beardsley Ruml of the Rockefeller Foundation. Wilson used reprints of this section extensively to publicize the plan. Elmer Starch had been a classmate of Christgau's, whom he introduced to Wilson and the agricultural brain trust. Christgau understood their ideas and warmly supported them.

The Christgau Bill called for "an agricultural adjustment program designed to lead to highly efficient and profitable farms," says Richard S. Kirkendall in his *Social Scientists and Farm Politics in the Age of Roosevelt.* The idea was to help farmers in various regions determine the commodities that they could produce to the greatest advantage wihout creating a surplus. Very poor farm land, where possible, would be acquired by the federal government and retired from production. Costs, prices, techniques of production, farm organization, land-use practices, present and prospective supply, demand markets and other aspects of what was just becoming recognized as the highly complex business of agriculture would be explored and reported. The secretary of agriculture was to divide the country into major areas related to production or types of farming in which the Great Plains would, of course, be designated as a wheat and livestock area. Recommended systems of farming were to be tested under actual farming conditions. In addition, an enlarged Extension educational program was proposed to bring the results of these studies to the practicing farmer and rancher.[17]

When the point was raised that such a program would require an army of people, Wilson replied that the army was already in the field in the form of the Experiment Station, Extension Service, Forest Service, and other related agencies, covering almost every county in the nation.

Such a program as that outlined in the Christgau Bill called for extensive organization and planning. The term was offensive at the time because of its association with Russia's various plans. Wilson had seen some of these rigid programs and did not care for them. But, he insisted, planning, both long and short term, was as important in agriculture as in industry, and he had seen Montana farmers plan in a democratic fashion. Many of his critics in Montana and elsewhere were labeling him "red" when he was standing almost alone in his insistence upon democratic controls. When his colleagues in the Department of Agriculture insisted that such a complex program could not possibly succeed unless the decisions were made at the top and controls extended throughout the operation, he was equally insistent that planning could be decentralized. The voluntary features of the New Deal programs were incorporated because of his unrelenting conviction, although this required taking the issue directly to President Roosevelt, who supported Wilson's point of view.

Wilson realized, of course, that a regional and national organization giving direction and supervision was necessary since the farmers were basically unorganized. Edward J. Bell, Jr. recalls an evening meeting of wheat ranchers in Bozeman in the fall of 1929 when the appeal of Alexander Legge, chairman of the U. S. Farm Board, for curtailed planting was being discussed. Wilson proposed the question as to how many of the ranchers would be willing to reduce their acreages. Zales Ecton, a large dry-land wheat rancher, later United States senator from Montana 1946-1952, arose and said he would be willing to limit his planting substantially, but this would not be effective unless other wheat growers in the state and nation did the same thing. On the way home Wilson remarked with conviction that this was the crux of the problem, and only a national plan of cooperation would solve it.[18]

At about this same time Wilson began to urge greater consideration of the various types of land and an adaptation to the best use which could be made of each of the types. He was finding the Gieseker survey to be highly valuable. By 1930, when the Christgau Bill was introduced, Wilson had framed a companion bill. This bill proposed "a more efficient utilization of all of the lands in the United States, but especially those not adapted to private ownership.'' Items contained in the bill were: The federal government should give direction, and in cooperation with the states, develop a national land inventory, and conduct fundamental research on land use problems; the federal government should embark on a modest program of acquiring sub-marginal lands and those which might fit into special use; it should coordinate projects for both irrigation and drainage into a national land use program; and it should formulate broad programs of scientific research pointed toward providing the best use of the nation's land. Wilson's bill, of course, incorporated many of the functions now carried out by the Bureau of Land Management.

Wilson proposed that the place to start was with the repeal of the Homestead Act and other national land laws which enabled individuals to obtain agricultural lands without regard for capability for their management. ''Certainly the time has arrived,'' he insisted, ''when we must reverse our traditional historical land policy from that of expansion to one of adjustment and perhaps of contraction of the tilled area.''

Wilson could not keep from planning for the use of the marginal lands which would be taken out of production, and for the people who would be displaced when this was done. His first plan was that of the subsistence homesteads. The name was unfortunate since to many this was a situation in which the family eked out a bare subsistence although they did not depend upon public aid. This was far from Wilson's dream. He grew up in the day of ''the garden of the world'' theory, and the agricultural myth, if myth it be, that rural living offered many advantages over that in the congested cities. Wilson was not only a philosopher at heart, but he was also historically minded. He was well versed in the numerous utopian attempts to establish agricultural-industrial havens, and he had watched Elwood Mead's Australian experiment with great interest. His immediate inspiration came from the highly practical and successful Mormon villages in Utah where agriculture, small industry and desirable social conditions were combined effectively. Wilson's second assignment with the Roosevelt administration was to administer the subsistence homestead project. He was soon moved to another position, however, and the project was combined with a larger movement which was also mostly of Wilson's creation—the resettlement administration.

Wilson and F.D.R.'s 1932 Campaign

Wilson's stature as an agricultural philosopher grew rapidly in the national leadership group, and when the University of Chicago planned a series of nationwide broadcasts devoted to important problems, he was asked to present the subject of agriculture. Wilson chose to limit his discussion to land utilization, and he suggested six points of attack:[19]

1. Congress should immediately repeal the Homestead Act, together with other free-land laws, and enact a new national land policy bill. He conceded that repeal was not so important, but that he was seeking a proposal with enough shock effect to draw attention to the land policy situation, which he felt was serious.
2. Each state should immediately classify its lands and develop an active plan of state land use. He cited Governor Franklin D. Roosevelt's plan in New York.
3. Taxes should be based upon land productivity.
4. Poor lands should be removed from production.
5. Production on good lands should be placed on a flexible basis to meet market demands without producing a surplus. This was his domestic allotment plan.
6. Combine for as many families as feasible the better features of farm living with opportunities for industrial employment. This was his plan for subsistence homesteads.

As the 1932 political campaign began to take shape, Rexford Tugwell and Raymond Moley, who were assembling Roosevelt's "brain trust," often crossed Wilson's path. Tugwell and Wilson met in Chicago in the spring of 1932 and as a result, a reference to domestic allotment was included in Roosevelt's dramatic acceptance speech in Chicago on July 2. Immediately thereafter, on July 6, Wilson was asked to come to Hyde Park where his suggestions concerning agricultural ideas and the strategy involved in obtaining their approval made a deep impression upon Roosevelt and his advisors. "What we were looking for," said Tugwell, "was a way of controlling production which would be politically feasible." Wilson thought he had it. That was the contribution achieved by the domestic allotment plan.[20]

Wilson was asked to prepare an idea outline for the proposed agricultural policy speech which was given in Topeka on September 14. Wilson included his favorite programs of land use policy, balancing of production and consumption together with proper compensation for the producers of the staple commodities. He kept the strong group of McNary-Haugen supporters in mind, but maintained that our European customers must not be alienated by dumping practices, and that the tariff must be used to help direct production and not merely for price-fixing. The administration of agricultural programs by the government must be decentralized in form, they must be voluntary rather than coercive, they must be brought close to the people for democratic and cooperative action, and the programs should finance themselves. Raymond Moley, who wrote the final draft, acknowledged that "the very heart of the speech was contributed by Wilson." An astonishing amount of Wilson's phrasing was used, changed only for brevity and force for oral presentation.[21]

The Agricultural Adjustment Bill reached President Roosevelt on May 12, 1933. The domestic allotment program was first applied to wheat. Wilson was the logical choice to head this program although the president had first proposed him as under secretary of agriculture. In 1934 he was made director of the new division of Subsistence Homesteads, then rapidly he became assistant secretary and then under secretary. In 1940 he became director of the Federal Extension Service, where he served with distinction until his retirement in 1953. He then accepted a number of foreign assignments and advisory engagements and continued to be a focal center for original thinking in the whole field of agricultural and social relations until he died in December 1969.

The manner in which Wilson was able to extend his influence nationally has not been well understood. Russell Lord observes:[22]

> It was, indeed, an extraordinary job of quiet, open propaganda. One wonders sometimes how Wilson, away off there in distressed Montana, working almost alone on a $3,600 salary which had been cut one-fifth, got so much done. His travel time and money were extremely limited. He could never pay more than two dollars a day for a hotel room, however great the city. But letter mail was only two cents in those days, and Wilson had friends the country over.

Actually, relatively few Montanans are aware to this day of the contribution which Wilson and his associates made to national thinking. He became a national figure in 1933. In June 1935 he was given the honorary degree of Doctor of Science by Montana State College, and in the summer of 1966 he returned again when many of his former colleagues gathered in a colloquium to do him honor.

In the meantime scholars from other states have been studying the record and expressing wonder that Montana State was able to make its influence felt so far. The first book to concentrate exclusively upon Wilson's contribution, William D. Rowley's *M. L. Wilson and the Campaign for Domestic Allotment*, raises this question. And the domestic allotment issue was, of course, only one phase of the Wilson spectrum of plans for the new agriculture. Rowley observes:[23]

> The campaign for the domestic allotment plan raises several questions. One of the most obvious questions is why, in the light of the disastrous conditions of northern plains agriculture, there were not more leaders in the colleges and the communities of the northern Great Plains states pushing vigorously for national relief legislation and devising programs of significant reform. Of the land-grant colleges in the northern plains, Montana State College appeared to be the only institution which recognized that the problems of the wheat crop extended beyond the questions of efficient production and marketing. At Montana State College, Wilson was given a wide latitude of freedom, which few other state college professors enjoyed, to seek agricultural reform legislation from the federal government. This enabled him to span, in part, the gap between the world of academic economics and the world of politics, personalities, and interest groups. From the community of academia, Wilson brought and sold a plan to help agriculture's crumbling economic life; his freedom of action enabled him to gain access to Roosevelt's brain trust, where he received the general committment of the incoming administration to the principles of the domestic allotment plan.

Wilson would not have predicted in 1932-1933 the persistence of production difficulties. Since this has been true, the wisdom of Wilson's most unalterable conviction has been a priceless contribution:[24]

> Reform in agriculture must grow from the ground up and be built upon the solid rock of democratic opinion. It must answer the desires of farm people and they must determine its form. Its character must be shaped out of the social soil of the country. And agricultural leadership can do no better than provide devices whereby the rank and file may set their local problems into national perspective, help articulate the opinions that are formed on this basis and finally assist in turning ideas into action.

REFERENCES

Since many references in this chapter are found in other than regular college publications of the time, specific references are given below:

1. Wilson, M.L. to Alfred Atkinson, Montana State University Archives. Hereafter referred to as Archives. Agronomy 1, AW F5, letter, April 23, 1910.
2. Wilson, M.L., county agent leader's *Annual Report*, 1915, p. 4 Archives.
3. Wilson, M.L. to Alfred Atkinson, Agronomy 1, AW F5, letter, June 13, 1911, Archives.
4. M. L. Wilson Symposium, Montana State University, July 25-27, 1966. Manuscript draft of proceedings, p. IV. Archives.
5. *Saturday Evening Post*, Vol. 196, April 12, 1924, p. 3ff.
6. Wilson, M.L., Farm Management Demonstrator's *Annual Report*, 1923, p. 7, Archives.
7. Wilson, M.L., "The Fairway Farms Project," *Journal of Land and Public Utility Economics,* Vol. III (April 1926) pp. 156-171.
8. This sketch of the Fairway Farms owes much to the extensive summary of the project prepared by Miss Rita McDonald, when as archivist at Montana State College, she was arranging the papers of M.L. Wilson and of the Fairway Farms Corporation.
9. Report of Dr. Henry C. Taylor, executive director, the Farm Foundation, 1937, pp. 6-7. In the M.L. Wilson File, AH-14. Archives.
10. Experiment Station *Bulletin No. 278.*
11. Wilson, M.L. to Dr. Henry C. Taylor, Ag Econ 1, SK F4, letter, November 12, 1929. Archives.
12. Starch, Elmer to Dr. Dana Myrick, March 6, 1969. Starch file, Department of Agricultural Economics, Montana State University.
13. Lord, Russell, *The Wallaces of Iowa.* Boston, 1947, p. 301.
14. Wilson, M.L. to O.A.H. Herzer, Ag Econ 1, Sh F8, letter, December 19, 1929. Archives. Herzer was with the Land Development Company of the Canadian Pacific Railway.
15. Lord, *op cit.,* p. 300.
16. Wilson, M.L., "An Agricultural Program for Montana," Extension *Bulletin No. 84*, 1927.
17. Starch-Christgau correspondence. Wilson File E-77, Agricultural Policy and Conditions. Individual correspondence. Victor Christgau, U.S. House of Representatives, 1928-1932. Archives.
18. Bell, Edward J., Jr., M.L. Wilson Symposium, *op cit.,* " Putting Ideas to Work." p. 53.
19. Wilson, M.L., "Land Utilization," *Economic Series,* Lecture No. 23, April 16, 1932. University of Chicago Press, 1932.
20. Rowley, William D., *M.L. Wilson and the Campaign for the Domestic Allotment.*

Lincoln: University of Nebraska Press, 1970. p. 152.

21. See drafts in pencil of Wilson's notes for F.D.R.'s Topeka speech, Wilson File, Summer, 1932. Archives.
22. Lord, *op cit,* p. 308.
23. Rowley, *op cit,* p. 199.
24. Starch, Elmer A., ''An Agricultural Reform Produced through Democratic Action and Incorporation of Technological Achievement,'' p. 84, and Tootell, Robert B., ''Some Land Use, Tenure and Credit Aspects,'' in M.L. Wilson Symposium, *op cit.* Both of these associates quote this phrase.

Additional detail on many questions mentioned here is to be found in the Columbia University Oral History Interview with M.L. Wilson. Archives.

V. The Taylor Years: Drought, Depression and War, 1924-1945

John C. Taylor, the informal "Jack" was universally used (his signature was J.C.), became acting assistant director of the Montana Extension Service July 1, 1924, upon the resignation of Director F.S. Cooley. Taylor inherited a broadly based and progessive program built on Cooley's pioneering work beginning in 1907 as director of Farmers' Institutes, and then as the first Extension director following 1914. Cooley relinquished the position quickly in an administrative rearrangement initiated by President Alfred Atkinson. Neither man discussed this situation. President Atkinson carried the title of director until July 1, 1925, when Taylor was named director. Cooley left behind a brief but comprehensive and invaluable history of his years in the Service.

Jack Taylor

As has been noted, Taylor grew up on a sheep ranch near the slopes of the Bear's Paw Mountains in Blaine County, which his father represented in the state Senate. He graduated from Montana State College in 1912 and became the third county agent in the state when he went to Glendive April 1, 1914, serving Dawson, Wilbaux and Fallon counties. He was named assistant county agent leader January 1, 1917, when the stresses of the impending war arose. He carried the work of the leader's office during 1920-1921 when M.L. Wilson was on leave and assumed the office April 1, 1921, when Wilson became Extension economist.

By experience and temperament, Taylor was ideally suited to assume the leadership in Extension at this particular time. Naturally conservative, he gave comfort to that large group in the state. Judicious of mind and ambitious for Montana, he quickly recognized the need for the best methods even if those methods were new. Keenly aware that he was in the company of original, theoretical thinkers, he readily adapted to the role of an open minded administrator. He found ways to sift and apply in a practical manner the plethora of ideas, sometimes seemingly bizarre, which steadily swirled around him.

He organized well and delegated authority to carefully selected associates. He outworked almost everyone even in a day when working

J.C. Taylor, second director of the Montana Extension Service, 1925—1945.

around the clock was customary, and he consistently maintained a friendly but businesslike relationship with his staff. During the numerous periods of excessive stress, he had the priceless facility of retaining the full confidence of the numerous officers with whom he worked in the college, state and county staffs, as well as the state legislative committees. Taylor's two decades in office presented more crises than any other director before or since has faced. Drought, depression and war produced a kaleidoscopic series of rapid changes; action programs had to be designed to meet these sudden demands. He steadily kept in mind that these were emergency and temporary situations and attempted to envision what a long range and "normal" program should be.

Postwar Adjustment

Although the years before the stock market crash of October 1929 later appeared to have been relatively calm, this period was also one of jolting adjustment. Long years of what were then considered normal precipitation suddenly shifted into a drought in 1917 which continued intermittently into the early 1920s. Nevertheless, continued high war prices enabled the hard-hit wheatgrowers and stockmen to survive and even expand their operations. The withdrawal of the federal war funds from Extension in 1918 was a severe shock, but the Service had proved its worth. On June 30, 1918, county agents numbered 24 and home demonstration agents 15. In 1920 there were 25 county agricultural agents. The slight decline in home agents was causd by difficulty of replacement. The director credited the support of the strong Farm Bureau organizations in these counties for much of the smoothness of the transition. As drought conditions continued, the number of home demonstration agents dropped to six in 1925. The 24 county agents, however, served in the leading agricultural counties accounting for 62 percent of the farmers of the state.

Taylor made his first report at the end of six months on December 1, 1924, devoting considerable attention to public relations. Merchants, implement dealers and bankers, as well as farmers, had suffered during the drought and postwar deflation years. Cooley had developed a program to provide "Main Street" with helpful information and attention. This in turn secured support for the county agent and his agricultural work. Taylor planned to continue this program—even enlarge it.

Taylor was well aware of the competitive relations existing among the various state organizations. Each was important to the progress of agriculture; organizations included the Farm Bureau, Farmers Union, Grange, Stockgrowers, Wool Growers, Dairymen's Association, Bankers' Associations and Chambers of Commerce. He was adept in setting a firm course for Extension and then working with all of the organizations to mutual advantage. The State Department of Agriculture, Publicity and Labor, established in 1921, was still another organization where careful working relations were required to avoid overlapping activities. Extension worked with that department in cooperative arrangements such as a joint Dairy Division conference, research on the protein content of wheat with the Grain Marketing Division, and with the Publicity Division in promoting state and county fairs.

The Extension Service helped to bolster declining income by assisting in organizing cooperative marketing associations—34 by 1926. These produced an estimated increased income or savings of some $174,898 that year. Individual savings often were not high, but still helpful; the 206 members of the Hill County Marketing Association, for instance, sold products valued at $119,853, with a total savings of $21,613. Ten poultry cooperatives comprised the largest number for any one commodity.

The rapidly broadening scope of Extension work is indicated in Taylor's first annual report, June 30, 1925. He listed 69 projects in progress. In agronomy, 27 tests were being administered in grain seed testing, and attention was being given to tillage, tillage implements, irrigation and weed control. Animal husbandry emphasized feeding, culling, and 4-H projects in beef, swine and sheep. The dairy industry was concerned with testing clubs, milk and cream grading, and improving breeding stock through bull associations. Poultry raisers concentrated on housing, culling, and egg and turkey marketing association. Horticulture emphasized disease control, grading, marketing associations, potato improvement, shelter belts, home beautification, and 4-H garden projects.

An Agricultural Program for Montana

The 1925 report also outlined the progress of a major program which was completed in December 1926 with the publication of Extension *Bulletin No. 81,* "Basic Facts about Montana's Agriculture." Over a two-year period, nine committees worked in areas of Montana's major agricultural products, with a tenth devoted to land utilization. This was a resource inventory following traditional trends of emphasizing production, but its purpose marked a departure. A surplus of agricultural products had accumulated as a result of the war effort and continued increased production, but steadily falling prices were creating financial hardships. This study urged that since production was doing well, that the traditional emphasis on *methods* of increasing production be replaced by studies of the *economics* of production concentrating upon costs, marketing and profitable farm operation.

This approach led to broadening participation in the annual state Extension conference on January 5, 1927. In addition to Extension personnel, invitations went to farmers, stockmen and businessmen, and representatives of more than 20 agriculturally related organizations such as banks, insurance companies, implement dealers, chambers of commerce, and newspaper editors. Here a plan was agreed upon to enlist the people of the state in an intensive but rapid study to evaluate the resource inventory which *Bulletin No. 81* contained. The state was divided into six districts to secure as much homogeneity as possible and reduce travel costs. A steering committee was named for each district and in the conferences some 1,500 people participated in making recommendations for the wise use of Montana's resources.

The recommendations were summarized by M.L. Wilson in Extension *Bulletin No. 84,* "An Agricultural Program for Montana," published in May 1927. Comment has been made of the impact upon Wilson's thinking and of the hearty cooperation of the Montana farmers in the study. The background was, of course, the boom and bust of the war and postwar period and the several years of drought, balanced against increased production induced by war prices and improved methods, and the failure of the Outlook forecasts to influence supply. Wilson summarized the conclusions which were reached in the several districts, with hundreds of the most active Montana farmers participating. To repeat, a strong agricultural program for Montana could be obtained through: 1) national legislation; 2) cooperative marketing; 3) increased farm efficiency, and by what Wilson felt was most important, 4) adjustment in production.

Brief general recommendations were made regarding Montana's chief agricultural commodities, but the chief value of *Bulletin No. 84* was, no doubt, its listing in detail the conclusions reached in each of the six districts in which the studies were made. These gave specific practical suggestions to individual farms and definite commodities. The recommendations were later studied further in economic conferences held in the counties in which each of the most important commodities produced in that county was considered separately. One study was made with irrigated lands and another with non-irrigated, each emphasizing both production and economic management.

Traditional interest in farm crops prevented livestock from receiving important attention, although a livestock committee functioned in each of the six districts. However, in the general discussion, Wilson took time to point toward some of the livestock industry's important problems.

Full attention was given in *Bulletin No. 84* to the homestead filings which had been made between 1910 and 1922 on 42 percent of the state, and naturally on the best land. These fenced and plowed areas had completely altered the former pattern of unlimited and free range, which the livestock men had enjoyed before the homesteaders came. Particularly aggravating to the range men was the practice of the inexperienced farmers of fencing and plowing lands wholly unfit for farming, whereupon the lands then became unfit for grazing for years to come. The rapid settlement made it necessary for the stockmen to purchase large tracts of land in the 1930s, and the price of land had not declined in comparison with the price of products. The Extension personnel in management felt that private ownership of grazing lands was desirable, almost necessary under developing circumstances, but their studies showed it was still unprofitable. Use of large amounts of public land was recommended, but this land was often over grazed and not well located. The fears of the economists were fulfilled in the 1930s when drought and depression caused prices of agricultural products and land to drop precipitiously. This left those range men who had purchased in the 1920s and were still heavily indebted with little choice of action, but with strong emotions concerning the boom and bust cycles.

Wilson's urging that "adjustments in production" be kept in mind underwent an immediate test because as the bulletin was being published in May 1927, predictions were being made for a good production year. The result was a huge increase to 81,700,000 bushels of wheat in comparison with 34,998,000 in 1925. Barley doubled from 3,000,000 to 6,000,000 bushels.

Director Taylor felt that the intensive study of Montana's resources and the widespread consideration of their best use in preparing the bulletin, "An Agricutural Program for Montana" had united farmers and ranchers, state agencies and the business community to a point where all of them became "thoroughly familiar with the agricultural situation in the state and obtained a much more definite conception of the agricultural problems than was ever before had."

Development In The Twenties

The numerous meetings in 1927 which produced the agricultural program bulletin probably sparked the renewal of the Special Train Demonstrations which had not been used for a time. The Northern Pacific hosted a "Better Livestock, Better Feed" special which ran for 17 days in

Better Livestock special train sponsored by the Northern Pacific Railroad, 1927.

October, made 29 stops, wtih 22,000 in attendance. In addition, a series of "Better Feed" conferences were attended by 13,000 people. Since the train ran through counties which in 1926 had produced livestock valued at more than $35,000,000, the potential for improvement was considered very high.

In 1928 another livestock train, this one on the Great Northern, emphasizing sheep and wool, was sent across the state. This train made 31 stops in 22 counties and was visited by some 18,000 people. The exhibits again stressed improved feeds and more efficient feeding methods, the advantages of better breeding stock, and the principles of better management. The secretary of the Wool Growers' Association accompanied the train, and an unusually attractive exhibit of sheep and wool, and demonstrations of effective handling of wool, proved to be of great interest. Members of 4-H clubs handled special demonstrations at many stops, which led to increased emphasis on sheep and wool in the clubs.

The wide range of agricultural activities called for equally broad capabilities among county agents. In this period when the Extension Service was new and agents were under suspicion as being merely "book farmers," there was considerable pressure to undertake "service" activities. Ralph Mercer remarks that only rarely did a county agent embark on a day's journey over his county in his Model T without strapping on a transit and rod. He would certainly be called upon to assist in surveying a ditch or fence line somewhere in the course of the day. As new machinery came rapidly into wide use, the agent needed also to be a mechanic. This was particularly true with the combines, of which 500 were in use in 1927, and breakdowns were frequent. With the new marketing techniques, the growers often had difficulty in determining the desirable stage of ripeness in grain, which led agents to help farmers obtain tests for moisture content before combining.

Publicizing the early auction yards for feeder stock became another service activity. The first auction yard was set up in Billings in 1924, and the idea spread. A suitable number of buyers from the Midwest to provide spirited bidding could be attracted only if a large number of animals were available. The county agent often had to conduct a considerable education program before the ranchers made use of this new method of marketing.

The Montana Seed Growers Association was formed in 1912. Following the appointment of an Extension agronomist in 1917, he acted as secretary for many years. The Seed Growers arose out of county agent work during the war when good seed sold at a high premium. As an illustration, county agents took the lead as early as 1916 in bringing a small number of alfalfa growers together to form an association for certifying good seed. By 1927 there were 474 fields in production for seed, 357 of which were certified; and by 1929 two alfalfa associations were active. In the five-year period 1925-1930, alfalfa acreage and production increased three and one-half times. Grimm was the favorite variety with Cossack having some supporters. Certification increased the income of these producers by $250,000-$300,000.

A special project, "Know Your Wheat," was directed by Edward J. Bell, Jr., and Charles C. Conser from 1927 to 1932. In many meetings this project stressed the importance of high quality grain to offset the high shipping costs. Since the value of the 1928 wheat crop was estimated at $64,900,000, even minimum use of information concerning quality seed supplies, protein content, control of moisture content, and care in storage and shipment was estimated to add at least $15,000,000 to the value of the crop to the grower and distributor. Bell had previously, in 1925-1926, worked with wheat growers, elevators, and the State Grain Inspection Laboratory to secure quick determination of the protein content. Then he secured the cooperation of the Minneapolis Grain Exchange which agreed to send, through the Grain Market News of the USDA in Minneapolis, price quotations to the Extension editor by telegraph each Friday. The editor, in turn, got this information to the county agent by penny postcard by Monday. The farmers were then able, through cooperation with the elevators and the railroads, to take advantage of favorable prices. Conser took charge of the project after 1929 and served until going to Washington, D.C., with M.L. Wilson in 1933.

The first Extension agronomist, F.E. Fuller, began his work in May 1917. As the pressures of the war expanded, he proved the importance of good seed in getting high quality production. In December 1919 he took the lead in calling a national meeting of seed growers in Chicago "for the purpose of standardizing pedigreed seed work and gathering information for a more effective national association."

In 1920 A.J. Ogaard succeeded Fuller as agronomist and secretary of the

Montana Seed Growers. He is given a great deal of credit for stimulating the improved production which brought Montana seed growers the World's Sweepstakes Cup at the International Hay and Grain Show in Chicago in December 1922, after only a decade of growing certified seed. This recognition gave Montana seed an added premium in national markets. The minimum standards for certification in Montana became the standards in Idaho and Utah, and these three states influenced the adoption of higher standards by the International Crop Improvement Association.

Ogaard became a legend before his death of a heart condition in 1928. By the simple methods of slowing his pace he might have extended his life by many years, but he chose to carry on with a crusader's zest. Miss Cushman recalls:

> The pattern set by Ogaard in certification work has been followed with few changes down through the years. He made the inspection himself until he had trained reliable men to assist him. First he made the field inspection. Usually accompanied by the county agent, he went over every foot of the field to be inspected to see that there were no noxious weeds, such as fan weed or mustard, present. Also, the field was checked for any other varieties other than the specified grain. After the harvest there was the bin inspection. If the grower's grain measured up, it was then sacked and sealed with the Montana Seed Growers label of identification. All sealed grain was fully accounted for and careful records were kept.
>
> Seed certification work was a full-time job in itself. Walking miles and miles in the baking Montana sun was an arduous task. Yet it had to be done, and Ogaard did it, as have all of those who have followed him.

A.J. Ogaard (right) consulting with an alfalfa seed grower, 1924.

Montana Seed Growers' display of registered seed.

The troubled twenties continued in character to the very end. The state Legislature cut the Extension budget by about $7,500 a year for the 1929-1931 biennium. Opportunities in business and other fields were still good and early in the year all of the home economics specialists resigned, as well as the agronomy specialist and the 4-H leader. Seven county agents followed suit, for what the director termed "the highest turnover we have ever experienced in one year. . . we hope it may not occur soon again." The financial crash of 1929 caused less difficulty in Montana at the time than in the East, but it was still a major calamity.

The "Dirty Thirties"

The terrible thirties opened in Montana with a severe drought which extended over three-fourths of the state. Decreasing price trends resulting from the 1929 crash added to the troubles of those who sought some relief by selling livestock, grain or land. The Extension Service, except for its changing staff, was in a basically good position to give help since it had been able to expand into new territory through the use of the Capper-Ketcham funds. Eighteen agents had been added in three years, increasing the total to 48. Home demonstration agents had also increased from seven in 1928 to 13 in 1931. This was in line with the provisions of the Capper-Ketcham Act which stated that "agents appointed under its provisions shall be men and women in fair and just proportion." Three associate agents, working specifically on reclamation projects, and a marketing specialist were included in the expanded staff, and they were particularly useful in the dry years.

One bright spot in the period was the passage of the 10-year millage tax in 1930. A special levy of two mills on property had been laid in an election in 1920 for the university system. Ballots in the November 1930 election proposed expanding this to three mills and added a $4,000,000 bond issue for buildings. To the surprise of everyone both measures passed. However, the Legislature levied only an additional one-half mill in 1931, but did use the three mills in 1933. Even then the decreasing land values and high tax delinquency prevented the raising of revenues required for the additional demands made upon the Service.

The federal government responded to the credit crisis of 1929. The Federal Farm Board was established in that year with $500,000,000 to support cooperative marketing associations which attempted to stabilize prices of wheat, wool, cotton and other commodities. In 1930 the Outlook program was geared to the Farm Board plans. In Montana, 85 meetings were held from which a number of cooperative organizations emerged.

Another special train was operated over the Northern Pacific in 1930, "The Low Cost Farming Special." This made use of many of the lessons learned from the Fairway Farms experience. Where previous emphasis had been largely on increased production, this exhibit pointed up possibilities of lower labor costs, sources of capital, marketing outlets, and an ultimate margin of profit. Flat cars carried the newest machinery available for dry land farming. Exhibits and illustrated lectures showed what could be done with them. The train stopped in 18 communities from June 23 to July 2, with over 12,000 people attending. The slogan, "Save Labor—Your Biggest Cost," was effectively demonstrated.

By July 1931, Montanans were fully aware that they were experiencing an extremely severe drought. Extension activities were increased in every way to give relief to people and livestock in almost all of the eastern half of the state. Governor Erickson appointed a State Drought Relief Committee, and appeals were made to the USDA and to the Red Cross, which had been active in 1930. Director Taylor reported that in many localities there were no cash crops and little feed for livestock.

A conference in July with merchants, bankers and railroad representatives resulted in assistance to 480 ranchers to sell and ship out their stock, or to ship in feed for their use. Ralph Mercer commented, however, that "the producer was dangling at both ends of calamity. He didn't produce much and what he did produce wasn't worth much." Extension hastened to make a county by county survey to determine the extent of the disaster and ascertain the number of families in need of relief assistance.

Unexpected rains fell in some of the driest areas in early July, which led to a quickly organized information program on replanting corn, oats and other crops which might produce some forage. A bulletin, "The Late Planted Garden," was also hastily issued. Both of these efforts proved to be

A late planted garden in the fall of 1930.

remarkably successful. Feed and seed loans were secured on a large scale. Grain and forage shipments at reduced rates totaled 1,192 carloads, and 1,500 cattle were shipped to Wibaux County to take advantage of the revived range following the July rains. Seed loans numbering 8,030 were approved at a value of $2,062,395 and 5,549 feed loans were granted at a value of $1,040,472.

The home economics specialists and the home demonstration agents prepared a crash program of canning and preserving food, particularly vegetables and fruits. Emphasis was also placed on the greater care and use of poultry and eggs. Bulletins and demonstrations were prepared on renovating, dyeing, dry cleaning, and, in general, prolonging the use of clothing. Special attention was given to 4-H groups in these emergency activities which merged into two programs for the family, "Live at Home" and "Buy Wisely." These proved to be so useful that they were broadened into a basic planning program and continued for a number of years.

In the midst of this strenuous and effective effort to provide relief and establish the soundest possible basis for assistance to rural people, the Extension Service was faced with attacks which threatened its survival. The 1931 Legislature again reduced the budget by $7,500 as it had done in 1929. In addition, the federal support, which was based on farm population, was reduced by $3,800 since a heavy out-migration of farm population had started in 1930. Counties were also distraught and every county Extension budget was reduced from 10 to 20 percent.

Taxpayers' associations formed in almost every county. Many farmers joined in the movement to reduce funds for Extension. Their discouragement was complete and they felt that farm programs urging more and better crops were useless when their present crops were often sold at a loss. Movements started in 11 counties to eliminate the entire Extension program. Agents were dropped in Fallon, Daniels and Judith Basin counties, and one home demonstration agent was eliminated in the Fallon-Carter district. However, at the peak of the movement, Dawson County, which had been without an agent for eight years, restored the position. A federally supported marketing specialist added in 1930 was continued, and in 1931 Robert B. Tootell was employed as Extension land economist, the first position of its kind in the nation. Both of these positions fitted admirably into the work of relieving the drought and depression stresses.

At the annual 1932 Extension conference, one full day was devoted to the subject of the drought, with Experiment Station personnel taking part. Ideas such as strip farming, soil blowing, mulching, stock water reservoirs, flood irrigation, trench silos, the best use of straw and range conservation practices were explored. Similar discussions were then held in Miles City, Havre, Huntley, Great Falls, and Glendive. Canadians were invited to attend and a profitable exchange of ideas ensued.

Hundreds of dramatic stories have been told of those hectic years. In the barren eastern counties a program for the use of roughage arose spontaneously, resulting in the harvesting of thousands of tons of Russian thistles. These were then fed with grain straw and a small portion of hay. A minimum of grain then carried the stock through the relatively mild winters of the early 1930s.

A whimsical movement came out of the deep drought areas. Extension personnel in these counties sensed that with both crops and income almost completely gone, the ranchers and their families had a great deal of time on their hands. A series of recreational programs were arranged, resulting in county recreational councils, which in many counties developed into year-round programs attended by thousands of rural people.

Miss Cushman relates a heartening reaction to calamity:

> On a dryland farm in Chouteau County the parents and five children had spent the entire spring putting in several sections of wheat with little outside help. Their two pushed-together homestead shacks stood in stark ugliness, but the prospects for a crop were fairly good. The grain had ripened and by some miracle neither rust nor hoppers had attacked it. They had purchased their binder twine and were to start harvesting the next day when a severe hail storm swept across the fields, flattening the heads to a pulp. The wife remarked, "Until that moment I had not known how wonderful my husband was. We stood at the door and watched our whole year's work go smash. The mister never cussed nor fussed. His eyes just looked sick until he noticed where the hail had piled up between the two shacks. It was about a foot deep. Then his eyes lit up and he said, "Mary, get the ice cream freezer out and let's make ice cream for the kids. They've never had none." What was one cream check more or less? The kids had a feast.

The home economics "Live at Home" project shifted strongly to the 4-H

Meat cutting and canning demonstration being conducted by men and women agents.

clubs in 1932, when adults were often too discouraged to take new chances. The adults did work with the youth, however, since a high percent of farm homes reported gardens and 18 percent had enough vegetables to last through the winter. The many county fairs, which were then in favor, centered almost exclusively around 4-H projects, which ranged from livestock, crops and gardens to food preservation and clothing, even recreation. The quantity and high quality of the exhibits gave renewed hope to the entire community.

The county agents were actually protected in 1932 when the State Board of Education asked other members of the Service to take an 8 percent salary cut. However, the agents had already taken a 10 percent reduction, and later did take an additional slight cut as counties tried to reduce expenses. High tax delinquencies caused state funds to be insufficient to cover salary warrants. These were then discounted by the banks from 2 to 5 percent, which was in effect a salary reduction.

Taxpayers' associations, with continued support from the farm population, remained active. To combat inaccurate statements, the Service found it necessary to take time from the urgent needs of people to develop its own studies of taxation, the costs and income of counties, and the benefits of agricultural programs. So skilled did the Service become in these statistical studies that Fred Bennion, county agent leader, who was in charge of the studies and their interpretation, was offered the position of executive secretary of the Montana Taxpayers' Association, which he felt he could not afford to decline. In addition to a salary increase it offered an opportunity to protect the Service. During the year the counties felt it necessary to part with the services of six agents—three men and three women.

The negotiation of crop production loans continued and a record number of 19,004 loans were processed in the spring of 1932. The usefulness of the county agents was again demonstrated in the approval of 14,441 loans in the 30 counties having county agents, for a total amount of $3,480,747. In the 24 counties not having county agents, only 4,563 loans were completed, to the amount of $919,747. It is true that agent counties comprised the majority of the farm population, but in many instances needs were really more severe in the non-agent counties where farmers gave up rather than search for help which seemed to be remote.

The new land economist, Robert B. Tootell, capitalized upon the needs of the dry years. To him the better use of land meant the best possible use of all the water available. He quickly assembled an inventory of irrigation projects over the state during the 1920-1931 period. He laid plans for a rapid expansion of the program and secured the support of Montanans, Incorporated, an organization which preceded the state chamber of commerce. The state engineer's office also assisted in obtaining desirable legislation and developing adequate organizational support. Blaine Ferguson's work as irrigation specialist, with the cry of "Dam the Coulee, Dike the Land, Save the Runoff Water," took on new vigor as counties and communities organized to utilize every drop of water.

Drought conditions in 1932 were less severe than for the two previous years, but farm prices for many commodities dropped to the lowest point ever known. Wheat sold for as low as 25 cents per bushel, and hogs as low as four cents per pound. Annual farm income was estimated at $46,900,000 in comparison with $160,000,000 in 1928. Director Taylor commented in his annual report that "with farm income wiped out. . . and pressure from tax discussions steadily increasing, it became one of the most difficult years ever experienced from the standpoint of Extension work and workers." However, after reviewing the frenetic efforts of the year, he sounded an encouraging note. "Greater interest," he said, "and more active participation now exists on the part of rural people than at any time since Extension work was established in Montana."

Drought and low prices continued through 1933. Montana had already made major adjustments and crop production loans were only $1,341,405 as against over $4,000,000 in 1932. The distressed economic conditions were reflected in the Legislature where, as Director Taylor noted in his annual report, "The most determined effort ever made to reduce state appropriations for Extension work was experienced in the legislative session of 1933." A less charitable view was that a determined effort took place to completely eliminate the Extension Service. Friends of the Service descended upon the Capitol in astonishing numbers, but the Extension Service and the Experiment Station both suffered a 33 percent cut, which together with a reduction of county support of up to 15 percent made a difficult situation, indeed. A reduction in salaries of 20 percent and a sharp curtailment of projects took up part of the decrease, and a threat of additional federal cuts was avoided by the severity of state action. Strangely, personnel changes were fewer than usual. Only two regular positions were terminated—one county and one home demonstration agent. Two permanent county agents were added, and Extension work was expanded into another two new counties on a part-time basis.

The legislative assembly adjourned with many of its members still convinced that assistance to Montana's agriculture was a lost cause. But within the same month the inauguration of the Roosevelt administration and the consequent New Deal farm legislation sent the Extension Service into another boom of instant expansion and intense activity similar to that of the previous war period. Again, Taylor remarked, "The changes were unprecedented."

As has been indicated, when the Agricultural Adjustment Administration Act was passed in 1933, M. L. Wilson took charge of the Wheat Section on May 1. Since plans for this program had originated in Montana, Wilson used the state as a project laboratory and gave it considerable personal attention. Seventeen emergency agents were appointed to the counties without Extension facilities. These people were hastily recruited from civil service rolls for a period of 90 days to assist with registration for the wheat production program. As one federal or relief program followed another, many of these emergency recruits remained for a lifetime of service.

State, regional and district committees were named and the state was divided into seven districts. Each had a supervisor chosen from the state staff. Continued drought spurred all sections of the highly organized campaign and insured the cooperation of the wheat farmers. At the close of the program a gratifying 95 percent of Montana's wheat acreage had been included.

In December 1933, Elmer Starch, who followed Wilson as head of the Department of Agricultural Economics at Montana State College, fortuitously took advantage of available labor time from civil works programs to summarize five years of records relating to 1) tax data; 2) A.A.A. wheat acreage and production data; 3) price trends for outlook purposes; 4) grazing land evaluations; 5) ownership maps for counties; 6) agricultural price and income data. This information, when meshed with the Gieseker County soil maps, provided invaluable data for land use planning in relation to irrigation, grazing districts, tax delinquency, land purchase, rural rehabilitation and other programs.

The Dry Summers of 1934 and 1936

The eastern half of the state experienced the most severe drought ever recorded in 1934, and the emergency measures of the previous years were continued. Regular projects looking toward increased production were dropped, and assistance was given to the new federal programs such as Land Purchase, Rural Rehabilitation, federally sponsored state and county planning, and the program of the new State Water Conservation Board which had been created early in 1934.

The benefit payments of the 1933 Wheat Allotment and other programs amounted to over one-half million dollars and was the only major source of income in the eastern counties. This would have bridged one severe year, but in 1934, in addition to a terrible insect outbreak, Paul Carpenter reported:

> There have been other drought seasons, but during 1934 occurred the worst in the history of the Northern Great Plains "cow country." . . . In the Miles City area, roughly a circle with a radius of 100 miles with this city at the center, the range did not become green in the spring. The grain crops failed. Very little hay was grown and by June scores of cattlemen had their stock on hay meadows. By July "never failing" water holes were drying up and cottonwood trees were being cut down along creeks to keep cattle and sheep alive. By August 1, the death rate of cattle and sheep was appreciable. This area temporarily was a desert in the truest sense of the word. Summer and fall losses would have been appalling without the emergency program.

Director Taylor blocked out the needs which "created additional emergency activities, such as removal of livestock, location of range, feed and water, and finally, feed surveys and county organizations for the distribution of feed."

The Emergency Livestock Purchase Program was undoubtedly the major enterprise of the summer. The Rural Rehabilitation Corporation adopted the organization of the State Drought Relief Commission of 1931 and obtained grass roots aid from local committees of the Montana Stockgrowers and the Wool Growers. The A.A.A. and relief agencies furnished the funds; the Bureau of Animal Industry made appraisals and condemnations and had general charge of arrangements for cattle movements to railroad points and shipments. The Federal Surplus Relief Corporation arranged facilities for the disposal of stock condemned as being too poor or diseased to sell and marketed the hides.

The original 10 counties declared as a disaster areas soon broadened into 38. Ranchers seeking aid listed cattle numbering 758,781. Of these, 349,824 were purchased by the federal government and shipped out of the area, and 9,785 head were condemned and disposed of. Powder River County reported the heaviest sales. Farms seeking assistance listed 61,682 cattle. Of these, 49,785 were sold, and 1,716 condemned, leaving just over 10,000 head for restocking. Many counties reported from one-half to two-thirds of their stock sold or destroyed. Sheep sales and condemnations amounted to 142,812 delivered for sale, together with 348,829 pelts.

A grasshopper control campaign in the spring of 1934 was also dramatic in its extent and the calamity which prompted it. A survey in the fall of 1933 predicted that in 18 counties from 50 to 90 percent of the total acreage of grain was so heavily infested with eggs that treatment would be necessary to obtain a crop. In another 11 counties, 25 to 50 percent infesta-

Operation Grasshoppers, 1934.

tion was reported. The area was divided into five districts, each with a supervisor and with a number of emergency agents assisting. A tremendous campaign of information was carried out involving 1,239 leaders in some 500 meetings. Bait used amounted to 26,543,255 pounds, covering 2,623,086 acres. Crop savings were estimated at $7,032,362.

What would have been a ruinously low income for the state was boosted substantially by the government programs. Payments for wheat control programs in 1934 totaled $8,360,107. In addition, a corn-hog control program was instituted on a small scale which provided payments amounting to $267,098. These, added to the emergency sales of cattle and sheep in the eastern counties, kept many Montana families from complete ruin.

In 1935 only slight improvement was seen in the eastern counties, and some western counties had their worst year. Better prices made an important difference. State agricultural income was estimated at $97 million as compared with $84 million in 1934, 58 million in 1933, and as low as $47 million in 1932. Since the "normal" income was closer to the $160 million of 1928 or even the $121 million of 1929, the federal assistance programs were welcome. The A.A.A. wheat payments amounted to $6,827,324; the corn-hog program brought in $490,704, and sugar beets $2,149,828, for a total of $9,467,958.

The successful work of the Extension Service during the drought years brought increased financial support from legislative appropriations. Of basic importance also was another federal grant, the Bankhead-Jones Act, effective July 1, 1935. This provided permanent funds for agents' salaries and services which had been supplied from emergency funds since 1933. Montana's allotment covered 11 county agents and three district home demonstration agents. This easing of pressure on regular funds allowed adding one county agricultural agent, three home demonstration agents, and a full-time specialist in irrigation, beginning January 1, 1936. Increased support from county funds started the slow process of increasing the severely depressed salaries. Nevertheless, the Service experienced the largest turnover of personnel in its history, as 13 agents with five to 17 years' experience left during 1934 and 1935. These people rather naturally went largely to the new federal agencies which offered attractive specialized fields.

Drought, with a malignancy exceeding that of 1934, descended upon an even larger section of the state in 1936. Director Taylor reported, "The entire area east of the Continental Divide has less grain and grass to fatten livestock than in any season for which records are available." Ralph Mercer added a touch of color:

> Dust covered the sun for hours at a time. The grasshoppers and Mormon crickets by the millions were with us. . . and after finishing the crops, they worked on hoe handles, fence posts and buildings. . . Grass was gone, water holes dried up and every morning the sun came up hotter and brighter than the day before.

Almost twice as much insect poison was used in 1936 as in any previous year and another livestock liquidation program was undertaken which was larger than that of 1934.

Federal funds for cattle purchases were not available in 1936, and the Extension Service set up a voluntary program in Miles City with County Agent N. A. Jacobson in charge. This was entirely a purchase program, and in addition to shipments to other counties in the state, Jacobson contacted buyers in California, Oregon and Idaho. Cow and calf pairs often brought as much as $40, which encouraged restocking in 1937 when the rains came. Since the best stock had been saved by the ranchers, recovery was rapid.

The summers of 1934 and 1936 were devastating in depleting the ranges of eastern Montana. The severity of the two summers and the usefulness of

A dust storm near Winnett, Mont., 1937.

the Extension Service in meeting these emergencies quieted its critics for a number of years. With an active organization which spanned the entire state, it had now proved several times that it could quickly, quietly and effectively give the necessary assistance by marshalling the resources of the state and also those of the national government. Taylor's 1936 report noted with satisfaction that the Service was now at "the highest level it has ever reached since Extension work was established in this state, a total of 45 counties out of the 56 having the services of an Extension agent, 13 of which now provide funds for two agents and 12 employing an agricultural agent on a joint basis." Federal funds assisted this expansion materially, however.

The drought had depleted state funds, and the hard-pressed 1937 legislative assembly reduced appropriations below those of 1935. However, as the rains came, the county commissioners raised their support. Appropriations for 1937-1938 totaled $126,474, as against $80,430 in 1933-1934. Taylor commented again:

> The attitude displayed by boards of commissioners in the face of the most severe drought ever experienced in Montana since weather reports have been kept, coupled with insect damage, not only to cultivated crops, but to range grasses as well, indicates a genuine appreciation for the service rendered through Extension agents and specialists.

Agricultural Planning

Overall planning for agricultural advancement in the state, which had generated great enthusiasm in 1926 and 1927, had necessarily been shelved during the years of stress. Improvement in 1935 led to plans for a two-day economic conference in each county in 1936. These conferences were to review the data in *Bulletins 81* and *84,* bring this information up-to-date, name county planning committees and make recommendations for further study and action programs. Planning committees were formed in 39 counties, and 10 counties were chosen for detailed study.

In the United States Department of Agriculture, M. L. Wilson was working vigorously to stimulate nationwide interest and action in planning and conservation. In 1936 A.A.A. introduced a new and important facet by outlining an agricultural conservation program, and range improvement practices were included. County agents and special county committees with federal support instituted improved range practices, stock water ponds and modified cropping practices. Over 5,000 applications for assistance in range programs came in, together with 35,000 in the crop division. A high percentage was approved.

A major state conservation agency was the Montana Water Conservation Board established by the special session of the legislative assembly in December-January, 1933-1934. It moved quickly to construct water storage facilities sufficient to guarantee ample water for 400,000 acres of land. The board constructed the dams and major ditches for water distribution. Then, by agreement, the Extension Service worked with the individual landholder in the preparation of his land, construction of ditches, the use of water and types of crops best suited to the farm.

In 1937 O.E. Baker, well-known social geographer of the Bureau of Agricultural Economics, came for a time to work with the two groups of committees—the county planning and the conservation committees. The county conservation committees served as action agencies supporting the Montana Water Conservation Board, then worked with the federal Soil Conservation Service and the Farm Security Administration when they were formed.

The County Planning Committees stimulated the formation of over 250 community committees with some 6,300 rural people attending meetings during the year. Taylor reported that planning work had "reached a very advanced position in its development in Montana." He noted:

> Community committees have already collected basic information about climate, water, and land resources, and the productivity of range and farming lands. They are ready and interested in going forward with the next step in working on problems affecting agriculture, such as soil erosion, range depletion, insect and rodent control, decreasing crop yields, schools, roads, size and tenure of farms, taxation and farm debts.

Wind erosion such as pictured above was a common occurrence during the "dirty thirties."

The Mount Weather Agreement

These numerous and active planning committees in Montana were among the first in the nation. Their success encouraged M. L. Wilson to continue his urgent desire to make the activities of the Department of Agriculture as democratic as possible. He promoted a meeting called in July 1938, at Mount Weather, Virginia, with representatives of the land grant colleges in attendance. The resulting Mount Weather report was an agreement in which the land grant colleges accepted greater responsibility for agricultural planning activities at the county and community level. Secretary Henry A. Wallace had this to say:

> In the Department of Agriculture we have begun a bold experiment in democracy. We are trying to put into practice the idea that in a democracy the economic decisions ought to be made by the people. We are slowly building a mechanism county by county in collaboration with the land grant colleges and the Extension Service by means of which the farmers themselves will determine the elements of their total agricultural program, will decide how these elements may be welded together in one effective program and finally will administer that program. We call this county planning. Actually the farmers in any given county immediately see that they must interest themselves in forces and problems which go far beyond the county line.

A major directive of the agreement was the establishment, by the Extension director in each state, of a State Land Use Advisory Committee made up of representatives of the state and federal agencies concerned and farmer representatives from the different types of farming areas in the state. This committee would form the link between federal, state and community activities, with particular emphasis upon individual farmer participation at the community and county levels. The Montana Advisory Committee illustrates the broad range of agencies:

Clyde McKee.................... Director, Montana Experiment Station
H. H. Curley.................................. U.S. Forest Service
E. H. Aicher............... State Coordinator, Soil Conservation Service
W. F. Bailey....... Chairman, State Agricultural Conservation Committee
Thomas Horsford............... Director, Farm Security Administration
Ray B. Haight........................ State Land Specialist, Bureau of Agricultural Economics
E. A. Starch....... Executive Secretary, Northern Great Plains Committee
H. C. Bolster...................... Montana Extension Land Economist
J. C. Taylor...................... Director, Montana Extension Service

The committee gave particular attention to the development of a water facilities program as an immediate aim. A coordinated land use program for the entire state was its major aim and long range purpose. This committee was one of the first to be organized in the country, and it took pride in the extent of its activities. Some 1,400 agricultural leaders assisted in directing about 500 meetings in 1938-1939, following the Mount Weather pattern. The central committee met for two-day sessions every two months to consider and establish policy positions, to consider reports of planning committees, to act upon requests for water projects, and to keep abreast of federal agricultural legislative and agency proposals. Ralph Mercer summarized the numerous problems of the time from the point of view of field operations:

> Let us remember that. . .there were such things in operation as the grass conservation district law, the soil conservation district law, an act providing for water conservation boards and programs and a weed law; there was a farm tenancy program, a land purchase program, crop and feed loan programs, and an agricultural conservation program. Yes, and there were Civilian Conservation Camps, and a thousand Extension problems to keep every person on the staff run ragged and spread thin.

The several planning committees worked closely with the various pioneering agencies which were beginning operation. The 1933 Grazing District Act and the 1934 Water Conservation Board have been mentioned. In 1935 a Grass Conservation Commission was established, and in 1937 this commission was authorized to work with the federal Taylor Grazing Act. The result was, Director Taylor felt, "to provide the most far-reaching control over lands yet achieved in the state."

Brief Normalcy

An important change in personnel took place September 1, 1937, when the director of the Experiment Station, F. B. Linfield, retired from that position. He had also been a major force in Extension since 1902 when he joined the staff, with one of his duties as that of director of the Farmers' Institutes. Professor Clyde McKee, dean of the College of Agriculture, vice-director of the Experiment Station and head of the Agronomy Department, was named director. Because of his familiarity with the work and the staff, no interruptions in the interrelationships took place.

In 1937 Montana production returned to normal yields, and continued stable prices allowed the rural community to relax somewhat. State and, particularly, federal agencies shifted from emergency crash programs to those with long-term goals. The Soil Conservation Service developed an educational program looking toward establishing soil conservation districts. Martin Hansmeier was named Montana Extension conservationist to assist with this program. Rural electrification expansion was meeting with wide approval, but brought many questions regarding use of equipment and wiring of existing structures. Armin Hill was appointed rural electrification specialist on a half-time basis in June 1928. He prepared helpful bulletins and conducted conferences and demonstrations. He also assisted with the 4-H Club project in rural electrification set up in 1939, which proved to be especially popular.

Emphasis upon production through improved methods rather naturally continued. In his 1938 report Director Taylor summarized the increased use of strip farming. Although this technique had been strongly recommended in Wilson's Triangle Bulletin in 1923, it was in use on only 108 farms in 1933, on 16,300 acres. The dry years, however, brought a sharp increase in 1934 to 339 farms on 52,390 acres. The continued dry years proved its usefulness, and by 1938 it was practiced on 9,272 farms on 1,151,442 acres. In 1939 Taylor noted that on 32 demonstration farms in 18 counties, strip farming, summer fallowing, and the use of new varieties and new machines were being used to advantage.

The organization of the Regional Junior Fat Stock Show in Billings in 1938 was the beginning of another successful program. FFA and 4-H members exhibited 807 head of stock, and sales were highly profitable. Extension personnel in 1939 assisted the Stockgrowers and the Wool Growers in sponsoring district and county livestock meetings, and by 1940, 32 such sessions were held. A range management school in June 1940 brought 20 county agents to assist in identifying important range grasses, demonstrate methods of range reconnaissance, and develop patterns of determining the carrying capacity of range lands. In 1940 eight Dairy Herd Improvement associations were operating, with the Extension Service providing the coordinating assistance. Flathead County had organized as early as 1924, Madison-Jefferson groups formed in 1938, and Yellowstone County organized in 1940.

Personnel was a major problem thoroughout Director Taylor's regime. The Service ballooned with the assistance of federal funding during periods of war and drought. Then the state encountered difficulty in maintaining

Strip cropping for wind erosion control had become a general practice by 1938.

the expanded staff. Fortunately, the federal government steadily responded to the pressures by increasing its assistance. Director Taylor listed in his 1936 report the sources of Extension funds:

Federal:	
Smith-Lever and	
Supplemental	$44,983.51
Capper-Ketcham	23,299.65
Additional Federal	26,850.00
Bankhead-Jones, 1935	73,610.76
Total	$168,743.92
State	52,000.00
County	122,430.00
Total	$343,173.92

However, Taylor frequently found that while funds were increasing, the demands for services were also increasing, and he repeatedly made a comment similar to that in his 1940 report, "There probably was never a time in the history of Extension work in the state when funds were so short for financing expanding field work."

The adequate training of new agents with the continuous rapid change of staff was always uppermost in the mind of Director Taylor and his staff. In 1927 County Agent Leader Fred Bennion went to California to observe its reportedly advanced system of organizing Extension programs and training its personnel. In 1929 a new landmark was set when two members of the staff were given leave of absence for advanced training. In 1959 President Renne succeeded in bringing Extension staff members into full equality with the teaching faculty in such matters as rank, leave for advanced study, cumulative leave and leave with pay.

This rising interest in broader preparation of agents was implemented by the "schools of philosophy" initiated by innovative leaders of the USDA, which included M. L. Wilson. This program was organized by Dr. Carl F. Taeusch, long-time dynamic professor of philosophy at the University of Iowa, where he developed popular seminars designed for business leaders. In the January 1939 annual Extension conference, Taeusch and his team led a discussion on "What is a Desirable National Agricultural Program?", treating the theme from the point of view of economics, sociology, government, philosophy and international considerations. Taylor acknowledged that the first reaction of his agents was "one of bewilderment," but upon further consideration they approved the program. In the 1940 annual conference special instruction was given in the use of visual aids and radio. In 1941 two training programs were carried on in district meetings, one on 4-H leadership and another on farm credit in cooperation with the Federal Land Bank of Spokane and the Farm Loan Association.

The nature of the office of county agent changed strikingly in the 1930s. Formerly the agent had answered his own telephone, personally participated in many of the rural activities such as walking miles through fields of grain to certify them for seed production, examining animals for disease symptoms, surveying a boundary line, and conducting demonstrations of a dozen kinds.

The development of many new state programs, and particularly the meshing of them with the numerous federal projects made the county agent a facilitator. He worked with a number of committees which were in charge of such projects as grazing districts, soil conservation districts, land acquisition programs, and irrigation developments. Training schools for agricultural planning often combined a number of these projects to provide information to a community concerning the rapid progress being made.

The director's 1941 report outlined the progress in land classification and mapping. Thirteen counties had completed area maps, and six others were well along toward completion. Forty-two county planning committees were in operation made up of 249 organized community committees comprised of 1,464 farm people. Elmer Starch, in the July 1966 symposium in honor of M. L. Wilson, commented, "The mainstays for democracy in action are knowledgeable participants." The county agricultural agent and the home demonstration agent provided order and direction to this extremely important cooperative movement which, Starch commented, "brought to our times a new concept of democratic economic action."

The Second World War

A warning of a cloud "no larger than a man's hand" appeared in the 1940 report as Director Taylor commented:

> At the September and December sessions of the State Planning Committee, consideration was given to the effect of the present war and the defense program on agriculture and the need for carrying to all farm operators the most complete information available along these lines.

War-oriented activities developed rapidly in 1941. In February a statewide series of 380 community meetings were held, attended by 15,000 people to urge increased production in view of the probable need. In June another 84 meetings were held.

In early autumn of 1941 a regional meeting in Salt Lake City was called by Secretary of Agriculture Claude Wickard. He was soon to coin the slogan "Food Will Win the War and Write the Peace," and he outlined production goals for the region. In October and November a production goals campaign was organized over Montana resulting in 392 meetings attended by 10,837 people. Committees undertook such activities as an aluminum drive, in which 4-H groups were used extensively, war bond purchase drives, and the collection of scrap iron, which was particularly successful in rural areas.

A meeting of the Nutrition Committee during World War II.

Agents and state specialists at the 1940 annual conference: (left to right, bottom row) Bernard Williams, Ralph Bodley, Booth Holker, Melvin Ostby, E.B. Duncan and Jack McGuire; (second row) John Bower, Gerald Beheimer, Elvin Duenst, Bob Rorvig, George Loomis, E.G. Ferguson, Oliver Lammers, Stan Halverson, George Gustafson and Ben Daggett; (third and fourth rows) Lambert Hruska, Francis Nevin, Bill Ross, George Severson, Walso Zirnstein, Ray Kelley, Glenn Baird, I.D. Zobell, Ralph Stucky, Ed Kempton, Kelley Hanson, George Wooley, Henry Oefstus, Fred Finch, Harold Dusenberry, N.A. Jacobson, Leonard Eliason and Gib Zeidler; (fourth and fifth rows) Paul Orcutt, Rusty Ralston, A.C. Peterson, Earl Bjork, O.P. Roberts, Jack Taylor, Carl Peterson, Charlie Jarrett, George Hart, Ted Fosse, M.P. Hansmeier, Bob Newell, Martin Peterson, Louis True, M.M. Oliphant, Horace Bolster, Bob Rasmusson and Maurice Zimmerman.

The 1941 annual Extension conference was held December 12-19, 1941, instead of in January, since this would give more time for firming up annual programs. The date was fortunate since the declaration of war one week earlier, on December 8th, determined the trend for the year ahead. The experience of Montana in World War I, and in the crises of drought and depression, enabled its administrators to form a State War Board quickly and to organize by communities, counties and districts. The organization of the Cooperative Extension Service again was exactly right for the action program at hand. Extension was particularly well "tooled up" because of the state-wide Agricultural Planning Committees and federal Agricultural Conservation Committees which had been operating for several years.

In the December meeting, the project "War Production Goals and their Attainment" had priority. A technical committee of seven members formulated plans which were taken to the State Agricultural Planning Committee in February. The state was divided into five districts, and an Extension specialist was placed in charge of each district. Early in February county meetings were held in 48 of the 56 counties, and the several committees and agencies were united in the common effort. Production had been good in Montana and production increases were set on a realistic basis:

Cattle	from	1,201,000	to	1,313,000	or	112,000	head increase
Hogs	from	103,000	to	147,000	or	44,000	head increase
Sheep	from	3,345,000	to	3,547,000	or	202,000	head increase
Chickens	from	1,926,000	to	2,194,000	or	268,000	head increase
Feed Grains	from	886,000	to	1,296,000	or	410,000	acres increase
Hay	from	1,827,000	to	1,941,000	or	114,000	acres increase
Flax	from	170,000	to	367,000	or	197,000	acres increase
Sugar Beets	from	64,000	to	81,000	or	17,200	acres increase
Vegatables & Potatoes	from	74,300	to	92,800	or	18,500	acres increase
Wheat	from	3,685,000	to	3,423,000	or	262,000	acres decrease

Reduction in wheat acreage was accepted with some regret since this was Montana's best crop. But recent good crop years had pushed Montana's production to a high level, and there was concern lest marginal lands be used after several years of expensive effort had been spent in taking them out of production. Uniformly favorable moisture and growing conditions in 1942 gave the state the best production it had experienced since 1923. The increased moisture and cooler weather almost eliminated the grasshopper and locust plagues which had been a problem since the mid-thirties. Since prices also increased, 1942 was the best all around agricultural year in the history of the state.

One of the most severe problems, again, was that of staff. In 1942 three county agents and three assistant agents were called into military service and another five left for other positions. The director lamented that new agents were sent into the field with as little as one month's preparation, but even so, two counties dropped the Extension program because agents could not be secured. This situation worsened and in 1945 the Service lost 16 agents and five home demonstration agents. Additional publications were issued to answer the more common questions. Most notable was a series of 44 leaflets prepared by the state entomologist and his staff covering such plagues as grasshoppers, wireworms, army cutworms, garden slugs, and the cherry fruit fly.

The labor shortage extended from the staff to the fields and ranches, particularly in such seasonal crops as hay and sugar beets. The university units made arrangements in class schedules to allow the men from the student body, and many faculty, to take part in the beet harvest and other urgent farm work. "Main Street" businessmen and their employees also assisted.

Community cooperation was greatly increased by the Neighbor Leader project supported by both the Extension Service and the Civilian Defense Commission. By the end of 1942, leaders and action committees existed in almost every rural community in the state, including towns of up to 2,500 in population. A total of 2,356 men and 1,238 women comprised the leaders, and the organization was capable of reaching 90 percent of the people in their communities within a day or two. The first project was an explanation of the price control and rationing program, the complexities of which caused many people real trouble.

The wide and quick reach of the Service was also effective in a new program which Secretary of Agriculture Wickard asked for specifically—that of a rural fire control project. The Forest Service was also particularly helpful in areas where it had personnel. Fire control boards were established in almost every county and although equipment was difficult to obtain, preventive measures of importance were carried out. Director Taylor remarked, "No program ever undertaken by the Extension Service met with more wholehearted interest and response on the part of local people than the plan outlined for fire protection."

Production goals in 1943 were readily attained with favorable weather conditions prevailing. Although there was an expansion of 6.6 percent in acreage, much of this was on marginal land, and the crop increase was only 4.4 percent. Higher prices resulted in a 32 percent increase in farm income in 1943.

A project which met with favor was that of repair and reconditioning of farm machinery, since replacement was almost impossible. Maintaining a balance between feed and livestock numbers was also a matter of concern as small areas of drought appeared.

The labor shortage deepened in 1943, and a program for the greater use of youth developed. In cooperation with the superintendent of public instruction and the vocational education departments of the public schools, special classes were quickly set up to train for special skills, and some 10,000 youth were placed in farm assistance jobs. The Federal Emergency Farm Labor

Act became operative May 1, 1943, and the Extension Service and the USFS worked closely together. Mexican labor was imported for the canning and seed pea harvest in the Gallatin Valley, and the sugar beet harvest in the Yellowstone, the Big Horn, and the Milk River valleys. The beet harvest program carried over for several years, and, in cooperation with the sugar companies, the Service instituted training programs. This included the recruitment of Spanish-speaking assistants who did continuous trouble shooting in matters of housing, transportation, illness, wage disputes and many other problems which arose. Extension also coordinated a program which brought German war prisoners in 1944 and 1945 to assist with the beet harvest.

In the midst of crisis activities, projects intended for the long term still received substantial attention. Soil Conservation Districts almost doubled in 1943 to a total of 14 comprising 8,715,710 acres. At the same time cooperative grazing districts increased to 18, covering 6,613,136 acres. Experiments in feeding animals resulted in an increase in barley production from 212,000 bushels in 1942 to 600,000 bushels in 1944.

De-escalation, 1944-1945

Almost from the first crash war-planning committee meeting, national and state leaders had urged that postwar planning be given consideration. Financial support for Extension services during the war was satisfactory. The Legislative Assembly in 1945 approved a record increase in support of $23,859 for 1945-1946, and $27,334 for 1946-1947. County governments, which were in a comfortable financial position, also increased support. State and county support the country over, however, was falling behind in continuing the new and expanded services developed during the war, for which there was continued demand. At this point Congress came to the rescue with the Bankhead-Flannagan Funds. The total support for Montana during 1945-1946 showed an encouraging gain over 1936, previously noted:

Federal:	
Smith-Lever	$34,918.54
Capper-Ketcham	23,030.42
Bankhead-Jones	83,243.96
Additional Cooperative	32,217.74
Bankhead-Flannagan	52,439.55
Total	$225,850.21
State	85,689.00
County	158,881.84
Total	$470,421.05

Director Taylor announced in his 1945 annual report that he would retire as director on January 1, 1946. This report, after noting the numerous activities of the year, reflected the planning which had been taking place with the incoming director, noting the many plans for increased services in the future. The Bankhead-Flannagan windfall was designed to persuade the states to place one or more agents in every agricultural county, which in Montana, the director felt, included every county except Silver Bow. Increased financial support had allowed raises of 10 percent or more in salaries both in 1944 and 1945, and enabled Taylor to comment:

> For agricultural agents this raised all salaries to a $3,000 level or higher, with a range from $3,000 to $4,000 per annum. For home demonstration agents, the range is from $2,500 to $3,000. . . This is the highest salary level that has been obtained during the history of Extension work for both state and county workers.

Nevertheless, the changes produced by the war opened many new and desirable positions in federal and commercial enterprises at good salaries, and the Extension people were attracted. Taylor commented ruefully that a record 21 agents left the service in 1945. Sixteen of these were men, five of whom did not return after war service; five transferred to other agencies, four had entered commercial fields, and two had returned to home farms attracted by increased prices and higher yields. Of the five women, one entered the armed services, two withdrew when their husbands returned from the service, one was married, and one entered other work. On July 1, there were 10 counties without agents, and by November 30 this number was 13. The director hoped to place home demonstration agents in as many as 30 to 35 counties, and this posed recruitment problems.

In 1944 a new postwar project was given to the Service—that of the Veterans' Resettlement program. Detailed work was done in preparation for the veterans' return. Only a few took advantage of the project since not many had the financial support necessary. However, the work was valuable in making leaders in state and county agencies aware of the limited amount of land available and the high costs of establishing a farming business.

The director regretted that 4-H work as such had been almost completely abandoned during the war, although youth in large numbers had been pressed into active work. Some projects were revived in 1944-1945, and Taylor recommended that the program be given a high priority in postwar planning.

Looking toward unfilled needs in the state, the director emphasized that a new position of agricultural engineering specialist had been authorized on July 1, 1945, but no one had been found to fill the post. The new technology brought about by the war and the increased prosperity of farm families provided many demands for such a position. An immediate need was to advise rural families who would be building new homes within a few years; the number was estimated at 5,000 to 7,000. Conveniences for the farm home and farm plant had increased rapidly; electric power, indoor plumbing, water under pressure for many purposes, and facilities for the protection and repair of motorized machinery were some of the desirable innovations. Faced with many requests, the home management specialist set up a series of meetings in November, billed as "House Planning Workshops," covering such subjects as approved designs, convenient floor arrangements, new materials and credit facilities.

The Shelter Belt program had continued with unabated enthusiasm and additional success with more favorable growing years. The value of woodlots was still high, and with the addition during the war of fire protection, the director recommended a specialist in farm forestry. A livestock specialist was authorized on July 1, and F. A. Ralston was appointed. This position had been vacant since E. Paul Orcutt had resigned in October 1944. The director also had plans for a dairy specialist and a beef production specialist immediately, and a sheep production specialist in the near future. A child welfare specialist was also recommended as an immediate need, as well as assistants to the specialists in the rapidly expanding areas of nutrition and clothing.

Although Director Taylor was restrained in his final report, he undoubtedly had great satisfaction in looking back over the 22 years of his directorship. All too real still were the dire circumstances of drought and depression, the pressures of war, of the lowest financial support on record, and violent attacks on his entire agency. But his final report listed the highest salary scale on record, an expansion which would finally cover the entire state, and new areas of service which would add to rural financial well-being and appreciably improve the quality of living statewide.

Jack Taylor retired as director January 1, 1946, but continued on a part-time basis to work with a favorite older farm youth project, finally retiring in 1949. He was awarded the honorary doctor of science degree in June 1952 by Montana State College. Until his death in December 1964, at age 77, he continued to be known affectionately by his former colleagues, indeed by the entire state, as "Mr. Extension Service."

VI. Mid-Century: Directors Tootell And Beers, 1946-1958

Robert B. Tootell became director of the Montana Extension Service on January 1, 1946. Tootell had grown up on a farm in Cascade County and graduated from Montana State College in 1927, majoring in agricultural education. He taught vocational agriculture one year in the Deer Lodge High School, and returned to MSC for two years, teaching economics and sociology. He took leave for further training and secured a master's degree at the University of California in 1931. He then returned to Montana State to take the post of extension land economist, the first position of its kind in the nation.

Caught up in the problem of credit in the depression, he went with the

Robert B. Tootell, third director of Extension, 1946-1953.

Federal Land Bank in Spokane in the spring of 1934. He served with the Bank's Cooperative Farm Credit System in various positions until the fall of 1943 when he returned to Montana State to become head of the Department of Agricultural Economics. He held this position until the end of 1945 when he was named director of Extension. He held that office until June 30, 1953, and then took a similar position in the state of Washington. Farm credit still interested him, and in 1954 he accepted appointment as governor of the Farm Credit Administration. Here he served with distinction until his retirement in 1961. His 15 years constituted the longest term as governor in that administration to date. Montana State College gave him the honorary degree of doctor of laws in June 1956.

Tootell made a number of changes in the Extension structure. He tried to 1) free the director from unnecessary detail, 2) provide more immediate assistance and supervision from his office to the County Extension personnel, and 3) coordinate more closely the work of the specialists. The office of county agent leader was discontinued, and the more comprehensive position of assistant extension director created in its place. Chosen for this post was Newell E. Beers, a 1941 graduate of South Dakota State College, who had extensive farm experience, and was a district supervisor of county agents in South Dakota at the time of his appointment. Supervision of county agents was his primary assignment, and he was given two assistants: William G. Stucky and Harry A. Cosgriffe. He also gave special attention to coordinating the programs of the state specialists and their work with the county agents.

The state home demonstration leader, Mrs. Frances Smith Patten, retired in the fall of 1946 to devote full time to her home. Bessie E. McClelland, who had been extension nutrition specialist for several years, succeeded her. Miss McClelland also assumed additional responsibility for supervising state specialists in home economics and for coordinating their work with the home demonstration agents. Soon afterward the position of assistant home demonstration leader was created to strengthen the work of supervision. Marjorie Downs, home agent from Richland County, was appointed. These new positions increased the cooperation between the agricultural and the home economics staffs.

T. Booth Holker, acting county agent leader, was named to another new post, that of administrative assistant. Here he was chief administrative officer and additionally, because of his broad Extension background, served as an effective trouble shooter for many special assignments. He took an active part in developing training programs for Extension agents. He was largely responsible for organizing two courses in the college curriculum on Extension methods, and he taught both of these until he left for a foreign assignment in Jordan in October 1949.

Conditions in the Extension Service at the close of the war, as has been indicated, were largely favorable. During the war it had shifted its emphasis to production needs, rather than expanding into new areas. As peace returned the organization required only moderate changes. Its finances were adequate, and its leaders were now making impressive plans for new programs, as has been indicated in Taylor's final report in 1945. The chief problem, indeed a perennial one, was that of securing capable replacements for vacant positions. In 1946 funds reverted because positions were unfilled and needed equipment could not be obtained. Employment opportunities at the close of the war were such that almost a complete change in staff at the state level took place in home economics and in 4-H.

The economic situation remained remarkably stable; the usual postwar decline in prices and demand did not occur. Both 1946 and 1947 were banner years. A 26 percent loss in European agricultural production in 1947 led to a movement to increase exports to prevent food shortages. The Montana Extension Service issued a special grain conservation edition of the revived *Outlook* publication called "Looking Ahead with Montana Farmers and Ranchers," urging the conservation of grain for shipment to Europe.

Postwar Expansion

One step in meeting the postwar requests for additional services was the appointment of Eric Wilson as rural engineering specialist in 1946. He remained until 1961 when the position was dropped. By this time the building boom had decreased and many standard patterns had been worked out for modern farm houses, functional farm buildings, electrical installations, and particularly for the problems of running water and sewage disposal in homes and barns.

Armin Hill, rural electrification consultant, worked closely with Wilson on problems of installation and equipment. Closer working relations between the Extension Service and the power agencies were promoted by a landmark conference at the College in March 1949, which was attended by representatives from 21 of the 22 units of the Association of Rural Electric Cooperatives.

During World War II the nation was forcefully reminded again, as had been true in World War I, that young men in rural areas came up with more physical deficiencies than those in the cities. Although starting modestly, Montana was one of the first states to employ an Extension specialist in health education. Beginning September 1, 1946, Mrs. Frances McDonald combined her background in nurses training with her work in the College Student Health Service, serving one-fourth time as health specialist until July 1947 when she began to serve one-half time. She devoted her efforts full time to this work from 1948 until September 1956, when the service was discontinued. Home agents particularly appreciated her assistance since they received many requests for health aid and advice, especially with children's illnesses. County leaders also sought advice on how they might secure a public health nurse, a health center or other community

County Agents Lois Knowlton Stevens, Frances Smith Patten and Lillian Stone Mikkleson demonstrate electrical appliances which become more and more commonplace in rural homes.

health facilities. A circular on "Planning Health Facilities for Montana People" was issued in 1947 by an Extension Health Committee, and a number of bulletins and circulars on specialized common health problems provided additional aid.

Shelterbelts had long been popular for farmstead beautification and wind protection, and for accumulating additional snow, which increased moisture for the family garden. Experience had taught that site selection, species sequence, proper planting and cultivation had much to do with the ability of trees to survive transplanting and periods of severe drought. The Extension horticulturist had given aid to this work, but by 1946 additional requests came in for field windbreaks to reduce erosion from wind and to provide protective cover for wildlife. The success of mechanical tree planters also added to the requests for skilled help. At the time there was also a growth of interest in better management and use of the farm woodlot in the timber growing areas. In December 1949, John P. Drummond was employed as Extension forester. This position continued intermittently on a full- or half-time basis until it was abolished in 1965. The Extension forester worked closely with the state forester, the United States Forest Service, the University of Montana School of Forestry, and the State Forestry Nursery in Missoula.

State Extension assistance with weed control problems had for years been one of the many responsibilities of the Extension agronomist and the Exten-

Farmstead shelterbelts a few years after planting.

sion horticulturist. Although weed problems continued to increase, especially in irrigated areas, chemicals for more effective weed control, such as 2-4D, made great strides during and after the war. To meet the many requests for help in using the chemicals, Donald J. Luebbe, county agent in Blaine County, was appointed weed control specialist in 1946. Agents and county commissioners gave immediate cooperation, and county weed districts were formed in 26 counties within the year. A combination of weed control specialist and irrigation development specialist was made the following year, but the position of weed specialist was later re-established as a full-time position and has been strongly supported. Another development arising from the new chemicals was the rapid rise of aerial dusting and spraying programs. Commercial pilots saw in the insecticide and weed control projects a source of much needed revenue to keep their new air services afloat. The county agents and state specialists were quickly drawn into the program to give advice and confirm the usefulness of what the dusters proposed to do.

The increase in farm income and rapid technological development during and after the war heightened the interest of rural people in improving the quality of living in farm homes. Some of this interest, no doubt, accompanied the numerous new and remodeled farm houses equal in appearance, comfort and convenience to those in the city. Repeated requests from home demonstration clubs for a family life specialist resulted in the appointment of Miss Amy Wold in 1949. She retired in 1963, and the position was broadened to human resources development under the leadership of Dean K. Hoffman and those who followed him.

The extent to which range activities and development had been slighted was recognized when Ray G. Johnson was named range management specialist in January 1950. Tootell remarked:

> Although Montana has nearly 60 million acres of range land, a specialist had never previously been employed for this work. The regular livestock specialists and the Extension agronomists had, however, done some work along this line. . . Although Johnson remained less than two years before returning to foreign service, he stimulated a great deal of interest in improved range management by a dynamic and imaginative approach to the problems involved.

Marketing problems had always concerned state Extension leaders, but they had received only intermittent attention. Bruce L. Brooks, who had been in the Department of Agricultural Economics, USDA, became Montana Extension marketing specialist in 1950, and five other specialists were asked to devote one-fifth time to marketing organization and education. These were: Harriette E. Cushman, poultry; J. O. Tretsven, dairy; E. P. Orcutt, lambs and wool; H. R. Stucky, livestock; and Mary E. Loughead, consumer education.

The farm labor shortage which was acute during the war continued for some time afterward. This was particularly true in the six sugar beet factory districts where seasonal hand labor was needed for spring hoeing and thinning and fall harvesting. A State Farm Labor Committee, under the chairmanship of the commissioner of agriculture, was created to meet this need. Large numbers of workers were brought in from Mexico and Jamaica under government contract, and a considerable number of German prisoners of war were also made available, working under the strict supervision of the U.S. Army. Extension had the responsibility of allocating the workers to the sugar beet districts. Former county agent leader Ralph Bodley took this assignment for a short time and was followed by Harold Dusenberry, county agent in Rosebud County. Anticipating the withdrawal of federal funds for assistance with labor, the Extension Service organized a Labor Saving Caravan in 1946, planned by Dusenberry and Eric Wilson. Exhibits carried in the caravan demonstrated many ideas and devices which could save labor both on the farm and in the home, and these were in demand for display at fairs for several years.

Extension Farm Management Specialist Ralph Stucky soon became aware of the need to link the assessment of farm and ranch lands to their productivity. He reviewed the work of Leonard Gieseker in classifying land by soil type, which indicated productivity, and assessment schedules based on production values in Teton County by Bert Clarkson, agent, and Otto Wagnold, assessor. When Wagnold was appointed to the State Board of Equalization, Stucky secured support to recommend revision of the state tax policy. There was considerable opposition, especially on the part of some stockmen. In 1955 the state Legislature passed legislation requiring systematic reclassification of agricultural lands for tax purposes taking productivity into account. This was a major accomplishment for which Extension can take considerable credit.

The ability of Extension to focus its resources quickly was again demonstrated in 1952 when an outbreak of brucellosis occurred creating a great deal of anxiety among the stockmen and dairy operators in the state. The Experiment Station geared quickly for research, and Extension carried on a program of advice and control which undoubtedly resulted in substantial savings.

Among the many frustrating incidents which regularly appeared was the onslaught of the "rain makers" in the early 1950s. Literally dozens of these people came into the state with their attractive promises to increase the rainfall during the growing season. They succeeded in many communities in signing up large numbers of farmers and ranchers at the relatively low charge of 10 cents per acre for crop land and one cent per acre for range. Almost uniformly, their method was cloud-seeding with silver iodide from simple ground stations. Naturally, these entrepreneurs undertook to enlist the support of the county agents in organizing the farmers and generally promoting their business. Fortunately, the state office had some advance warning on this invasion from the U.S. Weather Bureau, which counseled caution. The information was passed on to the agents who were instructed not to participate in the promotion, but to publicize some of the known facts about the limitations of cloud-seeding, especially by unknown promoters. Tootell received more than one phone call from local farm leaders and businessmen accusing him of maintaining an overly cautious and unprogressive policy in the matter.

The rapid increase of new programs, and requests for still more, added to the well-established activities, created an atmosphere of constant stress. Tootell met this head on, urging a system of priorities and a concentration on fewer and more basic programs. In the annual meeting in 1948 he counseled his agents, asking, "When you as an agent leave your county after having worked there five years or more, for what will you be remembered?"

The Hard Winter of 1948-1949

Continued favorable conditions in 1948 were broken in December when it became apparent that another severe Montana winter was in the making. Immediately following January 1, 1949, excessive snowfall and cold temperatures were joined by high winds and blizzard conditions, causing 17 southeastern counties to be declared emergency areas. The Governor named his head of the State Highway Department as state coordinator of the Snow Disaster Relief Program. Every available highway snowplow was used on and off the highways. This still was not enough.

Extension immediately entered the program, and its resources were of great value. It was able to assemble a large number of bulldozers in the stricken communities, or to have them shipped in. These were particularly maneuverable on the side roads leading to the ranches. An ingenious device, the "truck train," was developed where from two or three to 20 trucks followed a snowplow or bulldozer, leaving supplies at the stranded ranch homes or stock herds. Fuel oil, groceries and medicines were essential for people, and hay and water for stock. Many light planes were also of great value in taking supplies to ranch homes and in dropping bales of hay for small isolated bands of stock. Extension Livestock Specialists Ralston and Orcutt, together with Director Tootell, were in the field continually. Director of Publications Louis True, who had a light plane, flew Orcutt on frequent reconnaissance flights to help coordinate the relief work. The federal government supplied about one-half of the near one million dollar cost to the state of the relief efforts.

When the storms ceased in early March, the losses of cattle were estimated at about $3,000,000, and those for sheep at about $100,000. One cattleman lost 1,400 head, and a sheepman lost 800 head. Except for the alert and effective marshalling of assistance, losses would have been much heavier and human suffering would have been much greater.

Adjustment, 1949-1958

By the spring of 1949 the wheat region was alerted that a surplus was building up in the world markets. A drought in Montana in the summer of 1949 reduced the production somewhat, but prices continued to be firm, and an encouraging amount of cash reserves carried most of the operators over the severe winter and dry summer. A major grasshopper plague also attacked the Montana grain fields. Control measures were quickly put into effect, which reduced the impending losses considerably. Continued study and improved control methods together with unfavorable weather for grasshoppers have made this the last major outbreak to this time (1974).

The wheat surplus which built up nationally in 1949 brought into operation the government seeding allotments. In order to qualify for price supporting loans, Montana farmers took allotments which reduced plantings by about 700,000 acres, from 6 million in 1949 to 5.3 million in 1950. Extension personnel spent a busy season in educational work with farmers, first on the necessity for reducing their acreages, and then in determining the best use of the diverted acres. The Department of Agriculture's plans for reduced wheat production in Montana were thwarted by favorable weather conditions, which, together with the improved practices, caused the 1950 wheat crop to soar to 91 million bushels compared with 61 million bushels in 1949.

The opening of the Korean War in 1950 brought another quick reversal in demand for wheat. Acreage allotments were taken off and seedings for 1951 through 1953 were increased to 6.3 million acres. A break in prices in 1952 and increased drought marked the end of a period of almost continuous prosperity since 1939. Tootell warned that a period of adjustment was probably overdue, and Extension lent its efforts toward making this as easy as possible. Acreage allotments and marketing quotas reduced the seeding of wheat from 6.3 million acres in 1953 to 4.9 million acres in 1954, and this approximate acreage continued for a decade or more.

The decade of the 1950s brought a moderately stable situation. Income from farm marketings varied from $360.7 million in 1950 to $435.3 million in 1958. Drier conditions in the wheat areas were more than compensated for by increased yields from better methods and improved varieties. Average yields of 14.6 bushels per acre in 1952 had increased to 23.4 bushels by 1958. Receipts from the sale of livestock were less stable, dropping from $243.5 million in 1951 to $147.7 million in 1953, then up to $230 million in 1958.

During the 1950s also, the Extension people, particularly those in the field, began to receive a great deal of assistance from several sources. The newer federal agencies, the Production Credit Association and the Farmers' Home Administration, gave assistance on loans and payments which agents had often furnished. The Soil Conservation Service gave assistance in the area of soil and water conservation, surveying dams, field contours, and the like. Extension personnel had always depended on assistance from feed and seed companies, elevator marketing personnel, and small flour mills.

A new development during the war had been the growth in size and activity of the fertilizer and machinery companies. Many of their representatives were not only vigorous salesmen but highly competent agricultural advisors, often with previous county agent or specialist training and experience. County commissioners encouraged the farmers and ranchers to take advantage of these services since they "do not cost us." A few even considered discontinuing the Extension agent, but this was met by protests from local leaders saying that many important programs would be eliminated, particularly the 4-H program. However, the result was that Extension, which for years had a virtual monopoly on techical assistance and service programs, found itself in a somewhat competitive situation which was not without advantages. It availed itself of the friendly challenge to increase the excellence of its unique programs.

The several years of high prices stimulated a movement to increase the size of farms, which was facilitated by the use of the new large machinery. During the experimentation and the increasingly profitable production of wheat during the war years, confidence in a single crop had also increased. One result was that the farm family established a winter home in the nearby town where family members could take take advantage of schools, social contacts, recreation and health facilities. The Extension Service was soon astonished to find that about as many of their club members, cooperators and 4-H members could be reached in the towns as on the farms. A highly encouraging aspect of this movement, however, was actually an increased demand for Extension services. Residents who divided their time between town and country stimulated the requests for aid in both places for home services, club work, lawns, gardens, fruit and landscaping. Living in towns at least part of the year, the ranchers were also able to make more frequent use of the advice and services furnished at an increasing rate by agricultural products companies. Discussion, training and demonstration groups could be more quickly assembled than in an earlier day, and community leaders and organized groups could be more widely used. Extension methods and programs were modified to suit the new conditions and Extension activity increased steadily.

Director Tootell was somewhat startled, therefore, to be asked by a committee from the 1953 Legislature to assist in reducing the Extension budget since so many private and federal agencies were assisting people who were using their farms for exclusively business purposes. Tootell had a great deal of support in persuading the committee that the Extension Service was still a useful agency.

An Advisory Council and Changing Programs

Tootell placed a greal deal of emphasis upon long-term planning and thorough preparation. He immediately drew up a plan whereby the agricultural and home agents together with the specialists would be able to cooperate in preparing and evaluating the annual programs. Early in 1946 the state staff presented in four district conferences a draft titled "Situation and Problem Statement for Plans of Work." This was a suggested model for an analytical approach to building programs of work. The involvement of local leaders was urged, and an effort was made to reactivate the community and county agricultural planning committees developed during the depression period. The substantial increase in funds from the federal Bankhead-Jones Act, which became available in 1946, designated especially for use in the counties, was particularly useful.

The entire Extension staff felt the need for a statewide group which would be a sounding board for proposed ideas and would support those ideas which were approved. In 1944 President Renne created an Agricultural Advisory Council for Montana State College, consisting of 25 outstanding rural men and women from over the state. The council periodically spent a day at its quarterly meetings evaluating the Extension Service programs.

At such a meeting in June 1946, Director Tootell and his staff reported that the dividing line between town and country was disappearing at a rather rapid rate, and that both were in need of many similar services. Increasing farm efficiency was almost immediately reflected in larger incomes and better living for farm families. Emphasis was placed upon the mobility of young farm people, half of whom would soon seek employment in the cities. Hence the need for the same quality of education, health care, and citizenship attitudes and services for farm youth as for those youngsters in the towns and cities. The point was made that the householders in urban areas were also in need of services in home economics, gardens, lawns, trees, flowers, soils and insects, together with information on electrical equipment and home machines, home accounts and consumer problems. The Extension staff outlined its broad aim as:

1. To improve the quality of living of the American people
2. To encourage conservation of agricultural resources so that future generations may have the opportunity for better living

More specific aims were outlined as:

1. Increasing farm income through more efficient production
2. Increasing farm income by assisting with more efficient marketing
3. Helping people to become wiser consumers
4. Improving family living through better housing, rural electrification, and labor saving equipment
5. Improving health
6. Improving education
7. Improving recreational facilities
8. Development of leadership talents
9. Improving citizenship by helping people to be better informed and encouraging them to participate in public affairs
10. Promoting conservation methods in farming which maintain or improve resources without interfering with abundant production.

These objectives reflected a considerable departure from the traditional emphasis upon production which had long prevailed. Leadership training in every community to provide assistance in working with groups of people was stressed. These leaders would then assist Extension's small staff in reaching the rural people of the state. This clearly delineated view of the changing nature of rural America, and specific programming to meet its needs and wishes, placed Montana ahead, as it often had been before, of the national thinking of the time.

Tootell emphasized again in an overview of Extension activities in 1947, the need for creating in each county a planning committee which would advise in setting up programs of work for future years. The Extension field staff changed all too frequently and Tootell estimated that "only the exceptional agent develops a really strong program in less than about five years." A stable local committee could provide strength and continuity to the program. He urged that the committee be representative, including no less than 25 percent women, and at least one representative from the home demonstration council, the 4-H Club council, the county weed board, the Soil Conservation District supervisory board, and the county commissioners. Tootell pointed to 4-H as being one of the most successful divisions in Extension, observing that it had a desirable pattern of active advisory councils, strong community leadership and well-programmed activities.

Emphasis was placed on the hope that the county planning committees would work cooperatively with as many agencies as possible. Workshops were in vogue at the time, and a concept of "group dynamics" added a desirable effort to achieve action following the workshop discussions. The state staff stressed two aims: one, develop methods and procedures, and two, work toward the selection and training of leadership. A specific outline for workshops included:

1. Program determination—finding the needs of the people
2. Program development—finding solutions
3. Program execution—how to do it
4. Program evaluation—measuring progress

Two bulletins gave assistance to this program. One, "How to Decide What to Do," gave specific details on working with planning groups. The second was the well-known bulletin, updated with a new number, 1000, "Basic Facts About Montana's Agriculture."

Women's work received greatly increased momentum in June 1949 when many years of dreaming and planning were fulfilled by the organization of a State Home Demonstration Council. This will be discussed in Chapter Seven.

The 1952 Rural Progress Conferences

The College Advisory Council, in one of its sessions with the Extension leaders, while grappling with new ways and means of solving old problems in agriculture at mid-century, came up with an idea. The highly influential inventory and program projection conferences of 1927 have been mentioned. The council wondered if another series, which would survey the accomplishments of over 25 years and then look to the future, might be productive. The staff at the state level felt the idea was a good one, and in September 1951 the county agents came to Bozeman to organize the state for the conferences. Each of seven districts (instead of six as in 1927) would have a central committee and appropriate subject matter committees. Next, a statewide steering commttee would be set up to assemble information and provide overall direction. To prevent overlapping and add strength, the functioning county committees were asked to take the lead. The Extension Service discovered again that merely the planning process leading up to the conference had a great deal of value. Some 4,100 persons in more than 400 community meetings took part in preparing for the district conferences.

In the seven district conferences held during February and March 1952,

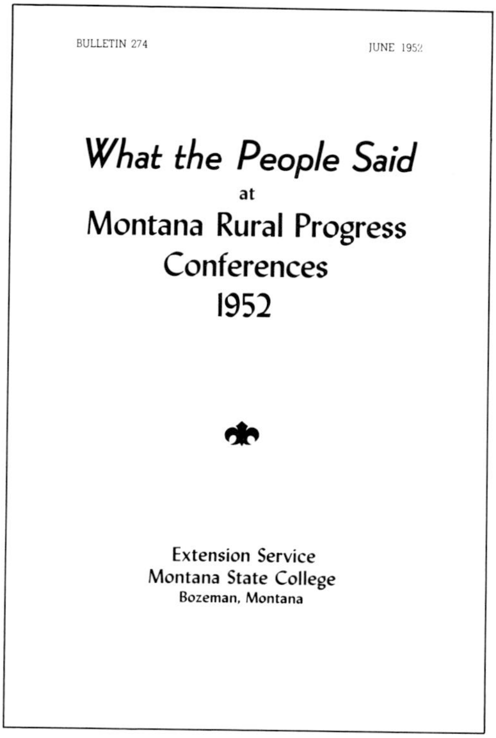
BULLETIN 274 JUNE 1952

What the People Said
at
Montana Rural Progress
Conferences
1952

Extension Service
Montana State College
Bozeman, Montana

Forty-five hundred Montanans attended the Rural Progress Conferences.

approximately 4,500 people participated, of whom about one-third were women and girls. The resulting report took the form of Extension *Bulletin No. 274,* "What the People Said at the Montana Rural Progress Conferences." Immediately apparent in the 25-year view was an increased emphasis upon people rather than things. The large number of women and youth who participated in 1952 underlined their activities in the home and community, whereas in 1927 men had completely dominated the conference, as did the problems of agricultural production.

New activities and services were emerging in 1952. They included such wide-ranging subjects as federal farm programs, farm credit, school consolidation, chemical, weed and pest control, rain making, and the youth programs of vocational guidance and military service. The need for broad educational and training experiences for youth and the broadest possible research programs dominated the scene.

Farm and Home Development

The Montana Rural Progress Conferences came at an especially opportune time. While their results were still being assessed in 1953, a national committee on Farm and Home Development and Public Policy was formed. Then Congress made a special appropriation for the program which was described as "a rather new field of coordination."

The new "unit" approach simply urged the agent and the farm family to sit down and consider the farm and home as a complete unit, and chart its overall development on a long-time basis. All of the specialists from the federal agencies, the banks, feed, seed, chemical and machinery companies should be called in. Planning considered the type of living which the family preferred, the type and extent of production, building plans, the children's education, and relationships with the community. A generous amount of consideration was also given to aspects of public policy such as taxes, credit, education, communication, price structure and marketing, national and international influences.

A series of articles appeared in the Montana *Farmer-Stockman* from October 1953 to February 1954, which publicized the approach taken. The public policy questions were emphasized, as the titles suggested: "Your Farm and Ranch Management Plan, Can It Be Improved?", "Records Aid Management, Income Tax Planning," "The Creative Use of Credit."

Director N. E. Beers, November 1, 1953

Newell E. Beers, assistant director, was advanced to director November 1, 1953, upon the resignation of Robert B. Tootell. His quiet, open-minded approach had made him highly effective on the state staff, and the state county agents' association requested that he be named director.

Beers was highly sympathetic toward the farm and home development "unit" program. He recognized that Extension had come full circle, since the earliest projects also considered the farmstead as a whole. He heartily subscribed to placing people first in consideration. The production aspects of acres, plants, animals, and silos and barns he rated as contributing factors to a good life on the land. He continued Tootell's enthusiasm for improving the quality of living among Montana's farm people, and he urged his staff to redouble their efforts. Beers commented in 1954 that the principle difficulty in making faster progress was a continuing negative "attitude toward change" on the part of both staff and people.

He appointed a state committee of nine specialists and supervisors to explore new methods and asked Acting Farm Management Specialist M. E. Quenomoen, Family Life Specialist Amy Wold and Home Management Specialist Eleanor Willson to work out the details of suggested approaches. New funds in 1954 provided additional personnel with time to aid families who requested help in making and carrying out plans for the farm and home as a unit. Beers summarized the Farm and Home Development project, "It is a personalized approach which has been tailored to suit the needs and desires of each of the many families with whom Extension has worked."

N.E. Beers, fourth director of Extension, 1953-1958.

The director recognized the need for maintaining a balance. He worked toward combining the results of the Rural Progress Conferences of 1952 with the Farm and Home Development program. In the fall of 1955 the county planning and advisory committees were asked:

> . . . to review agricultural conditions, income, loss or gain in soil and water conservation, and many other important factors affecting agriculture. They will be asked to recommend a progressive program in agricultural Extension, research, conservation, credit and other interests that they determine are important.

Director Beers planned for a similar report of progress each five years. By 1958, 51 counties were participating actively. In 25 counties, well-trained leaders were organized into advisory committees meeting three or four times each year. The remaining counties varied in their activities:

> Too often they are limiting their planning to commodities and to immediate problems of concern to the individuals, but not necessarily of importance to the county and to the community.

However, 122 active advisory councils and commodity groups with approximately 1,400 members were spending at least three days a year on rural progress movements.

In 1957 the director's annual report summarized the county agents' efforts in the farm and home management and development program. They were working on these problems with 327 families, 163 of which were in the low income group. One hundred seventy-eight new families had been added during the year. The plan of attack they reported included determination of goals, survey of resources, analyses of farm and home problems, development of plans, and evaluation of activities.

The year before, in 1956, the director's report had given considerable attention to the newer developments such as community organization, improved community services, health facilities, regional projects, national programs, world affairs, and a more effective citizenship. Of the dozen or so new programs, the report included detail on the four largest, which were:

Projects	Number	Communities	Voluntary Leaders	Persons Participating
Citizenship Activities	215	322	750	7,655
Health Facilities	72	262	595	9,516
Recreation Programs & Facilities	121	264	564	8,809
World Affairs	123	406	644	22,828

The director commented:

> Perhaps much interest is based upon a growing realization that the farm unit does not stand alone, but is part of the community, state and nation, and that success or failure of the unit may depend upon community activities or upon national policies and programs.

Inter-Agency Relationships

The New Deal period of mushrooming federal agricultural agencies, all exerting a strong influence upon the state structure, created a need for skillful coordination to keep all of the offices working together. When additional agencies were added as a result of the war, the director was often hard pressed in securing cooperation among them and avoiding duplication of effort. Merely a listing of the rich reserves of assistance available to the Extension personnel reveals the scope of the problem of keeping the working channels clear. Montana organizations, many of which Extension had assisted in forming and with which it had close working contacts, included:

1. Montana State College Advisory Council
2. State Home Demonstration Council
3. Montana Home Economics Association
4. State and County 4-H Leaders Association
5. State 4-H Advisory Committee
6. Montana Seedgrowers Association
7. Montana Potato Improvement Association
8. Montana Stockgrowers Association
9. Montana Wool Growers Association
10. Montana Turkey Growers Association
11. Montana Reclamation Association
12. Irrigation District Association
13. Society of Range Management
14. Montana Grain Growers Association
15. Rural Electrification Association
16. State Health Planning Council
17. Soil Conservation District Supervisors

Another series of agencies, many based in the federal or state government and some of a private nature, included:

1. USDA Council, state and county
2. State and County Production & Marketing Administration (PMA) committees, which later became Agricultural Stabilization and Conservation Service (ASCS) committees
3. Soil Conservation Service
4. Forest Service
5. Bureau of Indian Affairs
6. Bureau of Reclamation
7. Farmers' Home Administration
8. Federal Land Bank Associations and Production Credit Associations
9. State Farm Labor Committee and State Employment Agency, and eventually the U.S. Employment Service (USES)
10. The State Commissioner of Agriculture
11. State Forester
12. State Fish and Game Commission
13. State, regional and county fair associations
14. Farm organizations: Farm Bureau, Farmers' Union, and the Grange

The United States Department of Agriculture was also aware of the need for keeping a unified working relationship among its numerous agencies and those of the states. In 1945 the Secretary of Agriculture proposed that state and county USDA Councils be formed to coordinate the work of the federal and state agencies. In October the National Committee on Extension Organization and Policy recommended that such councils be formed in a selected representative manner except in counties where no more than three federal agencies were in operation, in which case all could affiliate.

The Montana USDA Council was organized in 1946 with Tootell as vice-chairman. Methods of cooperation were discussed and agricultural production goals for 1947 were agreed upon. The close of the war created immediate needs for readjustment, and the council worked intermittently. In 1947 it made a good sounding board for the discussion of the Missouri and Columbia Basin development programs. In 1948 Tootell was chairman of the state council which worked out long range agricultural programs for the state. These councils, made up of public employees, were only moderately successful and began to meet on call.

In 1955 Dean of Agriculture Maurice Kelso offered the services of the College by holding two meetings of all agency heads in order to broaden their knowledge of the activities of each and to provide better coordinated working relationships wherever possible. Another series of meetings took place in 1956, and in 1958 an effort was made to revive the county USDA Councils to full strength to avoid overlapping and misunderstanding between agency representatives.

Added Dimensions

Extension personnel were steadily concerned about the educational impact that they wished to make upon the state. They welcomed, therefore, the interest of the agricultural papers. The influence of the *Rocky Mountain Husbandman*, under the direction of the Sutherlin brothers, Robert and Will, has been mentioned. It gave considerable publicity to the work of the Farmers' Institutes and to early activities of the Experiment Station. No doubt the best single non-public outlet for Extension educational public relations has been furnished by the *Montana Farmer-Stockman*. This paper was established in 1914 and Chester Davis, its first publisher, his early editor, Harvey Griffin, and long-time editor, Lester Cole, have all been closely related to Montana State College and sympathetic with its purposes.

The Anaconda Copper Mining Company through its Fertilizer Department helped to usher in the new chemical age by cooperating closely with the College in its experimental and demonstration work. This was supervised for many years by Jesse Green, a graduate of Montana State, who was for a number of years a member of the Experiment Station chemistry staff, and also an experienced wheat farmer. Anaconda and a number of other commercial firms sponsored important 4-H projects—research and field demonstrations—the type of assistance which the railroads had sponsored so dramatically in earlier years.

The Montana conservation program expanded rapidly with the assistance of the New Deal Soil Conservation Act passed in 1935. The long background of concern for the best use of the soil in Montana went back at least to the extensive project along the Milk River under the Reclamation Act of 1902, one of the first such projects in the nation. The Montana Reclamation Association was also an active factor.

When Martin P. Hansmeier was named Extension conservationist in 1937, he instituted a vigorous program supported by the State Soil Conservation District Act and the state committees which were formed to promote it. To develop an educational program, former Director Taylor worked with Hansmeier during the summer of 1948 to map out plans for conservation education in each district.

Local programs observing a "Friends of the Land Week" were widely carried out with the assistance of the county agents. The education movement was continued in 1949 when at the third annual 4-H conservation camp, the Soil Conservation District supervisors entered into a firm agreement to assist county agents in a continuous educational effort with 4-H clubs. The Soil Conservation Service recognized the importance of close Extension cooperation and provided part of the salary of the Extension conservationist over a period of years.

Director Tootell was frustrated because state agencies and the general public failed to recognize the amount of work which many Extension activities contributed to the conservation effort. In 1952 he lamented that "unfortunately the Montana Extension Service is not generally considered by the public to be an agency which makes a vital contribution to the conservation movement." He then proceeded to assemble the story of its many conservation projects, which he presented to the Montana Association of Soil Conservation Districts. This gave Extension the image which Tootell felt it should have.

The conservation movement was greatly expanded in 1957 when Congress passed the Great Plains Act. This program was designed to coordinate the work in the several states which had drought conditions and wind and water erosion. It looked toward taking marginal crop land out of production and re-establishing a grass cover. A subcommittee of the advisory board worked effectively in diverting many cultivated acres into more acceptable use. Montana was slow in implementing this program because many leaders felt that conservation need not be taken seriously here. Favorable cropping years following 1957 confirmed this. Nevertheless, the secretary of agriculture declared the counties of Roosevelt, Judith Basin, Fallon, Rosebud and Hill eligible for long-term programs in taking marginal lands out of production. The study got under way, but the concensus was that "the going will be slow."

Rural Fire Control

Fire in the farm home had long been a major plague, but little in an organized way was done about it until 1941. Spurred by increased awareness of the possibility of war, a voluntary organization of several agencies administering public lands cooperated to form the Montana Rural Fire Fighters Service. The Extension Service was active in forming the organization. By 1946 the Fire Fighters were able to contribute $2,700 annually toward the salary of Ralph Bodley, who, along with other duties, assumed the work of Extension specialist in rural fire control. George Gustafson took over this work in 1947 and became state coordinator, with half of his salary paid by the Fire Fighters.

In 1950 with the outbreak of the Korean War, fire fighting was combined with the Civil Defense program. A separate status was obtained for a Rural Civil Defense program, which after 1953 incorporated the work of the rural fire control. The state forester acted as secretary for the Rural Fire Fighters, and this secured a sharing of the techniques and facilities available to the state and national forests.

Relationships with Farm Organizations

Extension leaders had long felt that if they could secure the active cooperation of the three general farm organizations, the effectiveness of both Extension and the organizations could be increased many times. The organizations made little attempt to work with each other, however, or with Extension. In 1947 "The Rural Health Program of The Montana Extension Service" was developed by a statewide Extension Health Committee, and all of the farm organizations supported it. This brought the remark that "Perhaps this is the first time these three have agreed on any sort of a program."

Many of the programs of the farm organizations were arranged at the national level, where opposition to some policies of the Department of Agriculture often prevailed. At the state level, then, there was considerable confusion as to which programs they should support. Director Taylor, through long years of personal acquaintance, had won the confidence of the leaders of all the groups, and this relationship has been retained by his successors. In 1947 the Farmers' Union stood well out ahead in numbers, with 18,000 members. The Grange had 2,000 members concentrated in the western one-fourth of the state. The Farm Bureau had only about 1,000 members with active groups in six or eight counties. More recently a fourth group, the National Farm Organization (NFO), with quite another approach to solving farm problems, developed a considerable membership.

The Extension Service had assisted in forming local Farm Bureau units, as has been noted. In fact, the enabling legislation in Montana recognizing the Extension Service and the Farm Bureau was contained in the same statute.

This statute still existed in spite of the fact that in Montana, as in most other states, an official relationship had not been maintained with the bureau because the units had joined to form the American Farm Bureau Federation shortly after the end of World War I.

Director Tootell felt that there was good reason for suspicion toward the Farm Bureau and Extension so long as this legislation existed. In 1949, therefore, he and the president of the State Farm Bureau proposed an amendment to the statute, removing the relationship. This was another measure to which all three organizations gave support. The amendment was introduced by a Farm Bureau member and supported by the Farmers' Union and the Grange; the measure passed readily. Extension worked regularly with each of the organizations, and Director Beers continued to search for projects in which all could cooperate.

Extension Finances, 1946-1958

Tootell voiced pleasure in his first report in 1946 that "The Montana Extension Service had more funds available for its work the past year than in any other year in its history." An expansion by Congress in the Bankhead-Flannagan Act helped to bring about this pleasant situation and allowed Extension to increase its activities in the county work. The war inflationary trend was no doubt responsible for an increase in salaries by 20 percent from 1940 until 1946. On July 1, 1946, the increase for the year averaged $200. This trend continued and the average salary for agricultural agents increased from $3,294 in 1946 to $6,400 in 1958. Each year most county commissioners resisted further increases, saying that county agents were the highest paid county employees. Director Beers countered with the statement that when the county employed doctors, lawyers, engineers and superintendents of schools, they were all paid more than agricultural agents. On one occasion Tootell removed both experienced agents from Chouteau County when the commissioners refused to approve an increase. Very soon the farm leaders became so vocal about the matter that new agents were employed.

This was a national problem, and actually Montana counties gave more than average support. Taylor, Tootell and Beers all remarked about the consistently high proportion of county support in Montana. Because of this, beginning in 1948, Tootell felt that it was important for him to meet personally with the county commissioners. He met with 19 boards in the 49 cooperating counties that year.

Again in 1952, Tootell noted that, "Unusually large county support is the most significant feature of the financing of the Montana Extension Service." He listed as the proportion for that year: counties, 34.3 percent; state, 33.6; federal, 32.1. Federal support was occasionally first in importance. By 1958, however, Director Beers could show that the state was assuming its reponsibilities to a greater degree. His listing for 1958-1959 showed the sources as:

Federal	$442,992	30%
State	518,117	35%
County	440,410	30%
Other	75,387	5%
(B.I.A., State Fish & Game, etc.)		

The total increase in funds from 1945-1946 to 1957-1958 was highly encouraging, rising from $422,200 to $1,476,906.

Recruitment, Training and Supervision

The expansion in the numbers of Extension personnel was also encouraging. In 1925-1926 the total Extension staff had numbered 45; 20 years later, in 1945-1946, this had reached 70. Fourteen years later, in 1958-1959, the number was 148, with 39 on the state staff and 109 working in all but three of the 56 counties.

The problem of replacement in a day of rapid technological advance was a severe one. As has been noted, academic courses were set up in 1946 for Extension training. The number of courses was small, but 13 enrolled in 1946. By 1949 the 17 who enrolled were more than could be employed in Montana. The situation changed quickly, however, and in 1951 there was again a shortage of replacement personnel.

Tootell was concerned that so few colleges were offering preparatory work for students interested in Extension work, and that subject matter departments did not urge their best students to go into Extension services. Speaking before the National Association of County Agents, December 1, 1952, on the subject, "The Future of Extension in the Land Grant College System," he stressed the need to "sell" Extension to the teaching faculty of the colleges, noting that the teaching, or so-called academic portion of the land grant institutions did not recognize either the Experiment Station or the Extension Service as "members of the Land Grant College team."

Tootell stressed even more than other directors that the work of Extension was a teaching job. He earned his master's degree in California, and took leave 1949-50 to complete the coursework for a doctor's degree in public administration at Harvard. He was sure that formal course work would be beneficial to his staff if he could find suitable programs and make satisfactory arrangements for them to attend. In this he had the full support of President Renne. By 1948 the State Board of Education was induced to place Extension personnel on a comparable basis with the teaching faculty by giving them the eighth quarter leave at full salary if engaged in activities leading to professional advancement. This landmark was emphasized by making the theme of the 1948 annual Extension conference "Effective Extension Teaching."

The staff in other areas of the state still had a disadvantage because county commissioners felt that they could not legally pay the county portion of the agents' salaries if they were on leave. To assist with this rather widespread situation, the Farm Foundation and the Horace A. Moses Foundation established a number of scholarships, and Montana agents were often recipients of these. In later years the agents on leave were paid from funds available to the state office.

Colorado State College sensed the need for short course programs for Extension personnel and set up special summer courses. Although the summer absence was often difficult, six Montana agents attended the 1948 three-week summer session. A similar number attended in 1949, and an additional six took a full year's work leading to advanced degrees. The state staff asked for advanced work in such numbers that limitations were established concerning the numbers to be absent at any one time.

The turnover in agents continued to be high, and the replacement personnel often had little training for the work. "New Workers' Conferences" were arranged periodically, and in 1953 an Advisory Committee on Professional Improvement for Extension Workers was formed to make recommendations on training and to develop a reclassification of professional rank. Harry A. Cosgriffe, former agent supervisor, became personnel and training leader in 1956. He coordinated the dual fields of training in the field with further academic study. By 1958 he had arranged an attractive series of short-term courses in several colleges, such as:

Communication workshops
Crops & Soils training schools
Junior Leader training schools
New Worker's training conference
Business Facts training schools
4-H Leader tractor schools
Livestock marketing schools on grading, shrinkage, and carcass judging
Farm and Range Management schools

Another important development in expanding personnel experience was the opportunity to assist with Extension programs in developing countries. Martin P. Hansmeier, extension conservationist, was granted leave March 1, 1950, to join a team of specialists going to Turkey under an FCA program. He remained for a second year, and was so impressed with the opportunities offered that after two months' home service he resigned and devoted a number of additional years to the Mideast development program. In 1950 also, Geraldine Fenn spent five months in Germany in educational youth programs. She made additional shorter visits to work with the International Farm Youth Enchange. In 1951 three specialists took foreign assignments, creating a condition in the state office which led to the policy that not more than three of the 15 specialists should be gone at one time.

Continued grants of leave for foreign service enriched the professional proficiency of the Montana staff through the contacts made and the experience obtained.

Rural Development Committees

In 1957 still another movement developed which encouraged a high degree of flexibility in county planning. This took the form of a State Rural Development Committee, which would in turn encourage county development committees. The plan was to form committees which would assist in carrying out projects adopted by the central development committee or by community groups.

The state committee was formed under the direction of Dr. Maurice Kelso, chairman. Kelso was dean of agriculture and also director of the Agricultural Experiment Station at Montana State. Other members were E. H. Sandberg of the Soil Conservation Service; John C. Bower, Extension economist; Perry Roys, director of the Montana State Planning Board; R. W. Arnold, assistant state forester; Dr. Carl Kraenzel, sociologist; and James S. Umber, president of the Montana AFL-CIO.

The plan had some background in an activity which took place in Ravalli County. A Ravalli County Improvement Association had been formed in 1957 and carried out several projects. This, in turn, had a background in a county evaluation study which had been made in 1945-1947 by Montana Study discussion groups. The resident county agent, Bill Thomas, was interested in the possibilities, had the confidence of his people, and took the lead in forming the county committees.

By 1958 Ravalli County had an active coordinating committee and subcommittees which had formulated extensive programs. These subcommittees were:

1. Small Industry Committee.Would offer aid to some small industries in operation, assist others in forming, and work toward a better marketing organization
2. Tourist and Recreation Committee. To publicize attractions of the valley, improve guide service for sportsmen, and increase available camping sites.
3. Agricultural Committee. Would improve irrigation systems, increase production of fruits and vegetables, improve hay production and marketing.
4. County and Community Improvement Committee. Secure aid of Army Engineers to control bank cutting on the Bitterroot River, extend and improve roads into scenic areas, study human resources in low income areas of the valley, encourage leadership training for various fields.

In December, Lake County requested support in forming a similar Rural

Development Committee, and Theodore Witzel was employed to give it direction. The program in Lake County was, however, much more traditional. Its subcommittees in the beginning focused on weed control, dairy marketing and land reclassification.

Broadening Objectives

The organization for the Rural Development Committees took place in the spring of 1957. As a result of the ideas which dominated the movement, John Bower, Extension economist and secretary for the State Committee, formulated an outline for a much broader Extension program. It was presented in July to the Montana State College Advisory Council in Havre. Objectives of the broader program were, he felt:

1. Increasing the understanding and vision of rural people
2. Developing leadership ability among rural people
3. Increasing the effectiveness of Montana State College in meeting the needs of rural people

These objectives might be reached by:

1. A conscientious effort to develop short and long-term programs which would help rural families reach their individual and group goals
2. Developing a method of helping people define their problems and solve those problems
3. Perfect a method of getting more depth and breadth in county Extension programs

At this same time a specific inquiry carried out among the farmers and ranchers showed that they, too, were including a broadened scope of activities in their recommended programs, but production activities still took first place. Their preferences were:

1. Livestock and Poultry Production
   breeding, health and nutrition
2. Soils and Crop Producton
   genetics, insects and disease, soils and fertility
3. Resources
   management, development, conservation
4. Economics
   marketing, farm organization and management, farm tenure, credit and financing
5. Government
   taxation and finance, roads, Social Security and welfare, land reclassification
6. Education
   schools, vocational training, libraries, citizenship

7. Youth
   leadership training, youth organizations—4-H, etc., pre-marriage counseling, military service orientation
8. Family Living
   health and safety, home management, recreation, religious and spiritual values

Probably the hallmark of the Extension Service at mid-century was the facility it had achieved in assisting rural people to group and re-group to consider problems and programs as diverse as those in the above listing. Improved production and disease control methods had made and were still making great strides. Questions of management, marketing, finances, education and family living were demanding increased attention. Extension had played a major part in creating an atmosphere in which the rural people had become adept in the art of community cooperation and in the business of planning and carrying out major enterprises. In 1958, 51 counties were cooperating fully with the Extension Service. The organization was more comprehensive and closely coordinated than ever before, better financed and more widely supported by a people who made less and less distinction between rural and urban interests.

VII. The Homemaker and the Extension Service

"Much attention was given to home economics instruction and recommendations were made for the permanent employment of an instructor in this subject," said Director F. S. Cooley in 1907, referring to the Farmers' Institute program. Proof of this lay in the employment of Miss Lucile Brewer for a short time, as well as the recruitment of several nationally known women for state conferences. Funds were in short supply, however, and priorities usually were given to agricultural production, with the expectation that increased profits would also mean increased benefits for the home.

Kathrine Jensen

Funds became more adequate with the passage of the Smith-Lever Act in 1914, and Director Cooley continued in his *History:*

> Home economics Extension work came into the state program at the beginning. Miss Kathrine Jensen of Fargo, North Dakota, was the first state home economics leader (appointed July 1, 1914). Few teachers are Miss Jensen's equal and none could surpass her in her hold upon the rural women of the state. Her splendid physique, fine training, genial personality, and prodigious industry stamped her work indelibly upon the minds of Montana people. She remained two years and was promoted to the head of the department in her alma mater, North Dakota, and later to the head of the Division of Home Economics in Idaho University.

Miss Jensen had only volunteer help. She found so much to do, and she so thoroughly enjoyed doing it in her semi-tornadic swirling about the state that she left few records and took time for very little formal organization work. But her energy and enthusiasm were so immense that her impact was permanent.

Miss Jensen's former public school teaching experience along with an immediate awareness of diet dieficiencies in Montana homes led to one of her first inquiries: What about the children in the rural schools? She found a distressing amount of malnutrition. Following a national trend, she began at once to promote a hot lunch program, and this proved to offer some hurdles in the small rural school on the plains. She persisted, however, and compiled recipes which would require a minimum of difficulty in stocking supplies and in preparation. In serving, she urged that the teachers enlist the cooperation of both the girls and the boys. She found ready cooperation among many of the teachers and the school boards. The results were so immediate and encouraging that the parents began to adopt improved practices in the home.

Her first tours over the state to work with homemakers were arranged with one- or two-day stops, but soon the women organized and insisted upon three- or four-day short courses in which a variety of lectures and demonstrations were given. With deficient nutrition in mind, she went into methods of preparing basic foods and set up demonstrations on cooking vegetables and meats and on the use of milk, eggs and cheese. She was particularly concerned with the miserable breads she found. Not only were there problems of poor stoves, unreliable ovens and unstable fuel, but there was also a great dearth of knowledge, skill and imagination in preparing the several types of bread which would attract children and adults alike. Simple dessert recipes were also in steady demand.

Interspersed with her lecture-demonstrations concerning specific foods, Miss Jensen also conducted informative and highly enjoyable general lecture-discussion sessions on such subjects as meal planning, homemaking efficiency and home convenience. The word soon got around that one of her liveliest discussions was that of "The Problem of the Daughter." The meeting room was usually crowded with the entire family, inluding the daughter, in attendance. She not only spoke to the regularly organized Farmers' Institute groups, but she used any free time available to speak to women's clubs of various types, and to church groups; she felt that she perhaps received the warmest support from men's groups.

She was also interested in boys' and girls' work, and since competition in local and regional fairs was strongly in vogue at the time, she spent a considerable amount of time helping them prepare exhibits. She joined in promoting fairs over the entire state, and judged extensively in them. During her vacation month, Miss Jensen noted that a considerable amount of her time was spent in work on bulletins, purchasing equipment and preparing lectures for the new season.

She was not averse to poking fun at herself if she could make a point related to homemaking. She sometimes used illustrations which came from her public school teaching days, one of which Miss Cushman recalled. It seems that Miss Jensen had a girl in her class who was utterly unable to develop an interest in geography.

> Miss Jensen sent a note home to Sally's mother suggesting that help be given at home. The next day a note came back from the mother stating, "Dear Miz Jensen, I don't mind if Sally can't do geogerfy. I never could do geogerfy, but I got married. Sally's Aunt Gertty couldn't do geogerfy and she got married. You can do geogerfy but you are an old maid. I don't want Sally to be an old maid.

Miss Jensen felt that the story illustrated a number of points in the social and educational structure of the time. Nevertheless, if the girls couldn't do "geogerfy," she was personally going to see to it that they could do home economics.

Her terse business-like diary revealed an impatient woman-on-the-go. For instance, the June 21, 1916, Ennis entry read: "Home Economics short course. Funeral in town. No meeting." June 26, Great Falls. "Great washouts prevented making Cut Bank." June 27th, Conrad. "Unable to meet appointments, mud-mud-mud." June 29th, Choteau. "No meeting. Roads impassable." This week was no doubt particularly exasperating, since it would be her last visit in this section of the state. Her work in Montana terminated on August 1, 1916. She would not have phrased a formal definition of home demonstration work, as did Blanche Lee in 1926:

> The ultimate aim of home demonstration work in Montana is to increase the efficiency, attractiveness and comfort of the rural home to the end that we may have a more healthful, happy, intelligent, efficient and contented rural population.

But Kathrine Jensen, as the first home economics specialist, did much to put this definition into effect.

Bess M. Rowe and World War I

Bess M. Rowe succeeded Miss Jensen on September 15, 1916, and she was fortunate in securing an assistant, Mina Ogilvie, on October 1. They followed the pattern of "holding schools" and prepared a considerable amount of compact equipment for demonstration use which could be moved with ease and speed. From October 1 until February 1, 1917, 20 schools or short courses and 72 single meetings were held.

By this time the idea of home demonstration agents at the county level was coming into use in the northern and western states. Miss Rowe recalled that in early May 1917, she and Director Cooley attended a conference of the 11 western states in Logan, Utah:

> Both Director Cooley and I were somewhat skeptical regarding this type of work but at this meeting Miss Ward assigned to Montana sufficient funds to put on the work in a demonstration county for two months—May and June.

Missoula County was chosen for this experiment since the county agent was highly cooperative, and a capable group of rural leaders promised assistance. Mrs. Clara Aust Bush, a graduate of the University of Minnesota, was appointed and did a successful job of stimulating interest in the new project.

Just as plans for the spring 1917 work were complete, World War I broke out and created the need for drastic changes. Comment has been made on the organization for war and the Flying Squadrons which swept through the state between June 16 and July 16, holding the enormous number of 387 Farmers' Defense Meetings. Each squadron was accompanied by at least one woman, well trained in home economics, who held special meetings with the women of the community, discussed the question of food preservation, and usually demonstrated the new pressure cooker recommended for canning meats and vegetables. These meetings were attended by 43,000 people. Mother and daughter canning clubs were formed in 61 communites.

In early summer 1917 the emergency bill "To Stimulate Production and Facilitate the Distribution of Products," passed Congress, and in August $16,000 was made available for home demonstration work in Montana. Eight of the 20 women who had been with the Flying Squadrons were chosen for these positions and placed in the following counties:

Fergus County—Mary Newell
Gallatin County—Mrs. R. G. Young
Missoula County—Mrs. Clara Aust Bush
Flathead County—Christine Wood
Sheridan County—Antoinette Olson
Helena District: Lewis & Clark and Broadwater counties—Lucile Reynolds
City of Great Falls—Nevon O. Betz
Yellowstone District: Prairie, Custer & Rosebud counties—H. Marie Barnard

In 1918 additional funds became available; new agents were employed and changes were made as follows:

Stillwater County—Jeannette Kelly
City of Butte—Mrs. R. G. Young from Gallatin, succeeded in Gallatin by Mrs. Georgia Roosevelt
Northern District: Phillips, Blaine, Valley and Richland counties—Gertrude Erickson
Musselshell—Elizabeth Fletcher, summer assignment
Wheatland—Kathleen Wilson, summer assignment
Ravalli County—Madge Switzer, succeeded by Ella Harmon
Blaine County—Gladys Dawson, when Blaine withdrew from the Northern District

The federal government gave extensive direction concerning the first program to be undertaken, that of producing, preserving and conserving food. Then the state organization outlined four major areas for study, organization and demonstration:

A. Food
   Proper feeding of the family, food conservation, food preservation and school lunch
B. Home Efficiency

C. Clothing
   Selection, care, and renovation

D. Health
   Maintaining health through right living; home care of the sick; household emergencies.

From War To Peace

The state staff recognized from the beginning that federal support for the entire county home demonstration personnel was of an "emergency" nature, related to the war. State officers moved as rapidly as possible, therefore, to put Extension work more largely on a state and county base of permanent financial support and leadership. The Farm Bureau, supported by the Extension Service as has been explained, gave major support. The 1917-1918 annual report of the Montana Farmers' Institutes commented on this organization:

> Twenty-three county farm bureau organizations have already been effected in Montana. In like manner, women's work organizations have been formed in 15 counties. These women's organizations, for the time being, have been termed "Women's Patriotic Councils."

The Farmers' Union, Grange and the Society of Equity also provided effective assistance, since all of these associations were organized more on the family base than was the Farm Bureau. An effective movement was begun in 1918 to add women's units to the bureau, which had started out largely as a men's organization.

The first and most important step was to secure an enabling act by the state Legislature to authorize county commissioners to budget home demonstration agent work, just as the 1913 act had allowed them to support an agriculturist. This action was taken in the 1919 session.

Miss Rowe felt that favorable action on this bill was strongly influenced by the women who came to Bozeman in February to the annual Farmers Week. Many of them stopped in Helena and interviewed their delegates in the legislature.

Only seven of the 15 emergency women agents survived the transition from the war period. These were in the counties of Blaine, Phillips, Ravalli, Flathead, Fergus, Cascade and Stillwater. In most cases the failure to continue was due to the lack of a supportive county agricultural agent, or of a strong women's organization. Miss Rowe wrote a famous report to the State Board of Education in December 1919 in which she reviewed the women's Extension program. She took an optimistic view in spite of the fact that half of her structure had toppled:

> When we consider that this work was unknown in Montana until about two years ago, that it was started as a war measure, and that the continuance of the work after June 30, 1919, necessitated the passage of new

laws by the state Legislature, and the action of county boards in each county, we appreciate more than ever the loyal support and hearty cooperation of the Farm Bureau women of Montana through whose efforts the action has been secured in every case.

She felt that the children's health program and the institution of hot lunches in the schools, for instance, should be credited with winning many enduring friends for the Extension program. Carefully selected statistics showed that during the war the women agents had stimulated the saving of $83,875.99 through home projects. Gardens had accounted for savings of $42,000, or over half of the amount, and clothing renovation for $10,163.

Mignon Quaw and Recreation

Mignon Quaw was named as a second assistant state leader January 15, 1918, on the basis of her ability to create and direct recreational programs, and because her out-going personality made her an effective trouble shooter in many of the new situations which arose. Her first project, a three-act play, "The Long Trail," has been described. This play, termed "the first truly Extension play produced in the country," and another titled "Nutrition," as well as her innovative entertainment programs attracted national attention and continued to be used through two wars and several periods of drought and depression.

Miss Quaw's programs for games, physical exercises, singing and dramatics arranged for do-it-yourself performances, especially for farm and small town groups, brought this comment from Miss Rowe in her December 1919 report:

> No work that has been done by this department has been more cordially received, nor has any apparently had more real influence in the rural life of Montana than has the rural recreation project. Although the value of this project cannot be estimated in dollars and cents, the statements we are constantly receiving from the people of Montana prove that the returns from this project are tremendously large, and that they must be figured in terms of better citizenship rather than in terms of money.

Miss Quaw also had a flair for display and the exhibits which she designed for state and county fairs are still remembered. Today they still help recall the progress which has been made in small but important essentials in everyday living. A state fair exhibit in 1919 on home conveniences took the form of a small cemetery with headstones bearing the following inscriptions:

MOTHER: Walked to death in her kitchen.
GRANDMA. Washed herself away.
COUSIN ESTHER: Gone where there are no more dull butcher knives.
SUSIE. Swept out of life by too heavy a broom.
Sacred to the memory of ____________. Scrubbed herself into eternity.

PRISCILLA. This is the first time she ever lay down to rest.

AUNT MARTHA. Too tired to twang a harp.

A larger sign at the back of the exhibit bore the legend: "Oh Death! Where is thy sting? A good washing machine costs about one-fifth the price of a first class funeral." Prominently displayed among the headstones were examples of fireless cookers, pressure cookers, iceless refrigerators, late model washing machines, light brooms, drop leaf tables, improved lamps, inexpensive easy chairs, effective mops, and other conveniences recommended by the home demonstration programs.

Dress Forms

Perhaps the most unusual project in clothing education, and a highly successful one, was the making of dress forms. Rural women, often isolated on farms and ranches, found the forms invaluable in making their own clothes. Myrtle MacDonald, home demonstration agent for Cascade County, introduced the method for making the forms in November 1918. The first forms were made of plaster of paris brushed on a tight garment worn by the member Stories were rife for years of the difficulty of making them, and of the exhaustion suffered, to say nothing of the abrasions received when the forms were finally cut from the victim. However, 422 women, stoically or otherwise, had sat through the process by 1920. By this time, a method of pasting gummed paper strips on a tight undergarment had been devised, which proved to be cleaner, lighter, faster and less expensive, as well as easier to manage when in use. This process usually took less than an hour in comparison with at least a half-day with the plaster of paris model. The dress forms remained popular well into the 1930s and the project of making new dresses or restyling old ones was often the top attraction for both mothers and daughters.

The first clothing specialist, Mrs. Georgia Roosevelt, assumed her responsibilities on July 15, 1919. But before she took over the duties, she took a two-month course at Teachers' College, Columbia University, for special work in dress construction and millinery. The making of hats was also a famous enterprise in this early period of adventures into home economics. Through the experimental work of the women in home demonstraton clubs, rural women soon became better informed than their city sisters, and whether homemade or wisely purchased, the differences in the dress of rural and city women disappeared within a few short years.

Blanche Lee and Home Demonstration Clubs

When Bess Rowe left on October 15, 1920, to take the position of field editor for *Farmers' Wife* magazine, Anna K. Turley, home demonstration leader from Oregon, came to take her place. Miss Turley resigned July 1, 1923, and the assistant leader, Blanche M. Lee, began a particularly produc-

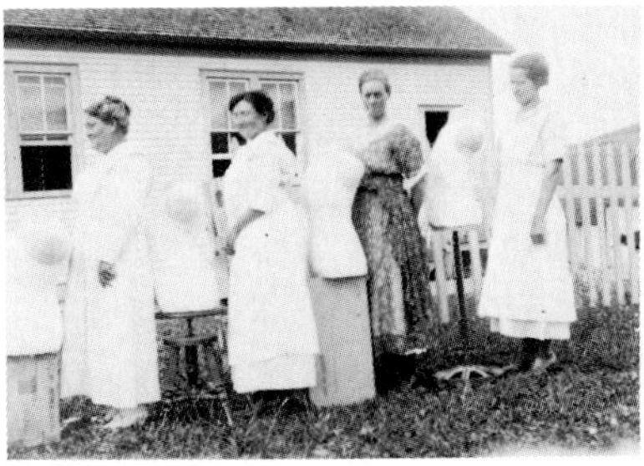

(left) Making dress forms was popular in the twenties and thirties.
(right) On their way home from an early Homemakers Club meeting.

tive period of 13 years as leader.

The county organization for home demonstration work was what Miss Lee called "rather loose." Her strength was in organizing, and the many clubs entered at once into a period of orderly arrangement and rapid growth. Many of the women's groups had formed as purely local and grass roots clubs, taking various names, and usually but not always aligned with Farm Bureau clubs. Many were not connected with the State Extension office. Miss Lee worked out a revised plan for a state organization, which she presented to the extremely active clothing and dress form clubs, and to the numerous Women's Camps in the summer of 1924, receiving firm support from them. She then prepared a new bulletin titled "Home Demonstration Work in Montana," and community groups were asked to take the name of Extension Home Demonstration Clubs. This was a landmark in the growth of women's work in Montana. A suggested constitution, bylaws and secretary's book were supplied to each club. Nineteen counties and 170 communities accepted the new format at once.

By 1926 only six home demonstration agents were on duty in the counties, but in 25 counties 152 home demonstration clubs with 2,509 members were functioning. The state staff consisting of the leader, the assistant leader and three specialists—for nutrition, home management, and clothing—assisted these agents and clubs.

The staff first made sure that the affiliated clubs were of a high standard. Then they gave every aid possible to the other counties where often the agricultural agent was doing an excellent job of "mothering" the Home Demonstration Clubs. A disturbing factor was the rapid change in this small staff. The average length of term for the state leader, 1914-1926, was only three years, and for the entire staff only two years and six months. Rapid growth continued to take place, and in 1927 the clubs numbered 251 and the membership 3,599. This membership still touched only a fraction of the 49,906 farm homes listed in that year.

County Councils and Project Leaders

A distinct improvement in organization came in 1927 when County Home

Demonstration Councils were formed in Valley, Carbon and Rosebud counties. Flathead followed in 1929. These councils were made up of the presidents and secretaries of each club, at least, and often project leaders came in numbers to promote their work. The periodic meetings of these representatives did wonders in generating ideas and enthusiasm, planning for unified work such as fairs, achievement days, tours, demonstrations and other activities.

In order to cover the vast areas of rural America and carry out the numerous activities, the Extension Service has always relied heavily upon its local "leader" system. Indeed, the development of this bedrock leadership is no doubt one of the crowning achievements of the Service. The officers of the many clubs, also bedrock leaders, were of great importance, but very early the project leader who developed the skills necessary to direct specific projects from making dress models to school lunches to culling chickens, became the strength of the movement. During 1928 five counties adopted the local project leader system. Training schools then became the pattern, and 75 were held during the year, with 1,012 women attending. The enthusiasm of the time was reflected by such incidents as two community directors coming by horse and buggy 65 miles to attend a leader's school. In another county, a woman came 36 miles on horseback in frigid weather. During 1928 the clubs increased to 421 and the membership more than doubled, to 7,359.

Foods and Nutrition

The two major fields of specialization which appeared first in women's work were those in clothing and home management. The first state leaders and their home agents began in 1914 to set up groups for study in these areas. The immediate coming of World War I raised an equally important consideration—that of food, its production, nutritive value, preparation and preservation. Miss Jensen had found that the children in the schools were suffering from malnutrition in the midst of plenty. When the young men from the rural areas appeared in the enlistment tests to be inferior in health and physical standards, rural leaders were ready for action programs.

As the close of the war approached, Red Cross personnel became available and the home economics staff again turned to a children's program in the schools as one which could be organized quickly and efficiently. Clinics on "Child Feeding" and others on "Child Welfare" were established. The response was enthusiastic but the problems of nutrition called for special skills and this led to the appointment of Miss Mary Ann Graber as foods and nutrition specialist, September 1, 1919.

Miss Graber, with great energy, set out to establish a broad and sound program, and in doing this sought the support of the State Board of Health and the State Tuberculosis Association. Tests in the schools brought to light a discouraging situation. One test revealed, for instance, that in three of the state's most prosperous counties 18 percent of the children were 10 percent or more underweight. Miss Graber presented her findings and her program to the January 1921 Extension conference. As a result, five test classes were formed by March, with 100 children in the program. These were not formed in rural schools, however, but in Billings, Glasgow, Park City, and two in Helena. The availability of physicians and nurses to assist with the program was of importance. For the school year 1921-1922 the number of classes expanded to 18 and the number of children to 360. The public health physicians who were maintained in Missoula, Cascade and Yellowstone counties also offered invaluable assistance with school programs and provided convincing evidence that the county Public Health Office was desirable. The home demonstration agents also gave strong support to the program, and all of them by various means undertook refresher work in foods and nutrition.

That weight and measurement did not reveal all of the problems of child health soon became apparent, and the program was broadened. Local physicians and nurses were enlisted, and clinics were held checking tonsils, teeth, eyes and ears. This led rapidly to the expansion of the county health units, and the Extension personnel were able to withdraw from the medical aspects of the program.

Miss Graber created a great deal of interest in the whole field of health, and became additionally enthusiastic about it herself, to the point that she resigned in September 1922, and continued study which led to an M.D. degree and a profession as a practicing pediatrician.

Miss Graber's work had been diagnostic and organizational to a con-

Weighing children at a rural school on a portable scale, 1920.

siderable extent. Her successor, Miss Martha Mae Hunter, was able to turn back to basic aspects of foods and nutrition and prescribe treatment. She wrote an attractive bulletin, "Good Meals for Children," which together with three popular leaflets, "Eggs," "Cheese," and "Cottage Cheese," changed the pattern of eating for many rural children and their parents as well. A "Nutrition Handbook," issued as Extension *Bulletin No. 62*, September 1923, also by Miss Hunter, provided a useful educational tool widely used by staff and the general public. Campaigns stressing the use of vegetables and "carry-milk-to-school" were national in scope at the time, and Montana was well organized to participate.

Miss Susanne Thompson, who succeeded Miss Hunter, continued to use the bulletin approach by writing one on hot school lunches. This was especially helpful in the small and isolated rural Montana schools where hot lunches were often difficult to manage. Milk was the key to relieving much of the malnutrition. When a survey in 1926 revealed that 27 percent of the children in three of Montana's most prosperous counties, Gallatin, Lewis and Clark, and Yellowstone, were highly deficient in the consumption of milk, specialist Jessie McQueen set up the largest single campaign of the early years. Her "Milk-for-Health" project in these same three counties was highly successful.

An Achievement Day in the schools toward the end of the school year proved to be a popular incentive for accomplishment. The event often took the form of a community festival with a program by the children and the awarding of ribbons to those who had made significant progress in nutrition practices. Since the awards could be proliferated almost indefinitely to include not only those who were improving their weight, but those who were cleaning teeth regularly, drinking milk regularly, and developing other desirable health practices, almost every student in the school participated and received recognition of some type.

This program attracted regional attention, and when Miss Frances Smith became nutrition specialist in August 1929, she was asked to present a paper at the regional meeting in Salt Lake City on "Montana's Blue Ribbon Children." She outlined the accomplishment in modest terms so far as its extent was concerned and regretted the limited staff which restricted the size of the project. She did, however, give high praise to the close cooperation in the schools by the local leaders and the homes which made the program a great success. It was largely this cordial unity of effort, which often appeared in Montana projects, which caused the undertaking to receive regional attention.

The vigor and widespread influence of the foods and nutrition movement is difficult to convey in words. Specialist Jessie McQueen's 1926 report reflected this interest by listing 319 nutrition meetings with 10,729 adults attending, and with 8,690 children in a separate program. In Richland County, to use only one example, the percentage of underweight children dropped in one year from 28 to 6 percent.

The nutrition program was quickly broadened from its emphasis upon children to include adults as well. Bulletins such as "Appetizing Vegetable Dishes" and "Recipes for the Use of the Whole Grain" were popular.

In 1929 a Home Health and Sanitation project received more support than any other offered that year. A plan for a series of meetings on "The Home Care of the Sick" was so widely requested that Mrs. Henrietta Crockett from the State Department of Health, and one of the best known of Montana's early public health nurses, was induced to assist.

The peak pressures on the Foods and Nutrition program came in the frequent periods of drought, depression and war in Extension's early years. Drought brought double difficulty because often home gardens, fruits and poultry were greatly diminished in supply. During periods of depression and war, the community cannery was rather widely used with state and federal aid providing the pressure machinery and the operator. The statistics of the food preserved and of family savings with and without the cannery were almost unbelievable.

The foods and nutrition movement developed concurrently with technological growth which was phenomenal in both speed and extent. In a summary of progress, 1919-1939, Miss Bess McClellan, who became nutrition specialist in September 1936, wrote:

> It has been a long step from the iceless refrigerators and fireless cookers of 20 years ago to the electric refrigerators and automatic ranges that are gradually going into hundreds of electrified rural homes today. Perhaps even more amazing is the progress from no canning of meat or vegetables, through the era of slow water bath method with its long hours of kitchen heat on hot summer days and its often inadequate processing, to the greater speed and efficiency of the pressure cooker, and newest of all, the freezer lockers with their products retained in garden-like freshness. Homemakers have had to learn new methods and readjust their thinking and their plans of work.

These technological changes also kept the Extension personnel busy. The revolution which came with the electrification of the rural home created the need for an extensive series of demonstration conferences. Proper wiring, proper lighting, effective equipment, the best ways of using refrigerators and electric stoves, and how to heat the kitchen when the old stove was replaced, all needed working out—sometimes on a house-to-house basis. The new freezers brought many new problems. To begin with, they were commercial locker-freezers, often on a cooperative community basis. The operator as well as the consumer asked for advice. Schools for cutting and wrapping meat, for instance, were extremely popular. The preparation of vegetables and fruit for freezing required study on the part of everyone.

Health officers were concerned about the operation of the freezer-lockers and Extension personnel took a large part in inspection and consultation. There were 37 locker plants in the state in 1939, the first year Miss McClellan noted the new development in her report. She recalled visiting one plant where considerable spoilage had taken place. Upon questioning, the manager admitted that this was all new to him, and that since the plant was not used much during the weekends, he had turned the power off on Saturdays and Sundays.

Both the commercial and the home freezers opened up opportunities for the preservation of wild game. Its proper cutting and care also brought many questions and additional demonstrations.

The Experiment Station scientific personnel and the Extension staff were squarely in the middle of the revolutionary changes. They were fortunate, indeed, that so few of the changes were of a controversial nature. The complex nature and value of vitamins and the proper treatment of foods to best preserve them resulted in more confusion than almost any other question, but the disputes usually took place only among research scientists.

The intricacies of high altitude cooking (baking particularly), and the value of enriched flour came in for extensive discussion. The highly volatile question of the chlorination of water was always promptly turned over to the Department of Health. The comparative merits of butter vs. oleomargarine caused much political controversy for a short time. In this, the Extension people usually kept out of trouble by leaning heavily upon the findings of the Experiment Station. Montana dairymen choke slightly at the recollection that it was a ranch woman—and a member of the Legislature—who persuaded the legislature to remove the troublesome law. She maintained that ranch women were perhaps most inconvenienced by anti-oleomargarine legislation since range catte were not dairy stock and mixing color into margarine was a waste of time.

Child Welfare

Child care, which was also going through revolutionary change in the second quarter of the 20th century, fell largely into the hands of the foods and nutrition people. Comment has been made of Miss Graber's work in the early 1920s. These programs assisted greatly in making parents aware of the services of the health units in pre- and postnatal clinics, and kept them abreast of the aggressive national movement to eradicate the numerous virulent children's diseases.

A continuing interest caused Director Taylor to seriously consider adding a child specialist, but he was unable to find a suitable applicant. The concern of everyone was expressed by increased effort. Frances Smith, incoming nutrition specialist in 1929, revived the earlier "Child Feeding" project and other child nutrition programs. The clothing specialist set up standards for the child's room and its furnishings. Miss Gladys Branegan, the first home management specialist, went to the Montana State College academic Department of Home Economics in 1921, and built up classes in child care for training teachers and home demonstration agents.

Extension personnel gave encouragement to a statewide movement by women's groups to direct more attention to children's problems. The American Association of University Women, with Gladys Branegan as chairman of the state committee in charge, took much of the leadership in promoting a Governor's Conference on Child Health and Protection in the Montana Capitol, February 5-6, 1931. Some 140 organizations were represented, and the conference has been considered of first importance in drawing attention to needs for public health and medical service, assistance for the handicapped and related matters. Considerable progressive legislation came out of this conference. A follow-up meeting on parent education was held in September 1932, on the Montana State College campus. County officials had been alerted several years before by a series of clinics; one in Rosebud County in September 1925, for example, listed 121 cases of infected tonsils among the 854 children examined. The Women's Camps, beginning in 1922, had become another focal point for child care support.

By coincidence, a new clothing specialist, Miss Josephine Pollock, began her work in September 1932. An important factor in choosing her for the position was that she had considerable training and experience in children's work. She gave about one-third of her time for several years to child care, taking advantage of the interest generated by the Governor's Conference.

Miss Pollock organized programs on "The Young Child" in six or eight counties at the time, and when these were operating effectively, she went on to other counties. As many as 100 communities carried on such a project simultaneously. In 1934, working with the State Department of Education, she set up a number of what were called "Emergency Nursery Schools," four weeks in length. Here an intensive program of checking for defective sight and hearing, throat problems and malnutrition was carried out. The program also included aspects of the more recent Head Start plan of adjusting children to school procedures.

During the period that Mrs. Frances McDonald served as rural health specialist, 1946-1956, she devoted considerable time to supporting the efforts of the nutrition specialist and the county health offices to improve the care of the child. In various ways, and somewhat intermittently, even without a children's specialist, the Extension Service has given leadership in the child health and care movement.

Drought and Depression

In 1931 another period of drought and depression, the "dirty thirties," settled upon the land. Increasing difficulty arose in inducing county commissioners to budget support for Extension work because of decreased agricultural income. In 1933 the Fergus County commissioners sent a questionnaire with the tax notices asking taxpayers to register their wishes concerning the continuance of Extension work. For the county agent, the vote was 860 for and 527 against. For the home demonstration agent, 600 taxpayers favored retaining the office, 714 were against, and that position was discontinued. Since Ferugus had been one of the best Extension counties, the shock of this reaction caused hard second looks at the entire program. At the following annual state conference, modifications were made in the favorite long-time subjects such as nutrition and clothing, and projects with social overtones were adopted. Some of these were "Aids to the Consumer," the "Outlook for Farm Family Living," Child Development," and "Family Relationships."

Two austerity projects originated in the early drought period and continued until the rains came again in 1937. These were the "Live at Home" and "Buy Wisely" programs. Both cut through every phase of rural living. The "Live at Home" activities extended, for instance, into home gardens which were nursed along with every precious drop of water available. Cheese making was encouraged and widely practiced. Raising more poultry was urged and methods of canning poultry and of preserving eggs were demonstrated at many meetings. Incredible amounts of food were canned, dried and stored. Houses and basements which had gone unimproved for years in relation to personal comfort were tightened to keep vegetables from freezing. Emphasis was placed upon economical and nutritious food, and in many homes through added effort and increased knowledge the diet was actually improved during the severe years. This program also extended to clothing, where homemade garments and renovation were the subject of many demonstrations. The "Buy Wisely" program extended to all phases of home living and the logic was enlarged to cover purchases for the farm as well as the home.

A district system of agents-at-large initiated in 1935 provided for economy and convenience in supervision in the face of the small state staff. Eleanor Harrer was assigned to northwestern Montana, and Hazel Thompson took the northeastern counties; Lillian Tubb was assigned to the southeast, while Lillian Stone worked half-time in Beaverhead County and half-time in the other southwestern counties. The bright spot of 1936, another year of disastrous drought, was an increase of 11 percent in the number of counties in which home economics work was carried on, 11 percent increase in the number of community clubs, and 6½ percent increase in enrollment in the 38 counties cooperating.

In 1936 Blanche Lee resigned to take a position in Wisconsin and Frances Smith became state leader on February 1. She had come to the state as home demonstration agent for Flathead County in 1927, then became nutrition specialist in 1929. Miss Smith stepped up the "Live at Home" program, following the summer of almost total crop failure in many Montana counties. Demonstrations were offered over the state on such subjects as "Your Money's Worth in Food," "Meat Cutting, Canning and Curing," and "The Year Around Food Supply." Other projects included "Feeding the Family

The Hathaway Homemakers Club enjoying a sewing class.

for Health and Efficiency," "Meal Planning and Serving," "Positive Health," and "Family Financial Planning." She recalls being strongly reprimanded by a visiting supervisor from Washington, D.C., for demonstrating a nicely appointed dining service in an almost destitute, drought-stricken plains community. She countered that she found that something like that occasionally did a great deal to boost the ranch women's morale.

The confidence which the farm women of Montana felt toward the Extension clubs is shown in the clubs' excellent growth which took place in the difficult years. The clubs dropped to a low in 1933 of 249 and a membership of 4,411. By 1938 this had climbed to 439 clubs and a membership of 8,122. Fergus County had reinstated its home demonstration agent in spite of continuing blowing soil, and the Fergus County *Democrat-News* devoted an entire section of its 1938 Christmas issue to a review of the 25 years of Extension work in the county. The 25th anniversary year of 1939 was widely observed over the state.

World War II

Just as the devastating effects of the depression were disappearing and a period of reconstruction and expansion was beginning, World War II descended with unprecedented demands upon the American farm and home. A summary of the effect upon agricultural production in Montana has already been made. The women and youth were caught with equal force. Miss Smith thoughtfully observed that the broken families where young men were suddenly taken for military service posed the most keenly disturbing element of these years. Young men who were deferred for work or health causes tended to withdraw in an inferiority reaction, and this was equally unsettling. The high school students were pushed precipitately into maturity as they took their places with their elders in the war effort.

The home demonstration program took the nationwide slogan of "Salvage, Save, Renovate, and Conserve." The war hit the homemaker directly as she dusted off the pressure cooker and vegetable storage bins of the recent drought and depression period. Women and youth often participated almost full time in the various salvage drives in search of metals, fat, paper and rags. In Fergus County an efficient organization shifted deftly from a metal salvage drive to a "Share the Meat" campaign, to victory gardens, and then to a labor survey. Numerous committees were formed to carry on essential work such as price administration, Red Cross, war savings, farm labor, civilian defense, public health and welfare, and postwar planning. Small reserves of workers made training schools necessary, and the neighborhood leader system proved its worth in the gasoline-short days. Statewide campaigns for victory gardens and food preservation were directed toward the home. The emphasis upon good health continued and 2,389 families were involved in home nursing and first aid courses.

The expansion in the number of agents and clubs after the war was assisted greatly by the passage of the federal Bankhead-Flannagan Bill of 1945. This furnished additional assistance to counties in supporting home demonstration work. Severe problems arose in securing qualified personnel and in obtaining wage adjustments and working hours which were somewhat comparable with the standards of the time. As has been noted, the building and furnishing of new homes was one of the most flourishing movements in postwar Montana. Extension specialists, Margaret Tuller in home management and Eric Wilson in rural engineering, took the lead in this rewarding advisory work.

Bessie E. McClelland and the Homemaker's Council

Miss McClelland became state home economics leader September 30, 1946, when Frances Smith, who had recently become Mrs. George Y. Patten, resigned to become a homemaker. Miss McClelland had served as nutrition specialist since September 15, 1936, and was thoroughly familiar with the Extension program. During Mrs. Patten's 10-year period, the state and nation had emerged from a severe depression and had gone through a major war. Home economics extension work in Montana had more than doubled in size, now having 15 single-county home demonstration agents, and one two-county agent. Twenty-two counties had full time county agricultural agents.

Plans were complete for rapid postwar expansion into every part of the state. The Bankhead-Flannagan funds were utilized as fast as personnel could be found. By 1950, 24 counties had home agents and there were two vacancies. A major addition to the service was made on October 1, 1949, when the position of family life specialist was added and Miss Amy Wold accepted it.

The organization of a State Home Demonstration Council in June 1949 was another major accomplishment. Blanche Lee had urged such a council in the early 1930s, and Frances Smith and Miss McClelland had continued to promote it. In June 1948 the Montana State College Agricultural Advisory Council recommended the Home Demonstration Council, and in November, eight women from across the state met at the College and made definite plans for a June organizational meeting. Counties responded heartily and by June 1949, 32 had formed an organization and 64 representatives were present. These women represented a record number of 518 clubs and 9,360 members. A rousing second annual meeting of the state council took place in June 1950, with 95 delegates from 41 counties present.

The county councils provided an excellent forum for generating ideas and enthusiasm. The new family life movement was given effective support and the annual Family Life Conferences were well attended. The women were disturbed that fewer than half of the counties had home demonstration agents, and to help decrease the shortage of trained agents, the state council awarded scholarships over a period of some 15 years to young women preparing for Extension work. The state council gave full support to a national campaign for safe driving among farm youth and farm women. Sheridan and Flathead counties won national awards in this program. The striking results obtained gave a great deal of support to the adoption of the automobile driver training courses by the high schools of Montana.

The annual meetings provided a sounding board for the presentation of proposed legislative measures of interest to rural Montana, to Extension and to education generally. The meetings usually also included the consideration of important aspects of international affairs. This in turn led to support for several phases of the International Farm Youth Exchange (IFYE) activities.

Extension women took part in the Rural Progress Conferences of 1952. The fourth annual session of the State Home Demonstration Council, with a record 168 attending, spent much of its meeting discussing the bulletin, "What the People Said."

In 1956 an extensive questionnaire was widely distributed to members of the home demonstration clubs which provided follow-up information from the 1952 Rural Progress Conferences. The shifting nature of the farm population became strikingly apparent when the questionnaire revealed that only 57 percent of the club women lived on farms and ranches. A full 16 percent, at least, appeared to live in towns above 2,500 population. Only 50 percent received all of their income from farming or ranching, while 36 percent confessed that none of their income came from farming or ranching. Sixty-eight percent had a high school education or more, and 11 percent were college graduates. An encouraging feature was that over 60 percent had been club members for five years or more. Seventy-nine percent owned their own homes, 89 percent had running water in the home, and 79 percent had complete bathrooms, while 33 percent had an automatic washer. Ninety percent had a radio, but only 67 percent had a telephone. The state office decided that the questionnaire revealed that this diverse membership required more than "just a farm woman's" program.

Woman's Week, 1965—

The Homemaker's Council steadily searched for new programs which would attract women not regularly reached by Extension work. Director Aasheim suggested a "Woman's Week," which one or two states were using with success. The council, Vivienne Kintz, home economics program leader, and President Johnson of Montana State College supported the idea, and the council set up an extensive planning committee which included major women's groups, and a prestigious campus committee.

The first Woman's Week was held on the Montana State College campus June 20-25, 1965. Extension affiliation was not required for those who came; they lived together in a women's dormitory, and 287 women came from all over the state for a four-day session in continuing education.

Instructors, largely from the College faculty, offered a wide variety of courses for the women to choose from. Courses offered the first year included: 1) Child Behavior; 2) Problems of Marriage; 3) Cultural Patterns of Families around the World; 4) Home Management; 5) Chemistry in the Home; 6) Green Thumb—Gardening Practices; 7) Women in Government; 8) Using Art in Everyday Life.

The council officers worked closely with the women who came to Woman's Week, with the council members, and Extension leaders to provide the type of organization, courses, entertainment, food and living conditions which the women desired. Women were encouraged to attend for four consecutive years, which resulted in these women providing considerable leadership as well as developing lasting friendships which were often beneficial in state Extension programs. These four-year participants were then "graduated" in a ceremony at the closing banquet, with appropriate color and humor. Since many of the women who came had rendered distinguished service to the state for many years, an idea for an additional annual award took form in 1973—that of recognizing three "Women of Action." At the 10th annual meeting in 1974, eight women who attended every year were presented with plaques designating them as "Wise Old Owls."

At the 1974 meeting, Director Aasheim, who was retiring June 30, commenting on the record attendance of 425, challenged the women "to make

(left) An exercise class held during Woman's Week.
(right) An assemblage of several program covers for Woman's Week.

it 500 in 1975." The attendance has been near 500 each year, reaching 600 in 1976. Woman's Week has been one of the most successful programs hosted by the Extension Service, and has helped establish greater understanding between the resident teaching faculty of the University and Extension personnel.

Family Life

State home economics leaders attempted from the beginning to bring the broadest possible consideration of the whole family to the Extension program. The technological changes which brought automobiles, improved roads, electricity and many conveniences to the farm home also brought additional changes such as mobility and enlarged educational advantages which have been noted. In 1946 a first annual Family Life and Education Training School was held, leading in 1949 to the appointment of a family life specialist. Family life discussions raised such questions as taxes, government organization, education costs, markets, price structure, management and similar subjects.

A series of discussion-based pamphlets was developed covering the whole scope of family living. One issued in 1953 was titled "Business Facts." Another, in 1959, concerned "Public Affairs and Responsibilities." By 1961 the more general subject of "Family Economics" was popular. These and other bulletins furnished the nucleus for a highly popular series of meetings extending over several years. The discussions followed the wishes of the discussion groups to a great extent, but usually included such matters as business records, credit, insurance, social security, estate settlement, state government organization and costs, and state and local school systems and their costs—all largely economic in nature. The annual Family Life Conferences, discussed elsewhere, considered these and other social issues as well.

Miss Wold resigned as family life specialist in 1962 and was succeeded by Dean K. Hoffman. The extensive revision of Extension organization which began in 1960 was extended to this position on July 1, 1963, when it took the title of human resource development specialist. The emphasis upon human resource and community development led to another type of program which was also extremely popular. Such subjects as leadership training, communications, understanding one's self and others, and an associated subject, transactional analysis, took much of the time of Vivienne Kintz, Burl Winchester, Rob Neils and Rex Campbell.

Better Kitchens

A demonstration program which caused a major revolution was "Home Improvement." This began in the mid-1920s as the "Better Kitchen" contests, adopted by almost every home demonstration club in the state, indeed in the United States. Dissatisfaction with the farm kitchen had mounted rapidly during World War I when farm women were urged to assist the war effort with farm gardens, can large quantities of food, undertake extensive sewing and Red Cross enterprises, assist in community projects, and also lend strong support to the general farm work. The traditional farm kitchen was, of course, also the family dining room, and usually the family living room. The room was large, no real thought was usually given to an arrangement for convenient working conditions, and all of the water had to be carried in and carried out. The wood or coal burning stoves were temperamental, uneven in temperature, and unbearably hot in the busy summer season. The customary priority was stated by one farmer: "I have built a good set of barns and sheds for my stock; now I expect them to build a house for me." Often this took a long time and the women became impatient. When the war prices brought larger incomes the women saw a chance to obtain some of the conveniences they had long wished for.

The state specialists tried to restrict the program to a few counties each year, enabling them to give direct advice and assistance. But in 1924 some 350 women entered the improvement contest. They undertook such changes as increased window space, bringing water into the house, and drain construction which took the waste water from the house. Sinks and work counters were common additions, as were renovated floors, either linoleum or hard wood, which were easier to clean than the customary soft wood, wide board floors. Iceless refrigerators and fireless cookers were both emerging rapidly, and both were as yet largely homemade. More convenient laundry facilities had a high priority. Cascade County was a pioneer in the effort to help women improve the "four walls" of the home, and make their work so efficient that they could get away from them more often. In the autumn of 1924 some 50 persons, men and women, visited five homes which had made significant changes without undue expense. Featured was not only the improved kitchen, but home power machinery, water systems and laundry facilities.

A 1927 survey showed that considerable progress was being made. Water had been piped into 18.7 percent of the homes surveyed. Septic tanks were in 5.2 percent; 12 percent had stationary wash bowls with running water, and 7.7 percent had stationary laundry tubs with running water. Flush toilets were in 6 percent of the homes. The availability of electricity was the factor which made the revolution in farm home conveniences possible, including water systems, home freezers, lighting, radio, and a host of other uses now long accepted as a matter of course.

A full-sized model kitchen was constructed and sent, in 1947, on the Labor Saving Caravan where it was shown in 41 counties to 35,594 people. Eight counties immediately requested that it be available for their county fairs.

The Mattress Project

Another indication of the changes which had taken place since the unbelievably "good old days" was the mattress program. This was a na-

Training session for home demonstration agents on how to make a mattress, fall 1940.

tional project designed to use surplus commodities, and was urged for Montana by observant local home demonstration leaders who knew, perhaps first hand, the kind of mattresses being used in many Montana homes. The project started in Montana in October 1940, when representatives from the Texas Extension Service interested in increasing the use of cotton demonstrated the methods at a state meeting. The project was further discussed at Outlook meetings in January and application blanks were distributed. When the demand proved to be encouraging, the program was organized. National Youth Administration girls' sewing projects sewed the ticks. A charge of $1.50 per mattress was made for materials and supervision at the several county work centers which were established. Following the annual Extension conference in December 1940, a training session on the making of the mattresses was held. When everything was in readiness, an exasperating delay in receiving the cotton prevented the rapid completion of the project which had been planned. In the meantime, as a means of getting rid of additional cotton and cloth, and providing additional people with productive jobs, a comforter program was added. A 1942 report tabulated 3,440 mattresses and 3,281 comforters made in the project. Inadequate as these mattresses must have been, they were most enthusiastically received. One home demonstration leader reported what seemed to be a majority reaction: "It is not unusual to hear women state that this is the first good mattress they have ever owned; straw ticks were the only kind of mattress they had previously had in their homes."

Women's Camps

A major program for the recognition of the homemaker and a testing ground for ideas for future programs, was found in the camps for women—often termed Mother's Camps—which began in 1922. They were so popular that even long after Extension ceased to sponsor them, they continued for 24 years until World War II brought them to an end.

Only two camps were held during the first summer. One was organized for Richland and Dawson counties with 22 delegates representing 12 communities. A second camp was held near Glasgow with 66 delegates from 30 communities in Valley and Phillips counties. Each met for four days and considered similar issues. Child care and nutrition, home management and community recreation were hardy perennials. "They were the high spot," said Director Cooley, "of Extension's work with women up to that time."

The camps increased to five in 1923, and the enrollment shot up to 224 women from 92 communities in 16 counties. In 1925 seven camps were held and the location of these furnished the chief centers as long as the camps continued. The camps were at Brush Lake near Dagmar, Michel Lodge near Red Lodge, Camp Tuffit on Lake Mary Ronan in the Flathead area, and others near Glasgow, Sidney, Forsyth and Helena. The pattern soon varied from four days to one day, and in 1926 four highly successful camps—really conferences—were held for only one day. During any camp many women came for only one day, and in some camps a day was designated as Visitor's Day. The peak of attendance was reached in 1929 when 13 camps were held with 1,469 delegates and a total of 2,038 attending.

The program was almost disbanded during the drought of the 1930s, and, the Extension Service dropped the camps as a formal project during that time. The loyalty of the women caused them to be revived in a number of communities following 1940. They suffered further during World War II and were disbanded in 1946.

The camps filled a need for enrichment of the social life in rural Montana at that time. Purposeful social contacts in the sparsely settled areas were so few and so difficult to maintain, that the situation is not easy to reconstruct after a half-century of progress in transportation and communication. When the camps originated, they represented to a number of women the first vacation they had experienced in their lives. The rugged physical aspects of the camps provided a desirable informal atmosphere and seemed to be appealing in themselves. These, and the companionship of kindred spirits, were exhilarating and the women immediately entered into the spirit of the camp.

They turned quickly, however, to serious thought on the projects which were under consideration, and their judgment on trial balloons for future state programs was accurate and valuable. The question of a Family Life Education program was refined in these camps, and support for a specialist was obtained. Flathead County added its home demonstration agent largely because of the enthusiasm generated at the camp on Lake Mary Ronan.

VIII. The 4-H Story in Montana

Evidence that the earliest planning for a more abundant rural life in Montana included concern for its youth is contained in the first printed report of the Montana Farmers' Institutes in 1902. This included an address given to a session of the institute by Professor W. F. Brewer of Montana State titled "Some Suggestions about Boys." Recalling the phrase, "A boy's will is the wind's will," Professor Brewer hazarded that the decade from 11 until 21 was the most impressionable in a person's life. He urged that the family and the community surround its youth with a multiplicity of interests and activities associated with real life situations from which they could make choices. He suggested participation in vocational activities, encouragement in starting hobby collections, provisions for sports and social life within similar age groups, and a flexibility in which the "boy's will" could gyrate but remain within constructive limits. The following year Professor M. J. Elrod of the University of Montana considered the problem of "How to Keep the Boys on the Farm." His recommendation was "to give him something which is his own and let him use his brains in making it a success."

The youth movement was progressing rapidly by 1902, and Montana was abreast of the times. Thomas and Marilyn Wessel of Montana State University, in their recent and excellent *4-H, An American Idea, 1900-1980,* recount how the idea grew from similar concerns in many places. In 1896 Liberty Hyde Bailey of Cornell University began to use New York Extension funds to prepare study materials and make them available to schools and clubs in rural communities to encourage greater awareness of agricultural problems and possibilities. In 1902 Albert B. Graham, superintendent of schools in Springfield township, Ohio, began forming Agricultural Experiment clubs for both boys and girls. In cooperation with the Experiment Station at Ohio State University, corn, garden, flower and other clubs were organized statewide enlisting several thousand youth. In 1905 Graham was asked to become the first superintendent of the state Extension type program.

At the same time in Illinois another county superintendent of schools, O. J. Kern, working with the Illinois Experiment Station, also developed youth corn and other clubs. The president of the Macoupin County Farmers' Institute, Will B. Otwell, observed in 1898 that working together in a group

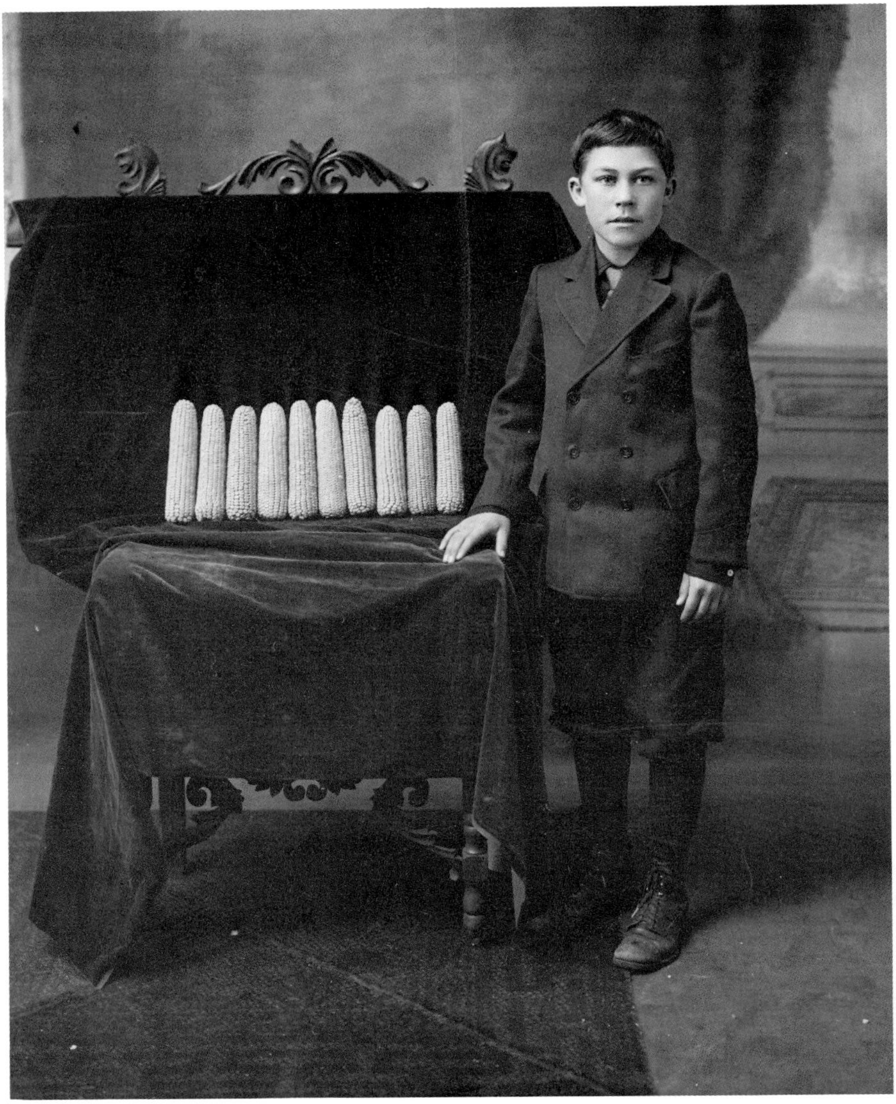

The winning 10 ears of corn at Montana's first corn show; exhibited by I.D. O'Donnell, Jr., Billings, Mont., 1910.

was one important factor in working with youth. He also added another feature which has remained a part of 4-H activities, that of recognition for superior work; an award or a prize, if even a small one. This idea spread like wildfire, and by 1904 as many as 50,000 youth were in friendly competition in corn clubs. An enormous pyramid of corn built at the 1904 St. Louis Exposition by the youth of Illinois received national attention. In Iowa two other educators, Jessie Field and O. H. Benson, received strong support from Henry C. Wallace and his *Wallace's Farmer.*

Montana Corn Clubs

Fred S. Cooley became superintendent of the Montana Farmers' Institute in 1907. He had a genuine concern for youth but lacking the finances and facilities for organizing and directing clubs, he attempted to initiate youth activity by securing a number of stimulating speakers to appear before high school assemblies to inform and interest the students in new developments in agriculture.

By 1910 a movement arose in some of Montana's eastern counties under the Farmers' Institute management to organize corn clubs. These received a powerful stimulus from a gift of $1,000 prize money from James J. Hill of the Great Northern Railway. The work of the corn clubs, "industrial contests" they were called, was displayed in a large exhibit in Billings in November. The effort was judged to be remarkably successful. Many of the farmers in eastern Montana had come from the Midwest and they had been unsuccessful in their continuing attempts to raise corn. They were as interested as their children in the experiments taking place.

The original directions in the corn contest called merely for a selection of the best 10 ears. When M. L. Wilson went into eastern Montana in 1912 as an agricultural supervisor, he instituted what Charles E. Potter called, "the first club project drafted by a Montana county agent." In his "Twenty Years of 4-H Club Work in Montana, 1914-1934," Potter quotes Wilson's simple rules:

> Plant a plot of 9 feet x 12 feet hills, dropping at least six kernels into each hill. After the corn is four inches above the ground thin to a perfect stand of three stocks per hill, and hoe frequently during the summer.

Each boy or girl was charged 10 cents for enough Northwestern Dent to plant the prescribed plot. Early enthusiasm brought 280 children into the contest but before fall the contest had simmered down to about 60. However, the entire project demonstrating new seed and more intensive care helped to establish corn, within climatic limitations, as an important auxiliary crop.

Dr. Hartman, Youth Specialist

The corn clubs really launched the organization of farm youth clubs, but pig, calf, canning, sewing and garden clubs followed closely. Another important segment of the youth movement had its origin in 1911. In that year, Dr. W. J. Hartman was employed as the first livestock specialist for the state. Since livestock was judged to be of special interest to youth clubs, Hartman was asked to give them particular attention. In the same year the new State Fair Association announced plans to "conduct short courses in stock judging and household arts at the State Fair." The fair was held in Helena and Hartman was asked to help in the preparation of the exhibits since one boy and one girl from each county was to be sent to the State Fair, the selection being based upon some industrial achievement.

The beginnings of wide public and private cooperation in the youth enterprises took place at the State Fair under the direction of Cooley and Hartman of the Farmers' Institutes, the State Department of Public Instruction, and I. D. O'Donnell of Billings for the Fair Board. The railroads willingly agreed to furnish transportation, and the Fair Board provided camping facilities and food for the youthful participants during the fair. By 1914 a short report was required, and the fair requirements for winning exhibits were listed as: perfect exhibit specimens, maximum yield, greatest profit, and the best account of the work.

Organization Under the Smith-Lever Act, 1914

Since boys' and girls' clubs were widely organized in almost all states by 1914, they were included in the Extension program under the Smith-Lever Act. Miss Augusts Evans from Illinois was chosen to direct the Montana clubs. She organized with whirlwind speed and by mid-1915 club enrollment was 2,939. Clubs participated in the State Fair in Helena, at the Western Montana Fair in Missoula, and in a new and important annual Corn Fair in Miles City. The first regional 4-H Club Fair was held in Wibaux in November 1915.

Miss Evans resigned July 1, 1916. Her abundant energy had enabled her to stimulate wide interest, but reports for her period of service are highly incomplete. Forty-eight leaders reported active work being done. Four of these were county agents, four were county superintendents of schools, 14 were teachers, and 26 were housewives and farmers.

M. J. Abbey and Charles E. Potter

In August 1916, M. J. Abbey from West Virginia became state club leader. In March 1917 Miss Mina Ogilvie, who had been an assistant specialist in home economics, became his assistant. And on April 1, 1917, Charles E. Potter also became an assistant and started a long and successful career. Director F. S. Cooley in his *History of the Montana Extension Service*, says of Abbey:

> Abbey brought to this field a very conspicuous quality of personal leadership and boundless energy and enthusiasm. His was about the best advertised work in the state. The numbers of children enrolled in the clubs were very large and the enthusiasm was very marked.

Abbey reported that the best work was being done where county agents could give direction to strong local leaders. This addition to their other work—especially as the war broke out—was not really possible, however, and by mid-1917 some 25 county club leaders had been recruited locally;

they in turn secured community leaders.

Early state leaders hoped that at least a county 4-H staff could be paid. A beginning was made with the employment of Harry N. Kauffman in Flathead Conty in 1917. He had begun working with the clubs while on the staff of the Flathead County High School. He found himself in the lowest-paid bracket on the Extension field staff, and soon accepted a position in Wisconsin. A short time later he returned to the high school in Kalispell and resumed his 4-H work. His success was outstanding, and his first Junior Fair easily surpassed the well established and highly rated adult Flathead County Fair.

Mrs. Mary Davey of Deer Lodge was paid for three months' service in

Charles E. Potter (right), state 4-H leader, visiting with a 4-H Corn Club member and his leader.

1916 by the Deer Lodge Commercial Club, and in 1921 Mrs. May Thomas in Fergus County, and Martha L. Eder in Big Horn County were also paid for short periods of club service. In the meantime, a certain amount of prestige appeared to develop around local leaders and they were easier to secure, although they were never in ample supply and the turnover was always high.

M. J. Abbey had been brought to the state primarily to direct the College academic program in agricultural education, and as this work increased, Abbey resigned from his club duties on January 1, 1920. His assistant, Charles E. Potter, then became state leader, remaining until March 31, 1935. Cooley commented again:

> Since taking up boys and girls club work, Mr. Potter has given it the best of his energy, pleasing personality, and analytical mind. Club work has been reduced to a science and the best system has been applied.

The 4-H Emblem and Pledge

The ceremonial features of the 4-H Clubs developed over several years. The three-leaf clover as an emblem was proposed for use as early as 1906. It was suggested by O. H. Benson of Wright County, Iowa, representing the "trinity of powers in education, the equal training of head, heart and hand." In 1911 a small conference of youth leaders in Washington, D. C., approved the fourth leaf, suggested for "health" by O. B. Martin of South Carolina, and the imprint of an "H" on each leaf was agreed upon.

In 1913 Martin further proposed that a "4" be placed before the "H," and the insignia "4-H" was created. Little attention was given to emblems and ceremony during the war years, but in the revival of youth work in the following years this type of emphasis was useful. Directing youth work in the USDA in the early 1920s were the original O. H. Benson and an innovative associate, Gertrude Warren. At least in part through their efforts, by 1924 the 4-H emblem had been adopted nationwide.

A creed and a pledge was also developed over a number of years but not until 1927, in the first National 4-H Club Camp in Washington, D.C., were they adopted. The 4-H Club Creed is inclusive:

> I believe in Boys' and Girls' 4-H Club work for the opportunity it gives me to become a useful citizen.
>
> I believe in the training of my HEAD for the power it will give me to THINK, PLAN and REASON.
>
> I believe in the training of my HEART for the nobleness it will give me to be KIND, SYMPATHETIC and TRUE.
>
> I believe in the training of my HANDS for the ability it will give me to be HELPFUL, SKILLFUL and USEFUL.
>
> I believe in the training of my HEALTH for the strength it will give me to ENJOY LIFE, RESIST DISEASE and WORK EFFICIENTLY.
>
> I believe in my Country, my State, my Community, and in my

responsibility for their development.

In all of these things I believe, and I am willing to dedicate my efforts to their fulfillment.

This rather comprehensive Creed has been concisely summarized in the Pledge:

I pledge my head to clearer thinking;
My heart to greater loyalty;
My hands to larger service;
My health to better living;
For my club, my community, my country, and my world.

The final expression "and my world," has been added in recent years as the 4-H idea has expanded to more than 80 countries and the International Farm Youth Exchange Program has become a major 4-H activity.

World War I and Postwar

The 4-H organization, indeed the entire Extension operation, found both frustration and high achievement during World War I, which followed so quickly upon its establishment. The leadership problem continued since there were insufficient funds to employ county leaders for youth, but considerable assistance came from the increased number of county agents who were employed under the "emergency" war measures of the federal government.

Fortunately, the federal Smith-Hughes Vocational program came into the schools in 1917, with its Future Homemakers of America in Home Economics for girls, and the Future Farmers of America in Agriculture for boys. School personnel had been called upon extensively to assist with 4-H Extension work since 1914, and they were even more widely used during the war. County agents also found that they were often able to reach parents more quickly on new and hurry-up projects when they started with the youth in the homes. The war demand for increased agricultural production coupled with greatly increased prices and a strong patriotic appeal was irresistible. The members and leaders of the 4-H Clubs and the vocational clubs found themselves in the middle of an important and satisfying effort to promote agricultural production and food preservation, and of nationwide drives to collect scrap iron, paper, and fats. By 1918 the enrollment in 4-H had shot up to 11,703 members in Montana.

In 1919 club enrollment dropped sharply to 5,251, only in part due to postwar relaxation. A supplement to the annual report of boys' and girls' club work for the year ending December 1, 1919, summarized conditions:

> The drought of the past year handicapped club work considerably. With only a 40 percent crop production throughout the state it is easily seen what the club members were encountering in carrying through projects. Much seed that was planted never came through the ground and the shortage of feed stuffs seriously influenced the returns on livestock production.

> The flu epidemic cut heavily on the bread and garment club work as the schools were closed down. Upon opening, the school work was so heavy in order to make up what was lost that many members had to give up the work. The number of working days was also cut short on account of this epidemic and many who had partially finished were unable to complete in-time for county contests.
>
> In spite of these conditions, the splendid spirit with which the youngsters accepted the inevitable is cause for much enthusiasm on the part of those connected with the work. "Give me a favorable year and I'll show what I can do," is the opinion of the club members of Montana.
>
> A club encampment was held at the Montana State Fair and the Midland Empire Fair with an enrollment of over 200 boys and girls who were county winners in their various projects. The railroad fare of the winners was paid by the county, while the Fair Association provided lodging, meals and entertainment. The club exhibits at these two fairs were remarkable considering the dry year. In 27 out of 52 counties, county club fairs were held at which splendid prizes were awarded.

In 1919 there were two full-time and 15 part-time paid club leaders who were salaried under the federal government "emergency" plan, plus 250 volunteer community leaders. Nevertheless, 80 percent of the clubs were supervised by school personnel. This was not a satisfactory arrangement since many of the teachers did not remain in the community during the summer season when 4-H projects were most active. The county agents retained their enthusiasm for the clubs, leaders were recruited from farms and homes, and by 1926 only about 8 percent of the local leaders came from the teaching profession. State Club Leader Charles Potter summarized the trend:

> The adoption of the policy that club work was extension work not school work, the selection of members out of school as well as in school, emphasis upon the voluntary membership in the 4-H Clubs, the selection of the slogan "every club with a local leader," and the emphasis placed upon the definite organization of a club group with a restricted number of projects whose requirements conformed to the four principles of project selection, together with the idea of using the Extension specialist as the consultant in 4-H Club subject matter, and the correlation of the club project with that of the specialist's program, were some of the adjustments which characterized the "after the war" period.

The rapid rise of the Farm Bureau and the deep interest of its members in all of Extension work proved to be a rich resource area for 4-H assistance. When the bureau took on political overtones, Extension ceased to use its facilities.

State leaders worked on the job of standardizing objectives and at the end of a decade, Charles Potter said:

> Leadership, cooperative endeavor, cultural, social and recreational opportunities, as well as community service, character building, and citizenship training, were now commonly spoken of as objectives in 4-H work.

New procedures emerged steadily, and shortly as a result, a new set of standards were designed in the form of a score card.

This provided for a program of work for local clubs, including well-planned club meetings, tours, exhibits, parents' meetings, health programs, achievement days, demonstrations and judging, picnics, publicity, well-kept records by secretary and club members, high percentage completion of club work, community service projects, and well-trained local leaders.

By 1923 a total of 2,720 members were enrolled in 264 clubs in 33 counties and 310 communities representing 1,500 homes. Eight camps were held in that year with an attendance of 431.

The completion of projects was less than 50 percent at the close of the war. A goal of 85 percent completion was projected at that time, and finally reached in 1926. Demonstration projects and judging contests proved to be effective methods, and county tours and interclub demonstrations were popular as well as excellent educational devices.

Camps and Fairs

During the first half of the century when travel was more restricted, the Extension camps were particularly attractive to adults and youth alike. The first 4-H camp was held in Blaine County in 1918, and the second camp in 1919 registered 81 boys for a four-day experience. Prairie County also had a camp in 1919 for 30 boys. By 1936 an increase to 16 camps enrolled 878 campers; in 1937 attendance was 1,434 representing 22 counties in 15 camps. Separate camps were originally held for boys and girls; then they were brought together in the same camp and program, and gradually the number of girls considerably exceeded that of the boys. By 1952 the number of camps had increased to 52, with 2,778 campers from 34 counties. In addition to a strong recreational program of a type which could be taken back to the home clubs, a stimulating educational exposure was obtained by using leaders from Washington, D.C., from the State College, from state and county government, and representatives from agricultural organizations.

Camp facilities were usually leased from a local organization, but the Extension Service assumed considerable responsibility for the Pines Youth Camp some 35 miles south of Glasgow on the shores of the Fort Peck Reservoir. It was built by the U.S. Army Corps of Engineers as a part of the Reservoir's accommodations, and the Extension Service operated it under lease through 1961. The Glasgow Kiwanis Club, the Elks Lodge and other agencies made desirable additions to it from time to time. Some 500 campers came each summer in the early years.

Picnics on a community or county basis were widely held during the summer months. These provided the opportunity for preliminary exhibits, a first round on competitive judging, and other contests related to project work, but the affairs were largely social and recreational. These were often combined with tours of inspection and demonstration of project work in the community or in the county.

The "show window" of the several methods used by 4-H Clubs for the display of their work was Achievement Day, which became a community event. These special days of displays, demonstrations, contest and awards, which still continue under various names, were often the trial run in preparation for the autumn fairs. A few such days were held in 1920 and proved to be highly successful. Mignon Quaw, recreation specialist, then prepared an outline of suggestions for a standard program and 28 Achievement Days were held in 1921, attracting 2,648 people. The community was usually the base for these occasions, although Big Horn County held a first countywide observance in 1922. Improved transportation following World War II made county cooperation more feasible and the single day display replaced many county fairs. The peak in the number of community Achievement Days appears to have been in 1936 when 404 were reported in 39 counties, attended by 18,756 persons. In the early 1960s over 50 counties were hosting these special days, although the countywide base greatly reduced the number to 87 in 1962, for example.

Youth participation became an inherent part of the Montana fairs following their spectacular contribution to the Billings Corn Show in 1910. This was followed by similar success in Miles City in 1911, and in the many county fairs which sprang up, 43 by 1939.

The official State Fair was held in Helena beginning in 1903. Its board gave a great deal of attention to incorporating the youth exhibits into the program. The Northern Montana Fair in Great Falls soon found favor, particularly after a special 4-H exhibit and demonstration building was constructed in 1928. This came to be considered the official 4-H fair following 1931. The Midland Fair at Billings also constructed a 4-H building in 1928, the largest on the grounds at the time. The Central Montana Fair Association in Lewistown also cooperated to the extent of building attractive dormitory facilities for a 4-H camp. The Western Montana Fair at Missoula and the Southeastern Montana Fair at Miles City as well attracted wide regional support from 4-H exhibitors. The Winter Fair in Bozeman, which began in 1946, has attracted outstanding exhibits and judging events from 4-H members.

The State and National 4-H Congress

The annual Montana 4-H Congress became the outstanding testing ground for competition in skills and leadership qualities. This was held on the Montana State College campus and began as the annual state convention in 1930 on a relatively small scale. Each county with a 4-H program could send two members and two leaders, with additional representatives in proportion to club membership. The first convention hosted 103 members and 43 leaders representing 31 counties. A strong leader training program was a continuing feature of the convention. Special dress was en-

Fifth annual 4-H Club Convention, 1934.

couraged, stressing green and white. The three major railroads of the state made transportation available at greatly reduced rates, and often local clubs and agencies helped finance other costs. Montana State made its then limited facilities for food and for lodging—a mattress on the gym floor as sleeping quarters for the boys—available at low cost, and nearly the entire faculty joined in providing an exceptional educational experience for the club members.

The annual conventions were not held during the World War II years, 1943-45, but opened again in 1946 taking the name of the Montana 4-H Club Congress. The proportion of leaders was smaller and the number of members was increased. A total of 278 members and 65 leaders from 39 counties attended in 1946. The annual attendance has been approximately 600-700 since the mid-1950s, with representatives from every county. An important educational program keyed to current youth interests is carried out. This includes extensive demonstration contests by individuals and teams, and judging in which both boys and girls participate. Dramatics and public speaking are also often emphasized. An elaborate dress revue contest has been a feature for the girls for many years, and more recently a few young men have intermittently displayed men's clothing as well.

The National 4-H Congress held annually in Chicago, beginning in 1922, has been without doubt the outstanding event of the year. Some 30 Montana delegates have been carefully selected each year. Montana representatives have measured up well. In 1949, for instance, eight were winners of national contests. During the years 1970-1976 Montana had 26 sectional and national winners. Included were two who received the silver tray awarded by the President of the United States. Warren Neyenhuis of Valier received the tray in 1968 for his outstanding work in junior leadership, and Theresa Whalen of Great Falls received it for all around achievement in 1969. The Milwaukee Railroad gave a great deal of encouragement to the Chicago Congress by providing special financial help for delegates and their advisors from the 13 counties through which the railroad operated. Beginning in 1925 Montgomery Ward provided one or more travel scholarships for a number of years.

4-H members giving demonstrations at an annual Montana 4-H Congress in Bozeman at Montana State University.

An important assisting agency is the privately organized National 4-H Service Committee, formed in the autumn of 1921 in Chicago. This is made up of nationally prominent business leaders and heads of farm organizations who had been assisting with a national 4-H tour to the International Livestock Exposition since 1919. The committee led in organizing the congress in 1922; it perfected a permanent staff and a steadily expanding program which, in its 50th anniversary year of 1971, had an annual budget of $2,279,383. Its major activity has been to promote the congress which brings some 1,600 4-H youth together each year to compete in the projects which the committee has arranged in cooperation with the national 4-H organization. Scholarships awarded in 1971 had a value of $180,000 and the total amount over the 50 years has exceeded three million dollars.

The assistant director of the committee in the 1950s was Tyrus "Ty" W. Thompson, Montana 4-H leader, 1946-1949. He initiated a number of committee services. The committee prepared numerous publications which promoted 4-H and supported the *National 4-H News*, which was the national 4-H publication. A series of television programs attracted wide attention such as "Living in a Nuclear Age," and one oriented toward nutrition, "Mulligan's Stew." The committee has developed numerous projects, prepared guide materials, and secured sponsors for project awards. It has also assisted in publicity and morale by merchandizing a wide array of materials such as award certificates, emblems, insignia, posters, banners, caps, shirts and other articles of educational and ceremonial use.

The National 4-H Club Camp which opened in Washington, D.C., in

1927, has also been a major attraction, particularly so since it provides an opportunity to see the Capitol city under favorable circumstances. Four delegates are usually chosen from Montana. An additional national event is National 4-H week, which has been a cementing event since it is observed simultaneously by every club in the country.

That 4-H provides more opportunities for travel and national participation than any other youth organization in the nation becomes evident by listing merely the journeys of one year—1957. The Pacific International Livestock Exposition in Portland drew four Montana delegates, as did the National Dairy Show in Columbus, Ohio. Twelve delegates traveled to the National Western 4-H Club Roundup in Denver, and four members from potato growing clubs attended the National Junior Vegetable Growers Association meeting in Springfield, Ill.; and a new Grain Marketing Tour drew three delegates to visit elevators and marketing facilities in Minneapolis and St. Paul at the invitation of the Minneapolis Grain Exchange.

A new note came with the National Junior Fact Finding Conference of the American Institute of Poultry Industries formed in 1953 in Kansas City through the efforts of Montana's poultry specialist, Harriette Cushman. Two delegates attended this meeting. Another three attended the Central States Regional Tractor Operators' Contest in Hutchinson, Kan.; 30 journeyed to the National 4-H Congress in Chicago; four went to the National 4-H Camp in Washington, D.C., and two 4-H and two FFA members attended the American Institute of Cooperation Conference at Fort Collins, Colo., sponsored by the State Department of Vocational Education.

An additional excursion was the annual journey each year of four

A county delegation to a Montana 4-H Club Congress in Bozeman at Montana State University.

delegates to the Junior Club Week observance in Vermilion, Alberta, to join with Canadian youth. In turn, four or more young persons from Alberta attended the Montana 4-H Congress each year.

Another conference in this period was distinctive. In 1940 William H. Danforth, founder of the Ralston-Purina Company, set up the American Youth Foundation Leadership Training Conference. One young man from agriculture and one young woman from home economics were chosen from each land grant college to attend the annual conference. They spent a week in St. Louis, where the women were taken through food service facilities, restaurants, tea rooms and food laboratories. The men were taken through the operations of the Ralston-Purina animal food factory, cereal mills and associated facilities. They were all then taken to Camp Miniwanca near Shelby, Mich., where they participated in an extraordinary program in leadership training. This was not directly 4-H related, but almost all of the participants had come through the 4-H program.

IFYE, The World View

The International Farm Youth Exchange (IFYE), which took the former 4-H member into the world situation, began in 1947. Montana entered the program in 1948. By 1973, with from one to three members going each year, 66 Montana youth had gone to 38 countries, and 131 youth from 50 other countries had come to Montana. This movement grew out of evidence at the end of World War II of the lack of understanding our people had of other nations. The program sends young men and women to live for three months or more in a farm home abroad, participating intimately in the work and social life of a rural community. A statement of policy for the IFYE maintained that "its most important contribution is the development of an informed youth leadership." Many individuals and organizations over the state contributed to the costs of sponsoring the young people in the early period. Since 1970 this program has been financed in the state largely by the Montana 4-H Foundation. The educational value of the returning exchangees was high, as the young people visited 4-H Clubs and community gatherings to present their reports, usually generously illustrated with slides of exciting and rewarding experiences in international living. In 1951, for example, Claire Vangelisti and Earl Peace made over 300 presentations, reaching an audience equal to about one-twentieth of the population of the state.

Awards and Support

The 4-H enterprise probably has been endowed with as many rewards as all of the other youth organizations of the nation put together. Parents have been proud of their children's accomplishments and have made livestock, plots of land, working materials and opportunities for information and travel available in great quantity and variety. The undergirding support which has come from the communities is difficult to report with fairness. The Farm Bureau, the Farmers' Union, chambers of commerce, and women's clubs have all given substantial encouragement. Widespread civic club support is well illustrated by the Helena Kiwanis Club, to use only one example, which has for over 40 years made the 4-H Achievement Day its major annual project and the day has become a notable event for the entire community.

The local banks have contributed generous assistance to the clubs and also entered into business arrangements with the young farmers to finance their enterprises under the most favorable terms. The railroads have made extensive travel possible through low cost group rates, and also have provided generous awards. State and national business corporations have made noteworthy contributions. The listing in the 1968 annual report is representative of annual contributions:

Contributors	**Amount**
The Railroads: N.P., G.N., U.P., Milwaukee and Burlington	$5,650 Total
Montana Power Company	2,000
Sears Roebuck Educational Foundation	5,900
Federal Cartridge	950
Montana Association of Mutual Insurance Companies	550
Danforth Foundation and Ralston-Purina Company	Scholarships
Minneapolis Grain Exchange	150
Peavey Company	900
Standard Oil Foundation	100
Montana Dairy Association	100
Montana Quarter Horse Association of America	250
Montana State University	1,200
Cram-Heywood Foundation, nine awards for work with sheep	900
Sperry-Hutchinson Scholarships	300
Union Pacific Scholarships, three at $200 each	600
National 4-H Scholarships	2,700
Club Awards	
Montana Power awards to 10 club electric projects	$250 each
Eli Lilly awards in health—10 clubs	
General Motors awards in safety—10 clubs	
County Awards	
3,738 awards (including 64 certificates in 55 counties participating in 31 National 4-H award programs)	

As the program has changed, from time to time, other assistance has been provided from the 12 Farm Credit Association districts, the State Association of Soil Conservation Districts, the Montana Medical Association, sugar beet companies, livestock sales companies and many others.

The several national foundations have made generous contributions to the states. The National 4-H Foundation has assisted with the annual National 4-H Club Camp in Washington, D.C., with the IFYE youth, and with programs which studied human relations, citizenship and character education. The National 4-H Service Committee in Chicago has assisted with the National 4-H Congress in Chicago and provided a large number of project awards, of which Montana delegates obtained as many as 27 in one year.

A list of the accomplishments of the Montana youth in state or national competition has not yet been compiled. Just a few examples taken at random from the annual reports illustrate the success which has been attained. In 1936, for instance, Daphne DeBruin of Chouteau County was awarded the distinction of having the best record of any girl in the nation in a Social Progress contest sponsored by the National Broadcasting Company and the Victor Radio Corporation. She was awarded a trip to Chicago and New York City and given a $500 scholarship. In 1937, at the National Dairy Show in San Francisco, Sam Lenz and Garth Blackburn of Cascade County teamed up in a demonstration on "Producing Clean Milk for the Baby," to win first place in the western states and fourth in the nation. Montana 4-H competitors presented the Grand Champion Fat Steer in the Junior Shows in Spokane, Wash., Salt Lake City, and Ogden, Utah, and a Montana judging team at the National Dairy Show in Columbus, Ohio, won first in the 11 western states and fourth in the nation.

In 1939 Clova Staehnke and Ruth Martin from a Yellowstone County 4-H Club won first in the United States at the National Dairy Show in San Francisco with a demonstration on "A Quart of Milk a Day." Also, in 1939, at the International Livestock Show in Chicago, Helen Monforton from Gallatin County won second in her class in beef, and with her sisters, Marie and Margaret, won first in the nation in the county group of three calves from any county; Harold Sitzman of Stillwater County won fourth place with an exhibit of lambs. The same year, two girls from Valley County won the highest award, that of "Superior," on their demonstrations at the World's Poultry Congress in Cleveland for "The Construction and Use of Egg Graders," and another team from Chouteau County won the next highest rating, "Excellent," for their "Stream-line Chicken." Other achievements are reported elsewhere in context.

Financial Returns

The work in 4-H Clubs was conceived as being educational, character building and of service to the community, but it was also intended to be financially rewarding and considerable attention was given to this aspect. An important part of the training received in 4-H projects was the keeping of records. Since the major projects reflected real life situations, these were not all profitable, of course. When the market started downward, and particularly in periods of drought and depression, the 4-H projects were in the same kind of difficulty which afflicted all of rural Montana. Fortunately, in many instances the intensive care which the 4-H member was prepared to give his project made it more successful than the larger operation on the same farm, and pointed toward some solutions in periods of stress, such as methods of culture, breed, variety, timing and other ingredients of the agricultural process which might be investigated on a larger scale. The annual reports were replete with financial detail. A couple of examples may serve to reflect the trend.

A summary of profits in the 1923 annual reports of the state leader notes that there were 264 clubs with a total of 2,720 members. The property of the club members had a value of $53,727.43, the costs of which were $22,012.90, leaving a profit of $31,714.53, or an average of $11.65 per member. In 1950 there were 766 clubs with 10,221 members enrolled. The holdings of various types were valued at $987,277.13. Expenses had been $502,874.42, leaving a net profit of $484,402.71, or a profit of $47.30 per member. Naturally, profits varied widely among projects; those generally regarded as most profitable included beef which made an average net return in this particular year of $129.75 per member reporting; potatoes averaged $105.81; small grains, $369.22; and sugar beets $344.07 per project.

At this time agricultural production of various sorts comprised the great majority of 4-H projects, and evaluation could be made in monetary terms to a large degree. Beginning in the 1950s, many other types of projects were adopted which are mentioned in "Objectives of 4-H Club Work" and most of which cannot be assessed in dollars and cents.

Length and Nature of Membership

Considering the practical and attractive nature of the 4-H program, state leaders were deeply disappointed in the early years when nearly 50 percent of those completing one year failed to re-enroll. This continued to be the national pattern, but Montana soon climbed above this average and by 1967 it held eighth place in the nation with a re-enrollment of over 70 percent. Local leaders also changed with discouraging frequency; two years of service was an average for a period of years.

This seeming lack of commitment led in part to a violent attack upon the entire 4-H program by a journalist, Ira Dietrich, in 1967. The book-length tirade was titled *Poor Damn Janeth*. Dietrich maintained that manifestly the projects in 4-H were not geared to the interests of the members, and that, in particular, the program did not expand with the age of the member since the average dropout age was 13. He went on to assert that those who did remain in 4-H did so largely because of pressure from parents. He insisted that the projects were geared more "to win" than to learn, and that both leaders and members were under intense pressure from parents and sponsoring organizations to win. In order to win, dishonest methods were used in preparing exhibits, animals were drugged and judges were bribed. 4-H leaders felt that the accusations were highly exaggerated, and even dishonest, since the book was written in a florid "expose" style and almost no evidence supported the sweeping statements. However, the book did raise, in a colorful context, enough of the well-known perplexities to cause a hard rethinking of many aspects of the 4-H methods of operation.

Concern for the high dropout rate led to a statewide study in 1931. Taking a positive approach, only a few simple questions were asked regarding why the member belonged to 4-H. Twenty-two percent were attracted to the general educational program, and 22 percent gave social contacts as their first reason for continued membership. Vocational motivation attracted 16 percent, and 10 percent admitted that they cherished the awards which were offered. The other 30 percent had other specific reasons.

In 1949 William G. Stucky made a study confined to Rosebud County which was incorporated in a "Western States Study." In this inquiry an encouraging 67 percent re-enrollment was found which compared with 50 percent nationally. The survey of the 25 clubs with 50 leaders and 280 members was interpreted through positive factors. Reasons for a favorable re-enrollment included good leaders who, as well as the county agent, remained in the background, leaving the conduct of affairs in the hands of the members who planned good programs, well managed business meetings and a variety of projects. The feeling that 4-H is more than a project, a progression in projects suitable to the age of the members, and a situation where the community gives approval to the organization, were additional features in the Rosebud County study.

Another study, made in 1955 in cooperation with the national office, took the negative approach. Here reasons given for dropout were early discouragement with projects which were too ambitious and slow in maturing, fear of failure, lack of parental support, distance from meetings, failure of the projects to appear relevant, and competition of school work and other activities with 4-H work.

Montana leaders felt that part of their success in securing a higher than average re-enrollment came from an early insistence upon completion of projects. During the first 10 years of club work, the rate of completion was only 53.7 percent. A goal of 85 percent was established in 1924 and was attained in 1926. In 1929 the Montana completion rate was 88 percent, the highest in the nation, and it has continued at a high level.

"What is 4-H experience worth?" was the quest of Weber H. Peterson, son of Carl Peterson, Montana's second county agent, as he researched a master's thesis at Montana State College on "The 4-H Student in College

Activities.'' He used student records for 1937-1938 and summarized the study in *Rural Sociology*, Vol. 3, No. 3, September 1938. Grade averages of 4-H and non-4-H students were almost identical. In participation and leadership he found that 4-H members averaged 59 percent above average. Only in journalistic and student senate participation were they less active. He found them more active in athletics by 24 percent; in church-related activities, 34 percent; in debate, dramatics and music, 36 percent; in honorary societies, 18 percent; in judging teams, rather naturally, 64 percent; and in fraternity activities, 70 percent. A significant figure was that in 1937-1938 some 82 percent of the 4-H members returned to college in comparison with 72 percent without 4-H experience.

Girls vs. Boys

4-H leaders have continuously been concerned by the larger number of girls than boys in the organization membership. This is probably a natural situation rising out of the greater mobility of boys, and as the older boys engaged in active farm work their time was less flexible. Some of the larger number of girl members came from home economics, where clothing and foods projects had high enrollments. However, in mixed activities girls often showed up in numbers, and in prize-winning positions in beef and sheep judging, fat stock and corn shows. In the early years the numbers were somewhat comparable; in 1924 boys numbered 1,500, girls 1,882. In eight camps held that year there were 201 boys and 244 girls; by 1928 the imbalance in six camps had changed to 64 boys and 203 girls, and by 1936 this had changed again to 140 boys and 594 girls. In 1936 the number of girls in the entire enrollment was about twice that of boys—4,839 to 2,137—and in 1964 this remained about the same—8,840 to 4,874.

Older Youth Programs

Together with every other state in the nation, Montana attempted to find ways to keep older youth in 4-H activities. In 1934 a new group was formed along the lines of the national pattern, ''The Montana 4-H Builders,'' comprised of 4-H members 16 years of age or older. A club of Builders usually drew its members from a number of regular clubs, the members maintained contact with their former leaders for aid with their projects, and they usually attended their old club, giving assistance and developing a movement which led into the Junior Leader program. One Builder's Club per county was recommended. These members were particularly useful in assisting with the community Achievement Days and with the county fair. An active social and recreational program was included in their activities.

The Builders grew slowly. By 1938 five counties had active organizations.

The Lewis and Clark County Club was the first, forming as an older club in February 1932; then as 4-H Builders in October 1933, and in turn sparking the state movement in 1934. By 1947 Builders Clubs were active in 19 counties. In that year three summer camps were held for these older members just before the regular 4-H camps. By 1952, 20 counties had Builders Clubs, with a total of 716 members. Instead of camps in that year, regional one-day "jamborees" were held. The Missoula jamboree attracted 85 members.

An idea which appealed to older members was that of conservation projects, and a summer camp centering on this project developed in 1947. Attendance was based upon winning in project and scholarship competition in each county. The first camp of 42 boys and girls from 20 counties was held at Camp Maiden. Charles L. Horn, president of the Federal Cartridge Corporation in Minneapolis, gave encouragement and funds, as did the State Association of Soil Conservation Districts. Agents reported that some 1,053 members carried on projects in competition for this first camp.

The location of the summer sessions changed from year to year and included, in addition to Camp Maiden near Lewistown, Beaver Creek Camp near Havre, the Little Bitterroot Youth Camp in Flathead County, the Lions' Camp near Red Lodge, the Bow and Arrow Ranch in Park County, and the Pines Youth Camp on Fort Peck Lake. By 1957 some 143 persons attended the camp, including representatives from Canada. Over a 10-year period, the subject matter emphasis shifted, covering such subjects as crops, livestock, grasses, soil, water, fire control, birds, fish, game and trees. Watershed problems had been considered at every one of the 11 camps which had been held up to that time; wildlife at 10; range management at six; timber management at seven; human conservation at five; insects at three; and weeds at two. The 1968 camp attracted 96 members from 40 counties to the Little Bitterroot Lake Camp. In 1971 the 25th annual camp was held at Camp Kiwanis near Havre.

The special attention which was given to the Builders produced encouraging results, and an additional refinement was made with the older members by designating "junior leaders" among the early teenagers, and "teen leaders" for those in the upper teens. Special training sessions were planned to develop the techniques of each group. By 1964 junior leaders numbered 2,201, almost enough to assist on a one-to-one basis the 2,966 adult local volunteer leaders. There were fewer teen leaders, and they often assisted the younger members on special projects. Camps for these groups also provided one of the best situations in which to recruit and train these young leaders, and by 1968 there were four camps attracting 400 campers.

4-H Alumni and Mu Beta Beta

As the State Congress met at Montana State College beginning in 1930, the desirability of having an organization of former 4-H members among the college students became apparent. A 4-H Alumni Association had, in fact, been formed in 1929, and found one of its first enterprises to be that of assisting with the convention in 1930. Even though the 4-H meeting was held when the College was not in session, the students returned to the campus and gave full assistance to the arrangements and operation. They provided the best possible contacts in recruiting 4-H members for college enrollment. In 1948 a chapter of Mu Beta Beta, a 4-H college honorary, was formed, which added stability and usefulness to the 4-H alumni organization. However, Mu Beta Beta has been inactive in recent years.

Leadership in the State Office

Direction to the state 4-H movement has been given by a very small staff in the state office; a leader, an assistant leader, and one or two secretaries. This has been made possible by the upward movement of older 4-H members into adult leadership positions. Fortunately, the directors of the Extension Service have been sympathetic to the program. Fortunately also, the agents, from whom much of the direction must come, discovered early, as has been frequently mentioned, that the 4-H Club was often the best place to publicize a new idea. The specialists also worked with the agents and with the club leaders directly in preparing projects, setting standards and judging performance.

The leadership in the state 4-H office has been of a superior nature. In the early years relatively short periods of service were compensated for by Charles E. Potter who was 4-H leader for 15 years, 1920-1935. In his "Twenty Years of 4-H Club Work in Montana," Potter commented upon the continuing need for organizational refinements and noted that the uniform method of writing programs, projects and progress reports adopted in 1927 had been a significant step. He watched the development of the numerous New Deal programs with apprehension, but was firm in his judgment that the 4-H personnel could become "social engineers" as readily as any other segment of the Extension Service.

R. E. "Scotty" Cameron, 1935-1946, through his genial personality and strong sense of showmanship, dramatized the movement and held it together in a remarkable way during the difficult period of the war. Tyrus W. "Ty" Thompson combined youth, an attractive personality, and an enthusiastic dedication to rebuild the clubs rapidly after the war. Following his short three-year period, 1946-1949, he later gave encouragement from his position in the national office.

Paul J. Moore, 1949-1966, has served the longest period to date (1974) as 4-H leader. His systematic planning and good judgment made his regime one of steady progress and close cooperation from his staff. Clifford W. "Brick" Vaughn was equally successful in promoting a strong cooperative effort and a steadily growing organization in a period of changing structure.

Vaughn came from a county agent position in 1956 to become associate state leader, and succeeded Moore. In the reorganization of the Extension Service, July 1, 1967, Mrs. Vivienne B. Kintz became coordinator of the new Division of Human Resource Development, while Vaughn continued to direct 4-H work as state 4-H specialist. Upon his retirement in 1973 the work continued in experienced hands as James F. Sargent, former assistant, became state 4-H specialist, and Harold Strobel, with wide experience over the state, became assistant. By this time Dr. E. Dean Vaughn, who had earned all of his degrees at Montana State, was national director.

Of the many capable women who were assistant and associate leaders, Miss Geraldine Fenn has, without doubt, made the most significant contribution. She organized well, planned imaginatively, and possessed physical stamina which allowed her to travel extensively over the state. In addition to regular procedures, she carried on a good deal of the tradition of Mignon Quaw in an emphasis upon music and recreation for the entire Extension program, state and national. Possessing a strong dramatic sense she provided leadership in ceremonial programs, illustrated by her presentation of a suggested form for a Rural Life Sunday service at the National Congress in 1947.

Her greatest impact upon national trends no doubt came from her pioneering work with individual 4-H members on self-determined projects, which crested in 1964. Together with agricultural production and consumption projects, her dedication to youth interests also included social, cultural and world relationships. Such projects had been surfacing for some time and usually caught local leaders unprepared and with insufficient time to guide them. Miss Fenn's encouragement greatly assisted the movement which will be discussed shortly. She also gave strong leadership to IFYE, to the Annual Encampment of Citizenship, which some Montana farm groups co-sponsored, and to cooperation with the Peace Corps. She had coordinated the preparation of the Montana delegation to the White House Conference on Children and Youth in 1950, and she was asked to do this again for the conference held in 1970. She was available in 1970 because she was drawn, soon after retirement, into a new statewide program directed by the Montana Advisory Council on Children and Youth. She had come to the position of assistant leader in 1946, continuing until 1967, giving her a longer period of service than any other member of the state staff.

The County 4-H Club Council

Given the limited state and county professional staff, and the many community leaders, a need arose to share information, to establish more uniformity in programs and standards, and to provide training for new leaders. The first county leaders association was formed in Cascade County in 1926, and the practice spread rapidly. In 1930 a refinement produced the County

4-H Councils, and a constitution was formulated. Montana was one of the first states to adopt the council idea. Since the activities, interests and need for support encompassed the entire community, some councils attempted to include not only the 4-H organization, but representatives from a wide variety of groups such as the farm organizations, service clubs, school boards, fair boards, bankers and other businessmen. The council meetings made arrangements in detail for fairs, achievement days, tours, demonstrations, picnics, dress revues, camp awards, and trips to conferences and conventions. Here aid was also enlisted for crash projects such as rodent, grasshopper and weed control. Finances, transportation facilities and other ever present needs, as well as general promotion, lay within the province of the Council.

Montana 4-H Leaders' Association

The perennial problem of securing and retaining leaders for the 4-H Clubs became crucial again following World War II, leading to the organization of a State Leaders' Association devoted to improving the situation. An advisory committee was formed in 1946 consisting of representative county leaders and several specialists whose fields were popular with 4-H members. This committee assisted in organizing active leaders in each county, and establishing four districts, each with its organization. In 1947 a State Executive Committee was formed consisting of the state president, state secretary-treasurer, and a president and vice president from each of the districts. This committee met four times each year to review county and state programs. The districts were increased to eight in 1948.

Leaders in the state office and those in long-time service over the state hoped that through this association of leaders in each county, new people could be recruited and encouraged to continue for a considerable period of time. Two years was a typical length of time of service, which made recruitment and training a continuous and time consuming task. To further encourage and recognize longer service, a ceremonial touch came into being in 1944. A "Clover" award was assigned distinctive designations at five-year periods of service: silver at five years; gold at 10 years; pearl at 15 years; diamond at 20; emerald at 25; ruby at 30; and sapphire at 35 years. Records were assembled and maintained, and the recognition given encouraged many leaders to continue their work. A considerable number have received the diamond award for 20 years of service, but very few have reached the 35-year mark. A search in 1972 reaching back only to 1967 revealed only four in the latter category. Mrs. Audrey DuPay of Great Falls, who was active in forming the Leaders' Association in 1947, received her award in 1967. Ramon Nile of Forsyth and James Nile of Hysham received similar awards in 1971. The record length of service at that time appeared to be held by Mrs. Martin Schultz of Dutton. With 40 years of continuous

Mrs. Martin Schultz of Dutton, Mont., with the silver tray awarded her for 40 years of continuous service as a 4-H leader, 1972.

service in 1972, she was given a silver tray in a public ceremony, since a formal award had not been designated beyond 35 years.

National Alumni Awards

The National 4-H office has worked with private donors to give suitable recognition for outstanding service by 4-H alumni at the annual National

Congress in Chicago. Montana has been recognized on two occasions for this award. In 1955 former Extension Director Robert B. Tootell, then Governor of the Farm Credit Administration, was given this award; and in 1962, Montana State College President Roland R. Renne was similarly honored.

Rural Life Sunday

During the war period the idea developed for 4-H members in a community to come together in a Sunday worship service. By 1946, at the close of the war, this idea was meeting with wide approval and 103 such services were reported in Montana in 1947. These were held in 38 counties with 4,175 members, leaders and friends attending. The movement received additional impetus in 1947 when Montana Assistant 4-H Leader Miss Geraldine Fenn prepared an effective 4-H Rural Life Sunday service, and using the entire Montana delegation, presented it before the National 4-H Congress in Chicago. One comment concerning the steadily growing function was that "It is the feeling that no event stimulates as much local initiative and feeling of worth with so little promotion." By 1962 Rural Life Sunday Services numbered 308 in 41 counties with 7,655 in attendance. The disappearance of many rural churches and greater intermingling of town and country people resulted in fewer services and the name was changed to Rural Church Sunday.

Work with Indians

The 4-H programs were taken to the Indian reservations at an early date, where they met with a cordial reception. An "Indianized 4-H program" was developed, which will be discussed later. In the early period the records from the Crow Reservation are quite complete and show that two Crow girls were included in a girls' team which went from Montana to the Interstate Fair in Sioux City as early as 1923. By 1926 a sustantial number of Crow girls were enrolled in an extensive sewing club program in Big Horn County, both on and off the reservation. There was interest among both boys and girls on the Crow and Cheyenne reservations in garden and potato clubs, but drought conditions brought severe discouragement. In 1932 some 14 garden clubs were operating on six reservations with 122 members cultivating 29 acres. By 1938 the Bureau of Indian Affairs and the 4-H organizations were working closely together and the clubs were reported as follows:

Reservations	Number of Clubs	Enrolled Boys—Girls		Total Enrolled	Total Completed
Blackfeet	3	11	12	23	Records incomplete
Crow	13	42	94	136	116

Flathead	5	18	30	48	38
Fort Belknap	3	6	21	27	22
Fort Peck — Agr.	6	57	10	67	62
— H. Ec.	9		83	83	51
Rocky Boy's	4	5	30	35	29
Cheyenne	4	29	33	62	58

Where the supervision was comparable, the percentage of Indian youth completing projects was close to that of other clubs in the state.

Calf clubs were without doubt the most popular of all projects among the Indian boys, although they encountered great difficulties in obtaining calves with promise. In 1939 the Crow Calf Club was outstanding, but the agent reported certain problems encounterd by the leaders in this accomplishment:

> Just how they induced the young Crows to start the baby beef club three years ago is not revealed. During the first two years it appeared difficult to get across the idea that club steers thrived better on plenty of grass rather than feeding them as cow ponies, and giving them daily exercise in bull dogging and roping.
>
> Somehow the interest of the boys was maintained and the patience of their leaders held out, and this year seven members of the Crow 4-H Beef Club produced 27 head of fat steers, many of them well finished and trained for the ring.
>
> Mr. Runyan reports that this carload was shipped direct to market and sold for $2,400, of which $300 was net profit to the boys.

In 1940 over 600 club members on the seven reservations represented a 20 percent increase in membership. Ninety percent completed their projects, which was three percent above that of non-Indian clubs. The value of their project properties was estimated at $6,044.79. The breakdown in membership in the various projects was as follows:

Projects	Number of Clubs	Enrolled Boys	Enrolled Girls	Total Enrolled
Clothing	22		198	198
Food Preservation	2		17	17
Food Preparation	11	14	93	107
Room Improvement	2		15	15
Handicraft	9	88	24	112
Potato	3	18	5	23
Garden	7	50	1	51
Beef	9	61	1	62
Dairy	1		1	1
Swine			5	5
Poultry	4	20	9	29
Totals	70	251	364	620

In 1943 a Blackfoot girl won membership in the annual delegation to the Chicago National 4-H Congress, and in 1944 a Blackfeet boy won. In 1946 six Indian youth attended the National 4-H Citizenship short course in Washington, D.C., another youth went to the Encampment for Citizenship in New York City, six participated in the Black Hills recreation Laboratory, and 176 attended six recreation workshops held in the state. In the past decade, national and state programs for minority groups have been greatly enlarged and work with Indians and Indian youth has increased rapidly since that time.

The 4-H Project

Without doubt, the heart of the early 4-H program lay in the project which the boy or girl undertook. 4-H leaders took special care, however, to emphasize that the organization was more than the project. Lest some members not participate in the other activities, the project was designed to provide the broadest learning experience possible. The choice of project must be made, the physical properties assembled, the plans perfected and patiently carried to completion, and then the profits or losses of a tangible or intangible nature assessed. Usually the young person worked largely alone in full charge of the project, but with the full cooperation of parents, club leaders, and youth engaged in the same type of project in the same club. The project furnished a real-life situation and gave the member a practical insight into the business aspects of agricultural life. It was education at its best and indeed the nucleus of one of the greatest educational programs the world has seen.

An Indian 4-H girl in Indian costume.

The questions of standards for 4-H projects in relation to the club member arose very early. One of the first listings of criteria was essentially as follows: The project should be productive and profitable; should be suited to the capability of the member; should extend over several months; should permit ownership; should offer worthwhile training, and be of a magnitude which would interest capable youngsters.

The early project subjects were relatively few in number. Corn, beef, gardens, potatoes and swine were highly practical subjects for the boys; clothing, foods and home furnishings attracted the girls. Departure from the standard pattern was discouraged and only four members were listed in a miscellaneous category among a total of 3,889 members in 1927. In that year the listing and number of projects included the following:

Beans	2	Food Preservation	99
Beets	55	Gardens	198
Beef	16	Home Furnishings	57
Clothing	1,341	Poultry	167
Corn	210	Potatoes	316
Dairy	86	Sheep	91
Food Preparation	85	Swine	162

A 4-H member gets some help with his tractor project.

Additional projects were incorporated with increasing frequency as new trends in technology appeared, and when leadership was available. In 1929 in 12 counties, 21 farm economics clubs were formed to stress keeping accounts on the farm enterprise. In 1940 clubs began to form to study the use of electricity on the farm unit. Continuous work was being carried on with the improvement of grains, and the challenge of experimenting with the new varieties often resulted in the older members setting up wheat clubs, oats clubs and small grains clubs. By 1958 tractor schools and tractor study and demonstration projects had become highly popular. In that year six training conferences were held over the state involving 76 leaders in all, 19 of whom were junior leaders. The establishment of a Western Regional 4-H Tractor Operators' Contest, often held in Hutchison, Kan., and the many prizes offered by oil and machinery companies, kept this type of training going in high gear.

Almost all of the projects have had a permanent influence upon production in Montana. The corn clubs were the first to bring the young people into cooperative work in 1910. These experiments on the small 4-H plots pointed the way toward the possible proper use of corn under Montana conditions without the larger losses which could certainly have been encountered. Potatoes were another case in point. Good seed was essential and trials on varied soils and different locations were made-to-order for 4-H projects. The 1930 annual Leader's Report noted that "the only certified seed in a community would be produced by a 4-H project." In the same year the report stated, "The bulk of seed corn produced comes from 4-H plots." Many such situations were no doubt in Director Tootell's mind when in a memorandum to all Montana Extension workers on March 28, 1952, he observed, "4-H Club work probably is the most effective *method* of doing *adult education* work."

Baby Beef and Fat Stock

Probably the most attractive project has been one called "beef," "calf," "baby beef," "heifer," and then "fat stock," but had to do with the raising of a calf. The 4-H Baby Beef project got under way in 1920 with an enrollment of only 44. The boys received a severe jolt when Eula and Hazel Thompson from Gallatin County ran away with both first and second place at the Montana State Fair with their twin Hereford heifers. By 1949 there were 1,452 project members with 1,569 animals valued at $246,584.98.

An important new dimension was added to the calf projects with the rise of the Montana Junior Fat Stock Show which was first held in November 1938 in Billings. Although lambs and hogs were shown and sold in larger numbers than calves, the emphasis and greater value lay with beef. The November date, which became traditional, provided a desirable time span from the earlier fairs for additional maturity for the animals and showing finesse on the part of both animal and owner. The Billings Chamber of Commerce gave strong support, and Keith Sime, Yellowstone county agent, developed the organization. The show was open to both 4-H and Future Farmers. The Danish system of judging whereby groups of animals were placed according to market grade was used for the first time in the state, and this was one of the earlier uses of that system in the nation.

At this 1938 show, 253 members from Montana and Wyoming exhibited and sold 161 head of beef, 347 lambs and 51 hogs. Three of the best calves were taken for exhibition to the Chicago National 4-H Congress. With these, in competition with 350 fat steers from over the nation, Marie Monforton of Gallatin Gateway and Arthur Jacobson of Clyde Park won first places, and Helen Monforton of Gallatin Gateway won 12th place. By the third annual show in 1940, the entries totaled 807 head of stock shown by 318 members. In 1947 a feeder calf sale was also held in connection with the fat stock sale, which added to its popularity and usefulness. The fat stock show was attacked as being artifical since fat stock had not been in the Montana tradition. However, the project gave the 4-H member a familiarity with all of the processes of stock raising and of marketing as well. The success of the Billings show led to the setting up of equally productive shows and sales in Missoula, Butte and Great Falls, and later at the annual Winter Fair in Bozeman.

Clothing Projects

The major project among the girls has been that of clothing. In fact, many clubs registered more students in this project in some years than in all other projects combined. In 1921, the first major year of organization, there were 64 clothing projects in as many clubs, with 865 girls enrolled. Leaders were always somewhat difficult to obtain. Mothers often felt they lacked the background and skills to provide leadership in this area, yet they participated in food projects with complete confidence. Fortunately, the teachers in the Smith-Hughes vocational homemaking courses, installed in the high schools in 1917, strongly supported the clothing project.

The project has stressed the study of cloth, colors, styles, prices and good taste. In 1928, for instance, the project was divided into three phases—the spring, summer and fall outfits. In that year 1,689 girls were enrolled from 179 clubs and 1,414 completed their projects. Many mothers remarked that the results were almost as beneficial for them as for the girls, since the projects also brought them into contact with the possibilities of better and more economical clothing. By 1949 there were 44 counties participating with 430 clothing projects and 3,332 members. The articles of clothing produced were valued at $41,116.29.

The demonstration of the clothing project has provided the touch of color and drama so essential in rounding out the achievement days, fairs and conventions. The Dress Revue has served as the major social event of these occasions, and the impact upon the entire community in upgrading an awareness of new styles, colors, fabrics, and costs has been massive and enduring. Since the judging at the revues soon came to include the poise of the girl as well as the fine points of dress, the girls and their entire families soon became more aware of certain social graces than their city cousins who for centuries were supposed to be particularly adept at these things. In more recent years a few young men have joined the young women in modeling varied types of men's clothing.

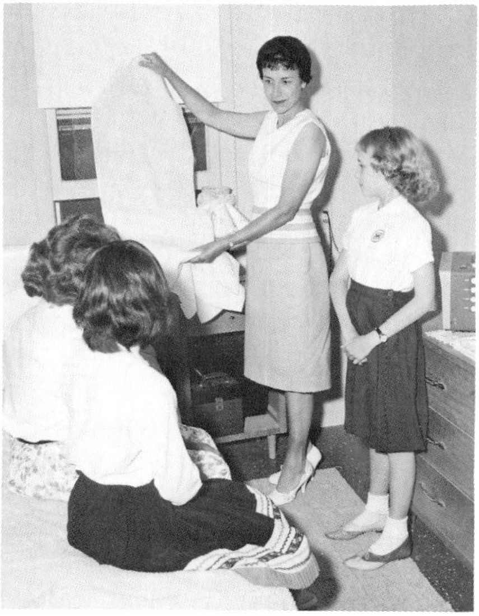

(left) 4-H member displaying a collection from her entomology project.
(right) 4-H leader with some of the girls in her 4-H sewing class.

4-H members exhibiting the three top 4-H steers at the state fair in Helena, Mont., in 1928. The three steers brought $11,355 at auction, top price on record for three top steers at any show in the United States.

The girls made equally great impressions upon farm home living with home furnishing projects, or "My Own Room" projects. Often the mother had been in a home demonstration club herself where similar subjects had been under study. In any event, not only the homes of the members, but many other homes in the community were strongly influenced by these projects. The food production and preservation projects have been popular, and during periods of drought, depression and war proved to be an invaluable feature of the national rural structure.

Self-Determined Projects

Although the numbers of projects expanded greatly, they remained for many years closely related to the immediate agricultural production problems; a "corn, hogs and manure" approach, is one interpretation. This pattern was, of course, much easier to administer in the earlier period of limited numbers of leaders and rapid growth, but slower travel and communication. Many states tried to maintain a standard of no fewer than five members working on similar projects in a club. This was difficult to maintain in the small Montana clubs, and an individual member often chose his own project within certain guidelines and usually still related rather closely to agriculture. Booth Holker notes in the 1938 annual report, projects such as that of the three Whitmer brothers of Bloomfield who built a small dam which provided irrigation for some 20 crucial acres, and of others who experimented with contour ditches. County Agent N. A. Jacobson developed 4-H work in Powder River County in the mid-1930s and worked with individuals on subjects of their choice in range problems, such as types of grasses and carrying capacity. In 1954 he outlined a 4-H horse project, which could be developed in a number of ways.

Gradually the range of subjects broadened and more freedom was given for individual or "self-determined" choice. The 4-H leaders discovered in 1964, with some astonishment, that 273 members were working singly on subjects of personal interest. In 1965 this number increased to 496. Miss Geraldine Fenn encouraged this movement, which was then little practiced in other states but has now been widely adopted. She traveled throughout the state conferring with local leaders who often felt quite inadequate in helping develop some of the new ideas.

A clear break was made, in many instances, with the older patterns which were tied to practical aspects of rural living. A number of projects in international understanding came out of IFYE contacts and visiting speakers from other countries at 4-H camps; "Understanding the Problems of the Congo" and "Contributions of Greece" are examples. Appreciation of art, music and poetry provided other subjects, but most were still related to agriculture. Patricia Heimbichner of Billings, who won a trip to the State Congress in 1966 on a dairy project, conducted an unusual study on "The Pecking Order of Dairy Cattle," which attracted the attention of the dairy association. A detailed study of "The Leaf-cutter Bee," by Peggy Christie of Townsend, was used by the state entomologist in his work. One of the most publicized projects was that by 14-year-old Jon Swenson of Shepherd. His "Study of the Black-Billed Magpie," won him trips to the 1966 Montana 4-H Congress and the 1967 National Roundup in Denver. The National Geographic Society then asked him to write up and illustrate his study for the *National Geographic School Bulletin.*

By 1966 "self-determined" projects were in 12th place in a list of 31 subject groupings. Reports to the 1968 State Congress noted such subjects as health, smoking, safety, careers, veterinary science, small engines, consumer education, fire protection and leisure time. Associated with individual interests, plans were also made for 4-H citizenship workshops which were held in Helena during January-February 1969, when the State Legislative Assembly was in session, and these continued to be popular during legislative sessions.

An Urban 4-H Program

Mention has been made of closing the gap between country and city dwellers, and the increased involvement of the Extension Service with people who lived much of the year in cities above 2,500 in population. On a national basis, considerable experimentation was carried on with youth groups in cities using 4-H organizational methods. When a study in 1969 showed extensive malnutrition among low income families, an Expanded Food and Nutrition Education program was established. It was broadened in the summer of 1970 when, to meet a need which school lunches had provided during the school year, a 4-H type organization was developed to upgrade the quality of food available for the children in these low income families. By June 1973, 31 paraprofessionals were working with youth in seven counties: Missoula, Cascade, Lewis and Clark, Silver Bow, Yellowstone, Hill and Fergus, and on the Rocky Boy's Indian Reservation. Some 6,551 youth in low income families were included in the program.

In 1973 funds were allocated for research programs in rather traditional type 4-H programs to determine modifications for city youth. Organizational work began in Great Falls and Billings. "The objective is to seek out low income youth, counsel with them, and provide them with a satisfying educational experience," said Mrs. Vivienne B. Kintz, coordinator for Human Resources Development at the time. She explained further, "We are preparing to offer classes in various 4-H projects, rap sessions on identified concerns, and recreational and educational events."

At the same time a nationally based Community Resources Development program was begun in several western counties centering around Missoula, and others in the east with Glendive as the center. The purpose here was to involve young people in community development. This movement worked through schools, 4-H Clubs, other youth groups and community development organizations. Mrs. Kintz explained, "This is being done with the idea that youth can serve and learn about their communities now and be prepared to fill leadership roles in various aspects of community development as adults."

These programs attracted several hundred young people at once, and under the direction of professional and paraprofessional advisors a distinctly new youth program was initiated in the autumn of 1973.

The Montana 4-H Foundation

The idea of a Montana 4-H Foundation came from the need and the dreams of the leaders for substantial funds which could be used at their discretion for encouraging or rewarding achievement, for initiating promising projects and for experimenting with long-time plans. Many gifts of individuals and organizations to the 4-H program have been previously listed, and in 1967-1968 $45,000 was generously contributed for use that year. Almost all of these funds were designated for a particular use, largely for awards and travel funds.

Still in need of funds which could be used more flexibly, Director Torlief Aasheim secured the support of his associates in the state Extension offices and key leaders over the state in incorporating the Montana 4-H Foundation on June 10, 1969. Its purposes as stated in the Articles of Incorporation:

> . . . are to aid and promote, by financial assistance and otherwise, all types of 4-H programs and other Extension youth programs. The objects and purposes shall also include provision for scholarship aid to qualified 4-H members attending accredited educational institutions.

With so many constructive programs in place, 4-H has continued to increase in numbers and influence. It has broadened its reach and has crossed lines freely with other organizations such as Future Farmers, Future Homemakers, and segments of the Scouting movement. The merging of lines between town and country and urban inclusion in a Division of Human Reources point toward the purpose of improving the quality of living among all people. Probably the most striking trend of 4-H in recent years has been its growth in urban areas.

During the entire span of 4-H Club activity centering largely in a distinctively rural America, there has been a continuing buoyant spirit in what has been widely recognized as the greatest voluntary educational and social youth program the world has ever known. The cooperative activities, carried out with considerable open philosophical discussion, made it easy for the youth to absorb what M. L. Wilson characterized as "a deeper appreciation of the spiritual values of living in the open country." The remarkable unity among the leaders also made it easy for them to heed Wilson's admonition that they be diligent in developing "the best crop the nation can grow, rural boys and girls with character, with self-confidence, with physical and intellectual strength, and with the courage and determination to insure that the democratic way of life is to survive."

In a discussion with former Director Robert Tootell in the summer of 1971, in response to a question as to what he felt was the most important contribution of the Montana Extension Service, he replied without hesitation, "The work in 4-H."

IX. People and Ideas

In a story as complex and innovative—and successful—as that of the Montana Extension Service, many persons stand out for their distinctive ideas and creative work. The regrettable situation in this story is that space allows the inclusion of only a few of these. Used as examples, these individuals may serve to illustrate the innumerable activities and contributions which were constantly expanding. Many of the activities show the land grant concept working at its best. First a need, then consideration by the Experiment Station and perhaps by the academic community to determine solutions, followed by the dissemination of information by the Extension Service.

Some County Agents

Directors and specialists often appear in the leading roles, but county agents were also in the center of the action.

Robert E. Clarkson and Tax Equalization

Robert E. Clarkson was raised in the Chinook area, and following his graduation from Montana State College in 1917, he went to Phillips County as its first county agent. He found a county as large as the state of Connecticut with one highway, and that not greatly improved. A frenetic homestead movement in the lush years of 1910-1916 had made a shambles of the open ranges of the larger cattle holdings and had produced over 7,000 small landholders.

The year of 1917 was one of drought, however, and among the numerous problems which faced Clarkson as county agent was that of credit. Banks were charging 10 to 12 percent interest, with short-term loans preferred. The new Federal Farm Loan Act met with great approval, and the county agent was inundated with requests for help in completing forms. Six loan organizations were soon formed in the county with applications totaling $250,000.

A touchy marketing problem also existed in that the smaller stockgrowers discovered that the railroads naturally found it to their advantage to work first with the larger stockmen. The small producer, then, often had to wait for cars, and in the meantime in many cases the market prices had fallen. The result was that several livestock shipping associations were formed and, thereby through cooperation, the smaller operator could market to equal advantage. Clarkson was somewhat embarrassed by being thrown immediately into a situation where he appeared to be in opposition to the established business institutions. He remarked in his first report in December 1917, "At the present time there is considerable opposition being shown by merchants and bankers in our larger cities against the work of the county agent in carrying on beneficial work for the farmers."

Clarkson served for a short time in the war and returned to the Extension Service as county agent for Teton County on October 1, 1918. A high degree of cooperation was characteristic of Teton County, and Clarkson used this to the advantage of everyone. A wartime campaign was initiated to eliminate waste through rodent control, and this was renewed in late 1918 and again in the spring of 1919. Some two million gophers were destroyed each year with a total estimated savings of $750,000.

Clarkson developed practices which he had used in Phillips County such as cooperative shipping associations, wool pools and turkey pools. As the drought continued and deepened, he urged experimentation with new methods of drilling wheat, the use of new varieties, the development of additional range waterholes and other devices.

Mention has been made of the therapeutic value of recreation programs in periods of drought and depression, and these Clarkson found to be particularly popular in Teton County. A county-wide recreation committee planned a coordinated program, and for several drought years beginning in 1933, a three-day "chautauqua" held in Choteau brought people of all ages from many communities to widely acclaimed performances, largely recreational, but also combined with serious discussions of agricultural and other problems from well known leaders of the state and region.

Clarkson will no doubt be best remembered for his tax reform plan. Teton County had also been homesteaded from border to border during the wet years and high prices of the war. As prices fell to the lowest levels in the century and the drought continued, Clarkson was greatly disturbed at the increasing tax delinquency and loss of farms and ranches by the score. The Montana Legislative Assembly had passed a statute in 1919, and revised it in 1921, providing for the classification of all property in the state into seven categories and setting up percentages for taxable and assessed value. Another statute placed in force in 1921 further detailed the classification of land for tax purposes. A constitutional amendment was adopted in 1922 establishing the State Board of Equalization to supervise classification and taxation. Many ranchers were sure that unfair evaluation of their lands and high taxation was the basic cause for their financial failure. Clarkson made a detailed study of the soils map painstakingly prepared by Leonard Gieseker, kept careful precipitation charts for the county, and studied the potential productivity of certain farms in comparison with the taxes levied.

He found many cases of inconsistency which the new laws and the new Board had not yet met.

The county assessor, Otto Wagnold, was equally concerned, and together with members from a county policy committee, they met with small community groups in every section of the county. Here the landowners went over each farm and ranch holding, comparing it with the classification in the new soil maps, and checked tax levies. The result was quick agreement upon several standard classes of land for assessment and taxable valuation, based largely upon productivity, which could be recommended to the assessor and to the State Board of Equalization. A distinctive feature of this method was the democratic manner in which neighbors worked with each other and with county officials, resulting in a marked reversal of attitude toward governmental processes.

This practice set up in Teton County by Robert Clarkson, similar approaches by Henry L. Lantz in Phillips County, and still another method developed in Stillwater County did much to cause M. L. Wilson to insist that national agricultural practices such as allotment could be mandated by the farmers themselves. When Fred S. Willson succeeded Clarkson as Teton County agent he continued to promote the productivity theory of taxation.

Clarkson was caught up in activities of New Deal agricultural programs, and at the time of his death in 1941, he was regional director for the Bureau of Agricultural Economics in Lincoln, Neb. He had worked closely with 4-H Clubs in his 15 years in Teton County, both in agricultural and recreational programs. As the World War II Liberty Ships were being built and named in 1944, his name was proposed for one through 4-H channels; the "Robert E. Clarkson" was dedicated August 31, at Houston Tex.

Launching of the Robert E. Clarkson at Houston, Tex., on Sept. 4, 1944.

Jones and Spaulding and Stillwater County Voluntary Tax Reports

One of the most noteworthy developments in democratic action in taxation methods came out of Stillwater County. In 1929 drought and depressed conditions were already limiting county government activities. Arthur C. Bishop campaigned for the office of assessor with the promise that if elected he could reduce the costs of the office substantially by eliminating some 50 part-time local deputies. Their duties were to call on each property holder, listing household items, personal property, livestock, machinery and other taxable property. These listings were usually made from the observation post of a comfortable chair, or the front gate, and were often incomplete and unsatisfactory. Bishop felt that if the property owner merely filled out a form which could be mailed, the result would be equally reliable. He proposed to reinforce his plan, however, by making the assessment lists a public record, and perhaps publishing them in the local newspaper. Grave doubts were raised concerning the credibility of returns prepared in this manner and considerable controversy arose, but Bishop was elected. County Agent William H. Jones, followed by agent Curtis J. Spaulding, encouraged this plan, assisted in the preparation of the forms, and worked extensively with rural communities and individuals in completing the returns. In so doing, they demonstrated again the ability of the Extension Service "to reach the last family up the last road," quickly and with effective rapport. Everyone was astonished when the returns were higher than before; about twice as many cattle were reported, and personal property listings were much higher. Among about 3,000 taxpayers' reports, some 150 were considered questionable. Upon investigation, about half of these were adjusted. The method was adopted statewide and is still in use. In his Columbia University Oral History report, M.L. Wilson recounts his visit to the county as the plan was being tested. He then comments that a few years later when the New Deal agricultural planners were faced with the enormous task of securing detailed information from millions of farmers for wheat allotment and other programs, he recalled the Stillwater experiment. Over the objection of everyone except President Roosevelt, he instituted the plan and found that voluntary, democratic processes were still effective.

Henry L. Lantz and the Malta Plan

Henry Lantz came to Phillips County as agricultural agent May 15, 1924. He found that many of the problems which Robert Clarkson had worked on still needed close attention as the drought continued. He was concerned about the general misery of the agricultural population reflected in increasing tax delinquency and a tremendous number of abandoned homesteads. The very large number of 7,000 land unit holdings which Clarkson had found in 1917 were reduced to 1,191 operating farms and ranches by 1925, and many of these were in trouble. The broad and level Milk River Valley with its meandering river offered natural conditions for federal projects.

But many of the farm units were operating with 100 tillable acres or less, and Lantz was critical of the allocation of land in the federal projects. Wheat, flax and oats were the popular crops, and even with the greatest skill the arid land production was not sufficient to support a family. Lantz stressed the approach of "production per man" to secure a realization that returns were at a poverty level.

He promoted all of the recommended devices with great enthusiasm and advocated diversification for the homesteaders, strip farming, summer fallow and new varieties. He endorsed the program of Blaine Ferguson in Pondera County of "Dam the Coulees" and use the water for flood irrigation, particularly for gardens. Lantz urged that money and energy should not be used in constructing these dams unless they promised real relief. Many, he felt, were so small and uncertain that they merely prolonged an already destitute condition. In 1928 he remarked, "The gross income of many farms in the county is less than $500 a year." Feed and seed loans were a stopgap, but in 1930 he noted that only 1 percent of the loans had been repaid. By 1933 he expanded his observations:

> A summary of the seed loans proves that the government has been paying farmers as much as $25.00 a bushel to produce wheat worth 25 cents to 50 cents per bushel. According to loan records some farmers have been paid the equivalent of $20 to $40 a month to live on submarginal dry farms trying to raise wheat on land that has not produced the seed in a five-year period.

With H.E. Murdock he continued to work on flood irrigation, and Extension Circular No. 17 was issued in 1931. When G.H. Bingham conducted especially productive work along the Milk River Irrigation Project in Phillips, Blaine and Valley counties, Lantz went further into combining units of irrigated and dry lands into what became "The Malta Plan." Available records do not show how much he influenced M.L. Wilson in devising the "resettlement" idea, but the two were thinking and working along similar lines. Lantz felt that revolutionary measures should be undertaken in returning arid lands to the public domain rather than to spend large sums of money and years of farmers' lives attempting to provide seed and feed with inadequate flood irrigation projects. Equally revolutionary measures should then be taken in purchasing lands where these farmers could be "resettled" on units which were stable. In order to utilize the limited irrigable lands along the Milk River, he proposed a small irrigated farm to assure subsistence joined to an area on the benchland which might be operated on a dryland speculative basis, this in turn adjoining a larger grazing area. This plan, of course, was at least as old as ancient Egypt. Lantz described the development of the plan in his 1933 report:

> The Malta Plan of land settlement was originated in 1929 by H.H. Johnson, superintendent of the Milk River Irrigation Project; Josef Sklower, president of the Malta Water Users Association; and H.L. Lantz, Phillips county agent. The Committee secures options on small tracts of irrigated lands and

resells them to deserving dry land farmers who have machinery and equipment to work their land, but no cash to make initial land payments. The buyer and seller appear before the committee and agree to such terms as can be fulfilled and are satisfactory to both parties.

> The purchaser agrees to pay taxes and water charges the first few years, one to four, depending on the development and improvements on the land, but no interest or principal is paid before the land is well enough developed to grow such crops as sugar beets with which to make these payments.
>
> Since the adoption of this plan, 45 dry land farmers have been helped to secure irrigated farms. Most of these have been able to hold their own and while they have not made much progress they have made a living without getting government aid.

Joseph Kinsey Howard, in his *Montana, High, Wide and Handsome,* called this "the first practical farm resettlement program in the country." The program got into high gear during the Roosevelt administration, and in his last report as county agent in 1934, Lantz reflected his pleasure with the progress being made:

> In the spring of 1934 plans were made by the government to purchase submarginal drylands in Phillips, Valley and Blaine counties to resettle 500 or more farmers from these areas to farms in the Milk River Valley. The old extension program of trying to make wheat grow where a little grass grew before has been gradually changed to one of land utilization or using the land for what it is best adapted.

Lantz left his county agent's position in 1934 to become chief of land appraisal in the Fort Peck Reservoir area, but he returned again to resettlement work in 1938 as a district conservationist for the Soil Conservation Service which incorporated some aspects of the Farm Security Administration. In this work he helped direct the resettlement of 147 families on some 17,500 acres of irrigable lands in the Milk River Valley.

The Mizpah-Pumpkin Creek Grazing Association

Another nationally important and original idea arose in the drought stricken range area southeast of Miles City. As early as 1926 a group of stockmen whose adjacent holdings had supported some 6,000 cattle and sheep in the wet lush years preceding 1916 found that now only 2,300 head of stock were losing the battle for the limited forage in competition with a million gophers. The land involved 108,000 acres in Custer and Rosebud counties, reaching from Pumpkin Creek to the west over the divide to Mizpah Creek on the east. In the "good old days" the ranchers maintained home ranches along the streams and together grazed their stock on the divide range which was largely public domain and Northern Pacific lands. During the good years many homesteaders filed on favorite locations, erected fences and plowed up fields. Eventually, most of these were abandoned, but remained fenced, and were tax delinquent in county hands. By 1928 the federal government owned 24.5 percent of the area, the Northern Pacific Railway 41 percent, private owners 29 percent, and the state and county governments 5.5 percent.

In 1926 the ranchers, led by Nic Monte, asked County Agent Paul Lewis for help. Alva Simpson, Custer National Forest supervisor, and Evan Hall, agricultural agent for the Milwaukee Railroad, gave close cooperation as did representatives of the Northern Pacific Railway and the U.S. Interior Department General Land Office. Working closely, they secured the aid of the Montana delegation in Congress, particularly Senator Thomas J. Walsh, and an Act of Congress was approved March 28, 1928, which permitted a leasing arrangement between the government of the United States and the Mizpah-Pumpkin Creek Grazing Association. The ranchers completed their organization on April 16 when a constitution and bylaws were drawn up and adopted by 15 stockmen. An executive committee consisting of a president, vice president, the long-time Secretary-Treasurer Nic Monte, and three directors, all elected annually, carried on many of the details of managing the pioneering enterprise.

The Executive Committee reached a particularly satisfactory rental arrangement with the General Land Office, which also agreed not to charge fees for three years in return for a pledge to install sufficient water holes to service the area, remove the division fences, and fence the entire area and upgrade the grazing quality of the range. The taxes on the numerous lands of the Northern Pacific averaged $20.00 to $25.00 per section, and the railway officials agreed to lease their lands for a 10-year period for the tax costs, since the contemplated improvements would increase the value of their holdings. Private owners also agreed to a similar arrangement. State and county officers were in complete agreement with the project, but by law were not permitted to lease school and tax delinquent lands for less than $40.00 per section, which was more than the association felt it could absorb. A feature of the federal bill, however, allowed federal lands elsewhere to be traded for state lands. Soon, all of the state lands had been eliminated from the Grazing Association area.

The association then proceeded year by year to improve the lands. The entire area was placed under satisfactory fencing, which involved the removal of most of the homesteaders' fences. Gopher eradication, when carried out cooperatively over the entire area at one time, was almost complete after two seasons of poisoning. This increased the carrying capacity of the land by 20 percent. Sixty-four water holes were added to the natural springs which were not numerous. The number of cow units per section was carefully regulated while the condition of the range was frequently checked. A charge of $1.25 per head of stock was made, and grazing was limited to an eight-month period in the early dry years of the experiment. The quality of the cows grazed on the lands was scrutinized, and soon the association purchased bulls on a cooperative basis in order to provide for uniformity of calves which could be sold in larger numbers on a cooperative basis. Cooperative buying and management of salt, cottonseed cake and other feeds, the management of the roundup, branding and dipping also cut the costs appreciably. During the first year, 1929-1930, the range carried 1500 sheep and about 100 horses together with 3,000 cattle. By 1932 only cattle were grazed, 2,800 of them, and the range was excessively dry again. The association hoped to eventually graze 6,000 head and this goal was reached when the rains came. The return of improved conditions resulted in a gradual dissolution of the association which had served so well in the disaster years.

The operation was carried out quietly, and County Agent Lewis reported in 1933 that the "Mizpah-Pumpkin Creek Area was better known in other places in the state than in this county." Those in charge, however, recognized the quality of their work very early. Lewis stated in his 1929 report following successful legislation in Congress and a first year of operation, "This is a new step as no other legislation similar to this has ever been enacted and upon the success of this venture will depend future legislation which will bring about proper use of the public domain."

The experiment did attract attention, though, and by 1931 two other associations were in the process of forming in the area, and another in northern Montana. This movement led Tootell to formulate a State Grazing Act which was approved by the Legislature in 1933. Similar pressures on federal and state lands grew up in other states, and in 1934 the Taylor Grazing Act was passed by Congress providing for the national use of some of the innovations first developed by the Mizpah ranchers.

Dr. R.R. Renne (left), president of Montana State College, presenting a plaque to Theodore Fosse, Cascade Count, Mont., Extension agent, designating him "All American County Agent."

Ted Fosse — "All-American County Agent."

For the all-around success of his excellent program, Theodore "Ted" Fosse, longtime Cascade county agent (1939-1969), was chosen "All-American County Agent" in 1961. He represented the United States at the International Agricultural Exhibition in Cairo, Egypt.

In 1968 the Montana Agricultural County Agents' Association accorded him the Silver Buffalo Award for excellence in performing his duties and serving Montana's farmers, ranchers and 4-H members. Fosse retired January 1, 1969.

The Beef Cattle Industry

Former Chancellor Edward C. Elliott, when speaking throughout the state, often quoted the reference of the psalmist to "the cattle on a thousand hills," commenting that in Montana "a thousand cattle on a single hill" was a common occurrence. This is supported by the State Department of Agriculture's figure of $559,000,000 as the value of all livestock produced in Montana in 1970, making it the state's largest merchantable resource. Of this amount, cattle and calves had the greatest value with $308,410,000, compared with wheat in second place valued at $115,505,000. The importance of livestock was recognized early when a veterinarian, Dr. W.L. Williams, was included in the first Experiment Station staff in 1893. He worked with all livestock, and the livestock census of 1900 listed the size of his job. Cattle numbered 1,200,000; sheep, 2,900,000; horses, 230,000; and swine, 86,000.

The Farmers' Institute Board, organized in 1901, included representatives from the Montana Stockgrowers Association, formed in 1885, and in 1903 a member was added from the newly formed Montana Registered Cattle Breeders' Association. Demands for assistance were so great that the board's first full-time employee was veterinarian Dr. W.J. Hartman, who worked in an Extension-type manner from 1911 until 1917. The federal government provided livestock personnel during World War I, but in 1921 an Extension specialist was added to the state staff.

From Iodine Deficiency to Bovine Vibriosis

The Extension livestock specialists have not been veterinarians, but working with the Experiment Station on disease problems has been one of their urgent jobs. The year 1922 was one of aggressive assault upon blackleg in cattle when 32,710 head were treated in community vaccinating "bees" organized by county agents. Federal veterinarians came in to assist in a similar attack upon tuberculosis, and some 61,039 cattle were tested in 15 counties.

The year 1922 is best remembered, however, for the discovery of the cause of hairlessness in livestock—chiefly in pigs—which was causing

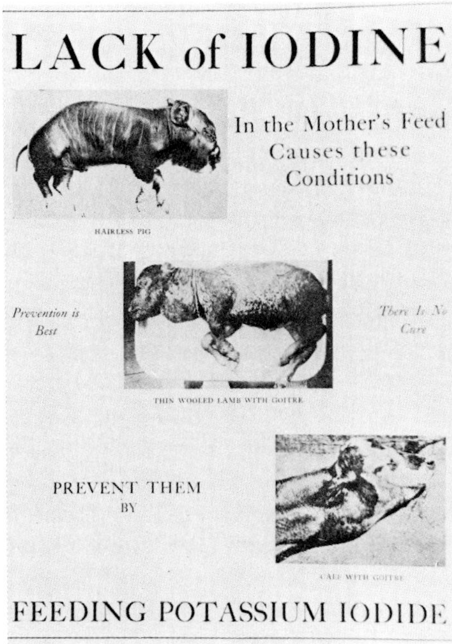

severe losses in several widely separated areas of the state. Dr. Howard Welch, Experiment Station veterinarian, through clues from other states and original research in Montana, diagnosed the disease as a thyroid malfunction caused by an iodine deficiency in the soils. He came up with the recommendation that potassium iodide be added to the animal feed in affected areas. This led to an increased examination of soils and grasses for deficiencies of other types, and the compensating use of minerals in feed supplements. National attention was given to this approach to animal health since the problem was widespread.

Montana has been fortunate in having relatively few difficult animal disease problems. Brucellosis was a steady threat, however, and the state was not declared free of it until early 1973. The staff of the State Livestock Sanitary Board, including longtime board member and Chairman Lyman Brewster, gave strong support to the Experiment Station and Extension Service work. Dr. Hadleigh Marsh, who directed veterinary activities in the Experiment Station from 1929 until 1950, had a worldwide reputation in animal diseases, particularly in sheep. He, together with Dr. Welch, Dr. Everett Tunnicliff and many others provided extraordinary leadership.

These men worked with the barest facilities until 1953 when a new veterinary research laboratory was made available. Then, in 1962 a combination of state and national resources provided one of the better animal health facilities in the nation. In 1964 another striking breakthrough at the veterinary laboratory was the production of a vacine which controlled bovine vibriosis, a disease which had been causing severe losses in cattle for many years on a national scale. The early day county agent was reported to be always equipped with surveying instruments; he was followed by one whose standard equipment was a vaccinating kit.

These accomplishments made the recollection of Lyman Brewster in the *Montana Farmer-Stockman,* September 5, 1968, seem remote indeed. Said Brewster:

> How many people today recall the slaughter and burial of a train load of hoof and mouth infected cattle at Glendive in 1918? How many remember the incredible early day work on tuberculosis? How many remember dourine and glanders in horses? The anthrax outbreak, cholera, lump jaw, bubonic plague, fistula and poll evil? Scabies? Foot rot in sheep? Paratuberculosis, and so on and on?

The Beef Performance Association

The Montana livestock industry recovered rapidly from the disastrous years of 1934 and 1936. The Livestock specialist's reports became almost cheerful when he discovered that the stock which had been retained were of high quality. He also reported that the stockmen were less independent and that, "they were now willing to cooperate to form grazing associations, build drift fences to control the range, and construct reservoirs and drill wells that will assure an ample supply of water in years to come." A statewide program of range improvement was developed and county agents took part in schools of instruction. Stockmen joined in associations for grazing, marketing and livestock improvement. Regional and state associations for Hereford and Angus cattle, for instance, and Montana organizations for the promotion of the new breeds of sheep, the Columbia and the Targhee, were of national importance.

The growth of the livestock industry is reflected in the increase in the number of Extension specialists. J. Oscar Tretsven had devoted part time to supervising the dairy industry since 1922, and he was strongly supported in teaching and research by Dr. John A. Nelson. In 1945 Tretsven was designated full-time dairy specialist, but was given the additional task of assisting the growing swine industry. Not until 1969 was Earl Peace named full-time swine specialist. In 1945 F.A. Ralston replaced E. Paul Orcutt as livestock specialist and in 1947 he worked full time with cattle, while Harold L. Dusenberry became the sheep specialist. In 1946 Fred S. Willson, who had served as county agent and as superintendent of the Northern Montana Branch Experiment Station in Havre, came to the Montana State campus in charge of animal industry, and he took an active part in experimental work. Orcutt returned as cattle specialist in 1948.

The Beef Performance project was one of the numerous programs to which these men gave full support. This had started as a record of performance experiment initiated by the U.S. Range Livestock Experiment Station at Miles City in the early 1930s. The Station urged ranchers to keep exact records of growth as a basis for the choice of young bulls and cows to be used for upgrading future herds. The promotion of this procedure by the county agents took the form of the ROM—Record of Merit—project for a number of years. In 1952 the Marias Hereford Association, centered in

Shelby, carried the project one step further by setting up young bull testing facilities. Bull calves which met traditional visual qualifications were evaluated in common feed lots for their ability to gain from weaning time to market weight. This led to a number of other bull testing centers around the state.

From this background the Montana Beef Performance Association was formed August 6, 1956, with Paul Orcutt, Extension livestock specialist, as the primary catalyst. Cattlemen, both purebred and commercial breeders, were urged to use careful records of calf weight at birth, at weaning and at one year of age, as proof of ability to gain weight rapidly. Estimates of a similar growth rate for the parents were added to the record. Based on heredity, rate of growth and the traditional visual qualities which had dominated the industry for so many years, certification of performance was issued. The association felt that this process promised to speed up and maintain the improvement of herds appreciably. The association was alert to new developments and made full use of artificial insemination.

COWS, CALVES AND COMPUTERS

by

E. Paul Orcutt

The First Twenty Years of the Montana Beef Performance Association

Cover page of Paul Orcutt's story about the Montana Beef Performance Association.

Paul Orcutt served as executive secretary of the association from 1956 until his retirement in 1967. Growth continued, and by 1970 the association had some 1,000 members and was certifying up to 2,000 young bulls and 15,000 heifers each year. This was another pioneering accomplishment which gained national recognition. In retirement Paul Orcutt traced this success story in his *Cows, Calves, and Computers, The First Twenty Years of the Montana Beef Performance Association,* (Bozeman, 1977).

New Standards of Quality

The Extension personnel in the mid-fifties found themselves in the middle of an economic and emotional hassle when the long-standing question of desirable standards for edible beef came into sharper focus. The taste for older beef—a firm, well flavored meat—had long been set by the beef-eating British people. In the days of the great cattle drives, at least four years were required for profitable marketing. As the day of the vast open ranges of free grass came to an end, the American producer discovered that beef animals took on weight much more slowly after one and one-half or two years of age, and marketing associations found that convincing the American meat-eaters that younger beef was much more tender and tasty was an easy task. In the case of older animals, well defined layers of fat appeared to decrease the toughness of the meat; therefore the most desirable profile for all beef for a long time was a short fat type. As younger beef became popular, large amounts of fat were found to be unneccessary, expensive to produce and to purchase, and perhaps cholesterol forming. A longer, lean animal soon became desirable.

The proponents of the age-old school of fat beef did not give up easily, however, in a day when the 4-H Fat Beef Show was at its peak. Demonstrations were given to prove the worth of new methods. Judging first proceeded under customary methods. Next, the animals were carefully photographed to keep recollections fresh, and then they were slaughtered. The same day or the following day, while the judges and the owners could still recall the merits of their favorite animal, the carcasses were carefully examined for the desirable "marbling" of fat and lean, which everyone accepted as high quality meat. This method proved to be convincing. Further refinements were made whereby the desirable degrees of marbling could be detected by electronic processes. The educational work on the part of the Extension personnel in this transition was extensive, to say the least.

Crossbreeding

The Extension livestock personnel found that their homework was increased considerably through the years, first with the search for purebred stock and then with the emphasis upon its modification, or crossbreeding, to secure better beef producing animals. Agents have spent endless hours in helping a stockman or his association develop herds of pedigreed Shorthorn, Angus or Hereford cattle. To these early favorites were later added the Galloway, Red Angus, Polled Hereford and others. The effort to produce the best of purebred stock still continues, but research has found that crossbreeding produces more rapid growth and a more desirable animal for beef purposes than the purebred. The experiments in crossbreeding have been greatly enlarged by the whole new array of "exotic" imports from Europe.

The Charolais, a French-bred import into Mexico and then into the United States in the 1930s, was one of the first of the exotics to make its appearance. Since its introduction into Montana in the late 1950s, it has continued to be popular. Other new breeds came with increasing frequency, such as the Maine Anjou and the Limousin from France, the Simmental—the most numerous breed in Europe—from Switzerland, the Chianina from Italy, and the Murray Gray from Australia. Although national and state associations were formed for a number of these breeds, Extension Service personnel continue to receive requests for assistance in acquiring animals from these new breeds, for crossbreeding purposes.

Reports from the specialists and county agents show many additional facets in the raising and marketing of cattle. The increase in the number of cattle fed and marketed in Montana is related to the increase in feed lots and sales rings which occurred following World War II. Feed lots, by 1974, totaled about 600, ranging from those feeding only a few hundred to the M & D lot at Shepherd with a capacity of 40,000. New management procedures such as custom winter feeding and custom calving also came under study.

Montana cattlemen, described by one observer as "among the best informed ranchers in the world," also expect their Extension agents to be as deeply interested as they in genetic advances such as artifical insemination, synthetic hormones and ova transplants. One prominent veterinarian remarked: "These are truly exciting times in the breeding of cattle. That which is probably not practical today may be commonplace tomorrow."

Oscar Tretsven and the Dairy Industry

A milk cow or two was standard equipment on Midwestern farms from which many of the Montana pioneers came, and they tried to continue the practice in the arid West. The early Farmers' Institutes devoted considerable time to dairy practices since including a milk cow in the self-contained farmsteads greatly improved the quality of living. Not until the federal government developed its crash war program in 1914 in the newly established Extension Service, did Montana have anything like a dairy specialist. F.M. Hillman, designated as a "field man," supported by the USDA from its Salt Lake City divisional point, served from 1914 until 1916. He spent much of his time in the dairy conscious valleys of the Flathead, the Missoula and the Bitterroot.

Dairying was developing generally by that time and M.L. Wilson in his first semiannual report June 30, 1915, as state leader commented:

> It is the concensus of opinion of the agricultural staff of the State College that this mountain region is particularly adapted to dairying and diversified farming while the eastern or plains region is particularly adapted to wheat as a cash crop and corn as a rotation and forage crop.

Dairy Associations

Hillman, his successor, W.E. Thomson, and others promoted cow testing associations, improved breeding practices, silos and cooperative community creameries, which grew up in large numbers. The cooperative creamery in Stevensville was one of the largest in the state. Cheese factories were also operated in Corvallis and Victor. The Bitterroot Cow Testing Association, organized in May 1916, was the first in the state. When its records showed the wide range from a high profit of $107.70 on one cow to a loss of $39.15 on another, greater attention was given to individual performance. In 1916 70 purebred dairy cattle were displayed at the Ravalli County Fair, a sharp contrast to the few which had previously been on exhibit. Similar advances were made in other valleys, including the Jefferson, where the Swiss Creamery in Whitehall continued a vigorous operation for many years.

A Dairy Herd Improvement Association had started on a national scale soon after the turn of the century, and several community associations

Cream can waiting to be picked up by the mailman for its ride to the town creamery.

began to affiliate in the 1912-1916 period. The Montana State Dairy Commission was formed in 1913, which encouraged the establishment of the State Dairyman's Association at the January 1917 short course meeting at Montana State College. A Montana Dairy Herd Improvement Association was formed in 1934 and Oscar Tretsven served as its secretary for many years. It was this period which Leland Cade, Montana editor for the *Montana Farmer-Stockman,* described accurately in the issue of June 1, 1972:

> At one time there was a major dairy industry in each county, a creamery in every village. The mailman hauled cream to town and hauled the empty cans back. Refrigeration was simple—the north side of a building, the cellar, the stock water tank, and while cans were waiting to be hauled to town they sat out in the hot sun under the shade of the mail box.

J. Oscar Tretsven became Extension dairy specialist on August 15, 1922, when the USDA field man was withdrawn in a retrenchment program. Tretsven had shown particular aptitude as an assistant in dairy work at the College and was familiar with the varied activities. He set out at once to organize a series of two-day schools over the state to reach as many communities as possible, and in 42 schools held in 1923 he contacted almost every active dairy community in the state. Reduced financial support made it necessary, beginning in 1925, for him to spend half-time on the State College campus where he was in charge of dairy production, and he served as principal of the large annual winter quarter Short Course in Agriculture.

John Nelson and Dairy Manufacturing

In compensation for Tretsven's assistance on the campus, Dr. John A. Nelson, who was in charge of dairy manufacturing, helped conduct meetings across the state in the summer. Both men sensed the need to improve the quality of dairy products, and this was Nelson's specialty. Milk and cream are highly susceptible to picking up distasteful odors, and under the conditions Cade described and others which he did not, Montana milk products picked up their share. Butter and cheese also often lacked the flavor, color and texture to qualify for top grade. Improved refrigeration was making the manufacture of ice cream more feasible, and this was a new field for most creameries.

The manufacture of dairy products increased strikingly under Extension supervision. The specialists' reports showed the following gains for the first years:

Year	*No. Dairy Cows*	*Lb. Creamery Butter*	*Lbs. Cheese*	*Lbs. Ice Cream*
1922	160,000	7,815,847	188,889	564,676
1924	194,000	14,178,938	934,065	761,638
1926	192,000	16,230,062	1,810,000	672,724

The increase in production of milk per cow was a major aim, of course. The estimate in 1926 was that the average annual production of Montana's dairy cows was 438 gallons or 3,502 pounds. This concerned the dairymen since the State College was taking particular pride at the time in its prize Konigen-Holstein, Grace Konigen, which had set a world's record in 1924 for a 10-year-old cow by producing 32,294.2 pounds of milk and 1,051.9 pounds of butter fat in one year.

Improved Facilities

Tretsven was highly practical in his approach to the care of dairy cattle, and he interested Harvey Murdock in agricultural engineering, whose inventive mind left enduring contributions in many areas. Together they developed new types of barns, sheds and equipment especially well suited to the area. Several stalls, for instance, variously described as a "model cow stall," a "free stall," and an "elevated cow stall" contained new features of convenience, economy and safety for both cow and owner.

Tretsven's "milking parlor" bore the brunt of many jokes, but was widely adopted, especially when milking machines became common. The parlors first appeared in the Bitterroot in 1916, and with some modifications, Tretsven promoted the pattern. The milking room in the barn contained a long elevated runway with a heavy railing. The cows willingly climbed the ramp to this platform because containers with feed and water were placed there for them. From this position, the milking machine cups could be easily attached to the udder which was about shoulder high from the floor. The feeding arrangement, while the cows were being milked, brought the cows in from pasture voluntarily at milking time. The parlor was economically constructed, extremely convenient and easily cleaned.

Tretsven also designed an outdoor shed-shelter which is still widely used. This was merely a shed of medium height with an extra wide projecting roof. A high board fence was built out from its open front, on the side from which the prevailing winds came. This provided protection of an area large enough for feeding of hay or silage, preventing undue drafts or blowing snow or rain during storms. The large dairy barns of the East and Midwest in which herds spent much of the winter, particularly the nights, was expensive to build and cumbersome to maintain. Moreover, the Montana cattle maintained better health and higher productivity if they remained in the moderate shelter of a well-bedded shed and were not subject to the extreme changes from a warm barn to the outside chill of a Montana winter day.

Industry Economics

The dairy industry in Montana found it difficult to adjust readily to the several economic crises which occurred. The low prices of the depression caused a drop in the production of butter from 16,791,660 pounds in 1930 to 11,586,093 pounds in 1936. Increasing production again was a slow task, and the specialist noted, "We not only lost five years time during the depression and drought, but we went back five years more in the development of the dairy industry in the state."

World War II also caused a strain, although the prices of dairy products were more satisfactory. Under the urgent needs of the government, the production of cheese jumped from 1,904,519 pounds in 1942 to 3,640,502 in 1944. In 1945 a readjustment of duties supposedly gave Tretsven a full-time dairy Extension assignment again, but he was expected to look after the swine industry too, which was also growing rapidly.

The reports of the specialist show a steady stream of varied activities. A movement to upgrade dairy stock resulted in importing several hundred choice heifer calves each year for several years from Tillamook, Oregon. A steady demand existed for acquiring high quality bulls. New dairy breed associations were constantly forming and sought assistance. In 1935 the Matanuska project in Alaska requested 100 purebred Guernsey cattle and these were collected from several counties in Montana.

A Dairy Industry Short Course of about one week's duration each winter made demands upon the entire staff. The movement toward artifical insemination led to the rise of a number of community and regional associations, each seeking information and assistance. Probably the best known early group was the Artificial Breeding Association in Western Montana, organized in 1947 in Missoula, with five Guernsey, five Holstein and three Jersey bulls.

The encroachment of oleomargarine also sent shock waves through the industry. The rapid adoption of the paper milk container produced a revolution of a type, since retailers were now relieved of the task of returning empty glass bottles, and the paper containers adapted more readily to pasteurization processes. Tretsven remarked about the changes taking place in his 1949 report:

> With the coming of paper milk containers, better roads and improved trucks for transporting milk, there is a gradual shifting of the industry into the larger and better irrigated valleys. . . . Through the introduction of paper containers it is possible for the consumers in many of the smaller towns to obtain pasteurized milk. Many of the operators in the smaller towns could not afford pasteurization plants.

These changes marked the end of many of the small dairies in the small towns which had always been accepted as essential. This also made the marketing of milk from the small farm herd more difficult.

By 1951 the total milk production in Montana was estimated at 530 million pounds, which at a price of $4.00 per 100 pounds yielded a gross value of $21,500,000. The value in terms of butterfat, cheese and ice cream was much higher. The value of dairy products was roughly estimated to be about 3.5 percent of the total agricultural production. Strong shifts began to appear in the statistics. By 1950 the average yearly production of milk per cow had increased to 4,960 pounds—up from 3,850 in 1935, but butterfat remained at 195 pounds. The number of dairy cattle in the state remained steady for the decade 1925-1935 at 180,000. By 1968 this number had declined drastically to 46,000. The number of creameries decreased by 50 percent between 1960 and 1970, but through improved breeding, better nutrition and improved management, half as much milk was obtained from one-fourth as many cows. The source of milk had also changed. In 1925 three out of five Montana farms possessed milk cows; in 1969 this number had decreased to three out of 10. About 350 licensed dairy producers were still in business in 1972.

The dairy industry, and this was true of the poultry industry as well, had served for many years to make the quality of living in many Montana homes and communities much richer than it would otherwise have been, but it was now adjusting to changing times.

Harriette E. Cushman and the Montana Poultry Industry

A crate of chickens was to be found on almost every covered wagon crossing the plains, and chickens and eggs were in urgent demand in the boom mining and lumber camps of the West. As the country developed, almost every home in town and country had its small flock of chickens which greatly enriched the family's food supply. Almost no attention was given to improving the stock, feeding or disease control, but when the Experiment Station was established many questions concerning poultry came to the staff.

The Station and the College academic programs gave extremely limited attention to poultry, so it was largely an Extension effort. In 1900 Station veterinarian, Dr. H.C. Gardiner, issued the first publication on poultry, *Bulletin No. 26,* "Poultry Raising." The subject was frequently discussed in short course offerings and in Farmers' Institutes, and in 1915 a Department of Poultry Industry was placed in the College curriculum. The department closed in 1925 because of lack of funds and was not reopened until 1945. A degree program was approved in 1952, but only three degrees were given, all in 1956. The degree program was withdrawn soon after and limited efforts were devoted largely to research.

Ralph H. Smith became the first poultry specialist, beginning service March 1, 1919. Everything needed doing, but since eggs were in short supply, he started by organizing egg marketing circles through which he promoted improvement in choice of breeds, standardization of flocks, disease control, housing, and care in the handling of eggs. Smith encouraged the growers to exhibit their birds and eggs in the county and regional fairs, which they did. They went on to display a surprising maturity in exhibits shown in the National Poultry Show in Chicago in December 1920. Sixty-one growers entered and won 34 prizes, including nine first places. Smith resigned in December 1921 and E.E. Weaver gave intermittent aid until a replacement arrived.

Harriette E. Cushman, Specialist

The specialist who was to direct the Extension poultry work of the state for 32 years, 1922-1954, was Miss Harriette E. Cushman. Educated in the

East at Mt. Holyoke College and Cornell University, she came to Montana from Idaho where she had been assistant poultry specialist.

Miss Cushman recalls that when she reported to Director Cooley the morning after arriving in Bozeman, she remarked that she would probably need a day at least to acquire housing and get settled at home and in the office. The director replied that she would have no need for either a home or an office for a time since he had arranged an itinerary which would take her over the state to get acquainted with the work. Her first stop was to be at Three Forks, 30 miles down the Gallatin Valley. Later the same day, after a whirlwind survey of the department and of Bozeman, she wound up in the railroad station asking for a ticket to "Two Forks," whereupon the agent replied, "You will have to add another fork to that." She contacted county agricultural agents, home demonstration agents, and poultry raisers in their farm homes, and "enough problems were unearthed to keep one busy for a lifetime," she remarked. Indeed, she spent the remaining 32 years of her professional career working on those problems.

She grouped the various questions concisely under two major categories—production and marketing. She worked closely with William F. Schoppe and then George P. Goodearl in their teaching and research, and found a valuable ally in Chester C. Davis, state commissioner of agriculture, who gave repeated assistance in statewide projects.

Demonstration Poultry Farms

Miss Cushman had the distinction of being one of the very few women specialists in agriculture in the nation, at times the only one, and then achieved the longest record of service. She was employed on the basis that the poultry industry was marginal, and was taken care of by women on the farm, and therefore she would be assisted by the home agents. She found that the agents were willing, but uninformed, and much more enthusiastic about clothing, nutrition and home improvement than mucking about in barnyards and chicken houses. The agricultural agents were also sympathetic, but they were extremely busy with what they felt were more important enterprises—grains, grasses, livestock and 4-H Clubs. Miss Cushman strongly supported the commonly held view of the day that "the family cow, the garden and the poultry flock are the greatest insurance for an adequate family living." She continued to be annoyed when agents did not include poultry in mainline planning because poultry "is not a major enterprise," they concluded.

In dire need of active assistance, the specialist came up with the idea of securing cooperating "demonstration farms." Her plan was to enlist at least three energetic poultry growers in each of the 26 counties which were considered best adapted to poultry growing. She felt that through careful instruction, experimentation and record keeping with these families, she could reach every inquiring poultry grower in those counties. She was never satisfied with the number but she had secured 22 cooperating farms by January 1923, and soon this number rose to 70. The number dropped in the drought period, but reached a peak of 80 in 1939.

Forty-eight of these farms soon had records showing that improved housing was immediately profitable, and shortly thereafter 71 sets of records proved that commensurate returns followed on the heels of improved feeding practices. Miss Cushman used the demonstration families somewhat as sub-agents, and she frequently held demonstrations on their farms discussing egg production and marketing, culling and disease control, as well as housing and feeding.

Housing, Marketing and Culling

Proper housing was a key factor in poultry production in such a place as Montana, and in the early 1920s much of it was completely inadequate. The houses were ramshackle, improperly built, sometimes too tight in construction and therefore disease producing. Not only did light and ventilation need to be considered, but properly built nests, roosts and drop boards were important.

The strategy in improving housing was to secure the cooperation of a poultry raiser who needed a new poultry house. A date was set for a building "bee" to which all of the neighbors would be invited to assist and to engage in a social occasion as well. The old building would be torn down and a new one constructed—usually in the same day. At intervals during the day many facets of poultry raising would be discussed by the specialist, the agent, the demonstrator, or all three. After considerable expermimentation, Miss Cushman came up with what she called the "Montana type" poultry house, with qualities which appeared to be most satisfactory for local conditions. In the decade 1922-1932 she estimated that she supervised 70 housing demonstrations resulting directly in the construction of 844 new and remodeled chicken houses. The movement grew steadily and in 1939 and 1940, agents in 33 counties reported over 500 building projects each year. As REA expanded, lighting systems began to be installed and in 1948 the construction of more than 150 such systems was reported.

When Extension entered the poultry field, large quantities of eggs were being imported. Miss Cushman followed the plan of Ralph Smith, which was to encourage egg-marketing circles which could guarantee a steady supply to major outlets and encourage them to pay a satisfactory price. She made contacts with the Butte wholesale houses which furnished the most useful markets for many years. Developing a reliable supply of suitable eggs for the dozen or more state chick hatcheries which existed at one time also took considerable planning.

The first annual State Egg Show was held in Butte in March 1924 with 109 entries from 25 counties. In a few years a standard was established for graded eggs, with the name of "Montana Express Select" to be used for the top grade. This provided the basis for the first law setting up standards for eggs for human consumption and for hatcheries. This law was passed in 1931. To cover the cost of the enforcement of the law, seals were designed showing the grade, and were to be affixed to each carton of graded eggs. The sale

Chicken culling demonstration, Richland County, Mont., 1920s.

of seals by the state soon became a sizable enterprise when 5,712,800 dozen eggs were marketed in 1953.

In the early 1920s, 104 eggs per hen per year were considered good production. By 1925 Clark Bumgarner, on a demonstration farm with a flock of 240 birds, ran his production to an average of 204.2 eggs with a labor income of $5.59 per hen. R.J. Streit of Kalispell set records over his entire 25 years of record keeping. For 13 years he maintained an average of 250 eggs per hen, and in 1950 this went to 295.7, a record at the time. In 1952 the average was 287.7 and his labor income per hen was $6.86 with a flock of 300 hens.

Poultry producers were slow in accepting the evidence that culling the flocks was a highly practical procedure, in spite of the fact that the accuracy of the practitioner could be checked immediately. Mrs. E.B. May of Richland County, for example, found it hard to believe that 23 hens, almost half of her small flock, were non-layers. She immediately shut them up to test the skill of the culler. The first week she got a few eggs and after that none at all. Mrs. J.N. Marquis at Bennett Lake felt that she was ruined when her extremely small flock of 19 was reduced to nine. She had obtained 67 eggs during the eight days before culling, and when she gathered 64 from her greatly reduced flock in the following eight days, she also was a believer.

Miss Cushman set out at once to teach her demonstrator flock owners and the agents the relatively simple art of culling, urging them to teach the poultry raisers, because culling the larger flocks was excessively time consuming. There was always a demand for her services, however. She estimated that in 89 demonstrations in 1923, with close to a 35 percent rejection, that the savings in feed and care amounted to $6,996—well above her salary and the costs of her office for the year. In 1928 she estimated that savings at her own culling demonstrations were at least $11,444, and by that time the practice was widespread. Non-laying hens were removed before reaching maturity, and savings were many times that amount.

Turkey Marketing

The foremost success story in Montana's poultry industry was that of the cooperative marketing of turkeys. The demand for turkeys rose along with that for chickens during World War I when red meat was rationed. Poultry was found to yield a desirable supplemental income, and by 1922 there was a surplus of turkeys in the state. Miss Cushman recalls that the director's first request was that she develop an organization of some kind which would move the turkeys out of Montana. A cooperative arrangement seemed to be the most feasible and she remembers that "Blaine Ferguson, county agent of Pondera County, was the first to respond to the specialist's plea. . . and in the fall of 1922 the first carload of cooperatively shipped turkeys in the United States moved from Conrad to Chicago."

The natural result was that with the promise of successful marketing, the interest in raising turkeys increased. In 1923 Miss Cushman was in great demand, and in addition to 11 one-and two-day schools, she held 80 other meetings for instruction in raising turkeys, and in preparing them for market with an attractive appearance. Very few now remember the days of the butcher shop window where the long, lean, blue, leather-skinned turkeys dangled from scaly feet to their long neck and feathery heads. Three new marketing associations were formed in 1923, and eight carloads of turkeys were shipped. An estimated $13,675 was added to the industry's profits by the cooperative effort.

In 1924 another three associations were formed and the number of turkeys marketed doubled that of 1923. A total of 108 training schools were held. To meet the questions which arose, since the Montana Experiment Station did not conduct turkey studies, the USDA opened a turkey research project at its Fort Keogh station near Miles City, which was of inestimable value to the new industry.

From the first turkey marketing association in the United States in Pondera County in 1922, the movement expanded rapidly. A Northern Montana Turkey Marketing Association was formed in which George Gustafson, county agent for Blaine County, became manager. Then a Southern Montana Association was formed and Harvey Griffin, a Bozeman turkey grower, became its manager. By 1929 the several associations met in Salt Lake City to consider mutual problems. In August 1930, in a Western States Extension Conference in Bozeman, the proposal was made for a Northwest Turkey Growers' Association. This was formally organized in 1931 with headquarters in Salt Lake City. George Gustafson became the first president, 1931-32, and Harvey Griffin became field manager. Miss Cushman was given the responsibility for grading the Montana turkeys in order to clearly mark and maintain the USDA and association standards.

This called for the rapid training of a number of hard-nosed graders, and the handling by Miss Cushman of thousands of turkeys in the final shipping stages. The Montana leaders gave strong direction to developing the "Norbest" trademark which became a symbol of the highest grade product. Association leaders had to be constantly alert to keep the growers from accepting what appeared to be attractive offers from "boomer buyers" who hoped to break the cooperative effort and then buy at their own prices. The success of the regional associations in direct bargaining with both grower and wholesaler was so great that other areas wanted to know how it was done. Miss Cushman was asked to narrate the success story at the meeting of the World's Poultry Congress in Ottawa, Canada.

The peak of the cooperative pool effort probably came in 1935. Thirty-five local pools were in operation with 2,614 members. Some 1,202,444 pounds, about 50 carloads of dressed turkeys, together with up to 8,000 capon chickens, which were in favor at the time, were marketed. The success of the Montana and the Northwest enterprise carried the seeds of its decline. Turkeys rose in national popularity, and by 1940 strong competition over the country made growing them less profitable in Montana. The number of pools dropped to 19, and the cooperative growers to 1,000. The Safeway Stores wholesale center in Butte contracted for a large portion of the Montana product.

4-H member with New Hampshire chickens used in his poultry project.

4-H Poultry Projects

Some phase of poultry production seemed to be a natural for 4-H projects, and indeed there was considerable interest in chickens, turkeys and egg production. By 1924 there were 25 clubs and attractive exhibits were displayed at local and regional fairs. When the participation did not increase rapidly, Miss Cushman analyzed the problem as the lack of a challenge. The "man-sized job" seemed to be a calf, pig, horse or sheep project. The poultry project usually started with chicks a few days old, with all of the liabilities of large losses due to weakness and disease in spite of the best care. Other animals were chosen when they were large enough to give promise of superior attainment. In profits, a dozen chickens, or the entire farm flock, did not begin to equal a single baby beef. Miss Cushman urged larger flocks, fine breeds which would display well at fairs, and turkeys or capons which would yield a more attractive income.

In regional and national poultry competition, Montana 4-H youth achieved notable results. Space permits only a few examples. In 1939 at the World's Poultry Congress in Cleveland, two of Montana's teams—from Valley and Chouteau counties—placed first in their class. In 1950 Marion Leuthold of Stillwater County was the top demonstrator at Camp Plummer, Portland, with a poultry dressing demonstration, "Slick Tricks with Chicks." In 1951 Kenneth Owen of Lake County won second place among all delegates at Camp Plummer. The most noteworthy performance in these years was by Elaine Sterland of Chinook in 1952. This was presented in Kansas City at the annual American Institute of Poultry Industries meeting termed a Fact Finding Conference. A regular feature of the meeting consisted of demonstrations by 4-H members. Miss Sterland's presentation was so effective and mature in its appeal that the president of the organization conceived the idea of a Junior Fact Finding Conference, made up entirely of young people who would share in some joint sessions with the senior meeting. Miss Cushman, who was Miss Sterland's sponsor, was asked to draw up the plans and organize the first Junior Fact Finding Conference, which held its first and highly successful meeting in 1953.

The Montana poultry industry declined rapidly following World War II. Higher grain prices made poultry raising less profitable; high prices for red meat and less time involved in raising it, led to the growing of beef. A change of taste to fryer chickens which could be raised most economically in warmer climates and closer to large markets became an additional factor.

When Miss Cushman retired in 1954, the director waited to check the demand for a new specialist before making an appointment. He decided that although there was a large and active interest in poultry, its problems could be managed by the professional associations and the state regulatory agencies, together with assistance from Experiment Station personnel, including Dr. George T. Davis who combined Experiment Station and academic course work during the period 1957-1972. He was assisted and succeeded by Allen F. Beckler, who had previously served as inspector of poultry for the State Poultry Improvement Board and had wide contacts over the state.

Thus, both the poultry and the dairy industries, after having served Montanans well through many years, fell prey to technological changes at about the same time. Both owed much to the Extension specialists who guided each to commercial success over a long period of time—Harriette E. Cushman and J. Oscar Tretsven.

Work with Montana's Indians

The Montana Extension Service had the welfare of the Indians of the state in mind from its earliest years. However, until the early 1920s the work of Extension was largely emergency and war oriented, and until 1924 the Indians were rather exclusively wards of the federal government. Agents appointed to counties adjacent to reservations were immediately aware of the situations created by the total of 25,829 Indians on the reservations, another 11,671 in nearby towns and cities, and of the 5,345,000 acres of Indian lands. The work of the agent was complicated by the operation of the Allotment Plan of 1887 which had apportioned lands to the Indians for family ownership, and opened additional lands for homesteading by non-Indian families. This situation is particularly well illustrated in northwestern Montana where the reservation of the Flatheads or the Confederated Salish and Kutenai Tribes had been almost completely broken up and Indian and non-Indian landholdings were intermingled as early as 1910.

Work with the Indians was greatly expanded in the early New Deal years. The Wheeler-Howard Indian Reorganization Act of 1934 stimulated the activities of the Bureau of Indian Affairs (BIA). Its dynamic commissioner, John Collier, presented a program to encourage the Indian to take pride in his heritage, and also to adopt those features of other cultures which would make a richer life for him and his family. He also realized that state agencies could best administer many of the programs he had in mind. The Johnson-O'Malley Act of 1936 provided that the federal government might contract with the states and their agencies or with private agencies to undertake educational work with Indian tribes.

Perhaps the active cooperation of Extension slowed down the formal operation of the Johnson-O'Malley Act in Montana. The agents worked closely with the federal agricultural representatives maintained on the reservations by the BIA. 4-H Clubs had been formed on the reservations in the late 1920s, and by 1938 a total of 47 clubs were in operation on all seven reservations, involving 168 boys and 313 girls.

The Johnson-O'Malley Act

Montana did not enter into the contract provided by the Johnson-O'Malley Act until 1955. Seven agricultural agents were to be assigned to the six Montana reservations; the Flatheads worked directly with the regular state-county program or under a separate contract. Full-time home agents were to be placed on the Crow and Fort Peck reservations, and part-time work was allocated to the other tribes. The budget grant for the year was $67,975. Each annual contract listed in considerable detail the areas of work to be covered. The customary general categories included:

A. Agricultural Production and Conservation
B. Marketing and Utilization of Agricultural Products
C. Home Economics
D. Human Resource Development
E. Public Affairs and Community Development

The selection of agents was carefully made and was not completed until 1957. Then the agents discovered that they needed to proceed with extreme caution. The change in agricultural and other assistance from the BIA to the State Extension Service had been announced abruptly and without consultation with the Indians. An important difference was that instead of dealing with officers whom they knew at the central Indian agency on the reservation, they must now go off of the reservation to the county seat, usually a considerable distance away. The change was also made in the context of extensive legislation by Congress to hasten the complete acculturation of the Indians. Consultation and amicable adjustments were made as rapidly as possible, but the Indians were justifiably apprehensive.

Verne Dusenberry, First Specialist, 1957

Verne Dusenberry, who pioneered Montana State's program in anthropology, was asked to evaluate and coordinate the program in Indian affairs for a six-month period, beginning in January 1957. His experience and interests lay largely with Indian youth and educational activities. Since the Extension agents, under the terms of the contract, needed to place the larger emphasis upon agricultural, home and technological programs, he withdrew, and the office of specialist was not continued. Dusenberry's chief recommendation was that the work of compatible home agents be emphasized since the highly important problems of health, education and morale lay in the home.

The new agents gradually broke down the distrust of the Indians. Howard J. Morton, working with the Crows, noted, in his first report covering 13 months following December 1, 1957, that in the first eight months he had received only 35 telephone calls from the reservation, and these were mostly from the Indian Service offices. In the following five months, however, he had 290 calls, largely from Indians who wanted information and assistance. The new agent for the Fort Belknap Reservation lived with an Indian family at Hays three days of each week during the summer in order to be more available and closer to his work. Dusenberry noted in 1957 that

Robert A. Roush had, in a two-year period, associated himself with some 30 progressive ranch families on the Rocky Boy's Reservation who, with his help, had reached almost complete self-sufficiency through their range cattle herds.

From these examples of success, Roush was able to secure the confidence of other tribal members. Robert E. Durham, longtime agent in Glacier County, already had the full confidence of the Blackfeet and continued progressive work. Elinor Clack, who was familiar with the Rocky Boy's situation, made striking progress and moved from half-time to full-time service. Even under the best circumstances, however, the agents often came face to face with the fact that the Indian culture carried distinctive concepts toward "time, saving, and the work ethic" which often got in the way of what the agents felt was productive effort.

Dr. Lloyd C. Pickett, Specialist, 1970

Dr. Lloyd C. Pickett undertook the work of specialist in January 1970, under the title of Indian program supervisor, and the record of accomplishment in Indian affairs is much clearer after that time. He was experienced in working with disadvantaged people and was tireless and resourceful in his efforts to coordinate the work on the scattered reservations and in bolstering the morale of his frequently discouraged agents. His ample newsletter, "The Reservation Agent," mailed monthly for several years, provided a great deal of useful information and also steady stimulation. Pickett's assessment of the problems of the Indians is reflected in a brief excerpt from a report of a discussion with his agents in a workshop on May 20, 1970. He commented upon the Indians' development as

> ...a social phenomenon—one which is culture bound. Here the change agent is seen as a human engineer. Technology is seen as a social rather than a physical phenomenon because it creates and specifies new social roles. A remote villager perhaps has only three major social roles, a family, village and religious role. If technology enters his village world, say in the form of a factory, these roles will become drastically changed. The new technology requires education, urbanization, new views of time, labor and leisure, cash markets, productivity, competition, impersonalization and centralized government. And the worker may now be paced by a machine, be required to move his home...no longer can he meet many social obligations demanded by his culture (such as dancing for three days at a wedding, because this will get him fired at his job), no longer can he blend work and leisure so that they are virtually inseparable....This does not mean that indigenous cultures must be completely destroyed in order for people to benefit from technology, but certainly they will be eventually modified to accommodate increasing technology.

Dr. Pickett then urged the agents to cultivate a sensitivity which would enable them to sort and evaluate the contributions of both the indigenous culture and that of the new technological age.

The evolution of a balanced program for the Indians proceeded slowly. Extension agents tried to avoid duplicating the work of the Bureau of Indian Affairs which maintained a superintendent on each reservation, together with a staff which carried on many traditional services. Its activities were gradually curtailed, however, until by 1973 it administered less than 40 percent of the total federal funding flowing into the reservations. Many of the other federal agencies which were also operating with low income people set up specific programs, as did a number of vigorous religious groups. The tribal councils also carried on economic and social enterprises. They often had substantial funds from their tribal land leases, coal and petroleum revenues, occasional favorable judgments in land claim suits, and with increasing frequency, income from tribally managed industries.

The home agents found a number of fields of activity clearly available. The need for better nutrition was glaringly apparent. An excess of starchy foods and sweets led to obesity, poor teeth and low resistance to diseases. The federal government provided foods which were in surplus at low cost or no cost, and the agents worked diligently to incorporate these into the daily diet. The use of vegetables was encouraged, but gardens were often difficult to grow on the arid reservations. Even meats were not always available. Indian families were extremely fond of visiting from community to community and often to other reservations. This made the care of chickens, or hogs and cattle, particularly milk cows, burdensome. The ownership of cattle was not feasible for the majority who lived in the villages where the parcels of land approximated the size of city lots.

The improvement of clothing for the family also offered another major area for the agents to concentrate on. The cost of new cloth was a limiting factor, and this led to investigations into appropriate types of cloth and lower cost materials, as well as the repair of older clothing, and alterations for the children.

Upholstering appealed especially to the women and often to the men as well. On many reservations, afternoon and evening upholstering projects were in almost continuous operation. Hazel Thompson, on the Northern Cheyenne Reservation, probably held the record for the amount of furniture repair which she supervised.

Agricultural and Industrial Development

The traditional approach in agricultural development with the Indians was to try to make them farmers in areas where very little land was suitable for cultivation, and irrigation possibilities were limited. Moreover, farming required machinery, buildings and day by day constraint. The range livestock programs met ready acceptance and made steady progress. A continuing problem was to secure suitable stock for the livestock enterprise. Even when the project was successfully under way, the universal problem of financial management which would maintain a favorable credit rating was always present. Another knotty problem for reservation agents was divided land holdings because of heirship claims which had accumulated over the years since allotments had been made. The agricultural agents devoted a great deal of time to assisting tribal councils as they attempted to secure for the tribe or for its members the widely scattered and mostly small parcels of land held by non-Indian men, or land greatly subdivided by heirship claims.

In addition to assisting individuals and families, the agents spent much time helping form organizations which might develop prosperous programs. The Northern Cheyenne Livestock Association was highly successful in a program which utilized almost all of their land. The Fort Belknap Reservation members organized two agricultural associations, and the Crows in 1973 successfully launched a vigorous Land and Livestock Association after four earlier failures. The Crows and the Northern Cheyennes formed special associations to investigate the intricate problems relating to the leasing of the extensive coal deposits which underlie both reservations. The poor attendance at these meetings in the 1950s perplexed the agents since the Indians were known to be fond of communal gatherings of all kinds. Later, when the meetings were called by their own organizations and they were familiar with the issues, the attendance and participation increased markedly, and the influence of their decisions became highly important.

Since the agricultural enterprises are not sufficient to care for the increasing population of the reservations, outlets in the arts and crafts and certain types of industrial work have long been sought for the well-known skills of the Indians. Funds were found to establish craft plants on the Northern Cheyenne and Rocky Boy's reservations. The Crows built a rug manufacturing plant, a motel and recreation complex, and formed the Big Horn Canyon Dam Recreation Area Association to operate the concession for the area of the huge Yellowtail Lake. At Fort Peck a Tribal Industries Association began operation with a military gun repair project. The Blackfeet formed the Great Western Industries and the B.F. Wood Products Plant at Browning. A pencil factory held promise for a time, and the Heart Butte Industries set up a small sewing enterprise. The Elmo-Dayton Association set up by the Flatheads secured the contract for the care of roads and bridges within the former reservation area, and secured additional employment through forest management and wood products projects. The Flathead tribe also gave strong support to the establishment of the federal Kicking Horse Job Training Center near Ronan. Many tribes established an industrial park site and a recreation project, and many of them set up a plan whereby their own people could participate in the extensive federal house building program. Increased employment opportunities and growing management skills were highly encouraging to Indians and agents alike. Extension was directly involved in the Heart Butte Industries and the Flathead road project and assisted in other enterprises as the opportunity arose.

New Houses, New Homes

A massive federal house building program developed just as the Extension Service expanded its work to the reservations. By 1970 some 1,242 new houses had been built and 450 had been renovated on the Montana reservations. An estimated 1,600 new homes and 700 renovations were still needed. Both men and women agents found a great need among many Indian families for aid in adjusting to what was for them a whole new way of life, since many were relocated from remote and desolate areas of the reservations to what seemed to be more desirable sites, sometimes to the major villages. The use of new stoves, new fuel and electricity was confusing and there were always problems in the use of water, particularly with indoor plumbing. There was also a need to organize home accounts, since the cash costs for utilities were the first regularly occurring imperative costs some families had ever met. The building of the house usually destroyed the grass in the yard area and produced a situation where in wet weather it was a sea of mud, much of which was tracked into the house. A great deal of effort went into the simple tasks of building a walk and seeding a lawn. The agents went on to encourage the planting of trees and shrubs. A Crow 4-H Club secured wide recognition by developing tree and shrub nurseries, and in making plantings and caring for them on the public areas of the reservation villages.

"Indianized" 4-H Programs

4-H programs received the cooperation of both agricultural and home agents, and here problems were also numerous. When the agents set out on the traditional projects such as calves, pigs, chickens and gardens for boys, and foods, clothing and home improvement for girls, they often found that the family resources did not include the basic materials for these programs, but tribal councils were often willing to provide funds for improved stock, seed and other materials. An additional problem was that very few adult Indians felt capable of assisting 4-H Club members, although Indian youth worked particularly well with their own leaders. Low attendance at club meetings was customary because of the distances between homes, unreliable automobiles and parental indifference. This situation was met in part by recruiting a number of school teachers, or by agents holding 4-H sessions in the schools during the school day or just after school was dismissed. The search for help was assisted for a time by the Aide Program whereby part- or full-time Indian assistants could by paid. The Neighborhood Youth Program (NYP) and VISTA also provided helpful assistance.

Lacking the facilities for engaging in some of the projects which the non-Indian youth pursued, the Indian students came up with an impressive list of subjects in which they had an interest. These included the Indian crafts of beading and leather work. The revival of Indian rituals led to interest in traditional dances and music. Crow youth and parents suggested studying Crow clan structure, Crow legends and Crow natural foods. Mothers remarked that they would like to have their daughters have some training in secretarial and office procedures as well as home duties. Horsemanship was popular where horses could be provided and leaders found. A few students chose the highly practical projects of home repair including electrical and carpentry skills, and water and plumbing appliances repair.

The success of 4-H work was best measured at the Indian fairs which provide one of the high points of the year on each reservation. The reports show striking progress on most reservations in the quantity and quality of exhibits, the steadily increasing responsibility of the Indians for the fairs, and the gradual increase in the number of exhibits which were taken on to county fairs. 4-H projects and similar programs were carried on at an increasing number of summer camps, where particulary good opportunities were offered for developing leadership qualities. Over a long period a number of Indian youth have participated in state, regional and national 4-H conferences. In 1966 Rosella Redwolf, a Crow young woman, was chosen for IFYE, but since she was in her first year of teaching at Haskell Institute, she was unable to go.

In 1973 a survey was made on the Crow Reservation to discover the interest in traditional Indian 4-H projects and the availability of Indian leaders. The response was encouraging and a request for a grant was made for a three-year program which would provide for a professional native youth leader and two half-time aides who would recruit and work with volunteer leaders. The grant was received and a program titled "Indianized 4-H Projects" was developed.

The 4-H program is probably one of the factors behind a rapidly increasing interest in college attendance. In 1973 some 150 Crow youth were college students, 122 from Fort Belknap, and 90 from the Rocky Boy's reservation. The enrollment of some 300 Indian students at Eastern Montana College allowed it to expand its Upward Bound program and offer courses and programs of especial interest to the Indians. Rocky Mountain College, also with a substantial enrollment, offered a number of academic courses and various short courses of value to Indian youth.

New Directions

At the University of Montana, Alonzo T. Spang directed the expanding program of Indian Studies until he received his Ph.D. degree. He was the first member of the Northern Cheyennes to receive this degree. He then returned to the reservation as BIA superintendent. At Montana State University, Barney Old Coyote, a Crow, who had held a number of positions in the Department of Interior, directed the University's Indian program, utilizing a unique system of tribal consultants to serve as liaison between the University and the reservations. In the Governor's office the position of state coordinator of Indian Affairs was created to direct the highly flexible program of assistance. The Montana United Indian Alliance coordinates the efforts of the off-reservation Indian people in the larger cities to provide job opportunities, housing and other assistance needed by Indians not affiliated with the reservations. The Montana Inter-Tribal Policies Board and its affiliated corporation, the Tri-State Tribes, coordinated and administered a number of multi-tribe activities.

Many success stories, in addition to those already mentioned, gave assurance that the work of Extension agents and other state and federal agencies were helping to alter the age-old pattern of Indian discouragement. Donald G. Petterson on the Fort Peck Reservation found that many otherwise indifferent older Indian youth acquired a changed personality when they were given the responsibility of assisting with younger students in a Youth Tutoring Youth project, and similar changes also took place in the tutored younger Indians. This idea succeeded so well that it was incorporated into the reservation schools. Petterson was later provided the opportunity to apply some of the same methods to a new urban 4-H program in Billings.

Mrs. Grace Miller on the Fort Belknap Reservation developed a tribal newspaper, *The Camp Crier*, which soon secured the active participation of the Indians and provided the important medium of communication required to promote many reservation activities, including one of the best Indian fairs in Montana. Robert A. Webert found that when he undertook conservation projects and tours with the Crow youth, the entire reservation became more environmentally conscious.

The home agents found that the Indians' concern for their elderly made it easy to form senior citizens groups, and often these groups opened doors for the introduction of new ideas and methods. They also found the North American Indian Women's Association to be a stimulating movement. The association was organized at a conference in Fort Collins, Col., in 1970, and the Blackfeet women secured the national convention for Browning in 1973. Local clubs existed on each reservation and the cooperation among the tribes in this successful event was something of a landmark.

Another instance of the new cooperation with off-reservation culture came with the ready response to the Kellogg Extension-Education Program (KEEP), which hosted an annual Tribal Leaders' Workshop in 1973, and another in 1974. The regular Kellogg program included 29 active Indian leaders in 1972, 34 in 1973, and 53 in 1974. The appeal of on-reservation workshops dealing with human relations was also exceptional. In 1973 about 230 Indian people participated in seven workshops on transactional analysis.

Director Aasheim's annual report in 1974 contained a section on "Work with Indians," which highlights many of the segments of the Extension Indian program up to that time:

> This program was funded by the Bureau of Indian Affairs and conducted by nine agents, one supervisor and four aides or Extension assistants. The agents conducted education in ranch management, livestock and crop production, human nutrition, housing, furniture renovation, consumer education, and community development.
>
> A pilot "Indianized 4-H" project was initiated through Special Needs Funds to adapt 4-H to Indian culture and to reduce the drop-out rate. The staff of three had 140 Crow youths enrolled by the end of the first year.
>
> A vigorous Land and Livestock Association was organized on the Crow Reservation where only 16 percent of the land is now in Indian use. The association made remarkable first-year progress with the assistance of the agent and the full-time help of a temporarily-assigned Extension agent. A land-policy study was made and a report was written. The association is fully organized (50 members) and tribal resolutions on land policy have been developed. A suit has been brought against the "competent lease

system" practiced on the Reservation.

A statewide land use, land policy meeting was called by Extension in February generating lively discussion and a request for a followup meeting planned in the fall.

A ranch management workshop was held on four reservations with mostly poor attendance.

At Fort Belknap, a team consisting of the agent and two BIA staff provided intensive support to tribal credit clients with very encouraging results.

Participation in the Kellogg Extension Education Program (KEEP) workshops by Indian people increased in FY 1974 to about 53.

Upholstery workshops were very successful on the Northern Cheyenne and Blackfeet reservations. An estimated 200 families participated there and on the Crow Reservation.

Landscaping and home improvement projects were conducted on all reservations to some extent, with satisfactory results.

Highly successful fairs were conducted on four reservations through Extension sponsorship with a total attendance of about 3,000 people. Two of the fairs were new.

A consumer event on the Blackfeet Reservation drew 200 people.

Extension sponsored a highly successful youth rodeo on the Northern Cheyenne Reservation.

Donald Petterson, after years of up and down work on the Fort Peck Reservation, in reviewing the shape of the future with considerable optimism, recalled that Chief Washakie, upon being pressed by a federal agent concerning the advantages of farming over the migrant life, exclaimed in exasperation, "Goddamn the potato." Petterson felt that this position, which he had frequently encountered, was changing for better or for worse. The Indians were thoughtfully surveying their culture and making choices which gave promise of blending some of the best features of both of the cultures involved.

X. New Directions: The Aasheim Period, 1958-1974

The retirement of Director Beers came at a time of normalcy when a transition could be made smoothly. Montana had enjoyed a stable and productive decade since the close of World War II, and the work of Extension was steadily expanding.

Director Torlief S. Aasheim

On September 1, 1958, Torlief S. Aasheim became the fifth director of Extension. He was born and raised on a farm in Sheridan County Mont., and received a B.S. and M.S. in agronomy at Montana State College. During his

Torlief S. Aasheim, fifth director of Extension, 1958-1974.

undergraduate years, he worked at the Experiment Stations at Moccasin and Bozeman in the summer months. After graduation in 1937 he served as range examiner under the Agricultural Conservation Program, and as assistant land use planning specialist with the Bureau of Agricultural Economics in 1938 and 1939. From 1940 to 1951 he was research project supervisor for the Soil Conservation Service, taking time out from 1942 to 1945 to serve during World War II in the Pacific theatre with the U.S. Coast Guard. In 1951 he became superintendent of the Huntley Branch Experiment Station from which position he was appointed director of Extension. He brought to the job, therefore, a broad experience, an understanding of Montana agriculture, and an awareness of the problems of Montana people.

Aasheim was appointed only after serious consideration was given to securing an out-of-state director who would give critical and objective appraisal of the Montana program. There was growing criticism that the Extension Service had made few adjustments to changing conditions, and was continuing 1914 programs with little question as to their suitability. President R. R. Renne and Dean of Agriculture Roy E. Huffman agreed that the adoption of desirable new directions should be given high priority.

They also agreed that a director chosen from in-state could secure these changes more smoothly than an outsider. Aasheim accepted the directorship with the clear understanding that he would be given freedom to administer the Service, but that he should consider broadening Extension's scope and tailoring it to the current needs of Montana people wherever changes seemed to be appropriate.

The SCOPE Report

The theme for the first annual Extension conference under Director Aasheim in October 1958 was "The Challenge of Change." Change was to characterize his entire period as director. The theme grew out of a major study conducted by a committee of the land grant colleges and Universities Association and the USDA Extension Committee on Organization and Policy. The study, titled "The Scope and Responsibility of the Cooperative Extension Service," was released April 1, 1958, and was popularly referred to as the SCOPE Report.

It described the rapidly changing conditions in the nation and the states. Adjustments in the family farm economy, for instance, were necessary as farms grew larger and fewer. With higher costs and greater financial risks, more efficient farm management became imperative. Increasingly powerful off-farm influences were developing around such a situation as surplus production. The Extension Service had indeed performed so well that perhaps it could be terminated. Urban congressmen were suggesting additional acreage controls, a more even balance between production and consumption and better marketing techniques in both foreign and domestic trade. The decline in farm population was noted. The rising educational levels in the farm populations were also observed, together with the resulting demand for the improved services which these people would expect. The closer relationship between rural and urban communities was commented upon, as well as the accelerating changes in family living in both. The increasing total population and the demands upon the natural resources, together with the resulting urgent need for conservation, were also considered.

In addition to making recommendations in the fields of production, marketing, management and conservation, the report made recommendations in five other areas, namely: family living, youth leadership development, community improvement, resource development, and education in public affairs.

The report then went on to emphasize the trend which was under careful scrutiny at the time, that of the increasing demands of the non-farm rural and urban residents. It recommended that these people should also be served by Extension. Even further, farm commodity and related organizations which handle farm products or which provide such services as credit, feed and fertilizers should also be able to call upon Extension for assistance. The American Farm Bureau Federation, which had for many years been closely associated with Extension, took exception to this broadening of activities.[1] The Farmers' Union, on the other hand, which had often been critical of Extension's work, supported the report. The challenge of change existed on every side.

The 1959 annual conference continued to probe the question of change of direction in Extension. The theme was "What Kind of an Extension Service for Montana?" It was structured around five panels which were made up of representatives from 1) producers and consumers, 2) agribusiness, 3) USDA agency heads, 4) MSC administrators, 5) the Extension staff.

The producers and consumers advised against radical change. The Service should perhaps give greater emphasis to the humanities, economics, sociology and related fields. The outstanding success in production now demanded more attention to marketing, and the relationship between these areas seemed to call for more emphasis upon developing civic leadership. This raised the question of competing farm groups. "The Extension Service must exercise a high type of courageous leadership," said the producers and consumers, "in helping these groups settle minor differences in order that they can cooperate in presenting a united front in solving their important common problems."

The agribusiness people, sensing that the future 25 years would be a period of immense technological and organizational change, urged county agents to keep one step ahead of these changes and thus insure the best possible guidance for their counties. In the recurring question of service vs. broader educational approach, they felt that the new technology (for instance, little was known about computers at the time) would require continued service activities.

The USDA Agency heads expressed the opinion that Extension was facing a period of even more strenuous effort in view of the trend to consolidate offices and activities. The agent, therefore, needed the broadest preparation possible, in order to work in areas of the specialties—some knowledge of forestry problems in certain counties, for example—and still remain flexible enough to be the practitioner in educational rather than chiefly in service activities.

The College administrative staff felt that Extension must take the lead in the increasingly important field of adult education. And adults should be expected to pay for these services. Additional work must also be done with children and youth, and if this imposed an uncomfortable load, then could some less important services be eliminated?

The Extension staff again debated vigorously whether they were part of a technical agency, rendering chiefly technical services, or should they be an agency emphasizing the more basic educational needs and functions? Some people asked, "Have staff members been doctors of farm failure, dealing mainly with problems of vocational skills resolved a generation ago? Have they been neglecting areas now demanding attention such as human attitudes, personality adjustments, and cooperative efforts from the smallest Montana community to international relationships?" Extension leaders looked toward their public and called attention to the action of the home economics people who long ago abandoned the narrow concept of a purely farm clientele. They began to look at the problems and let the problems determine the clientele. Other divisions should consider the same approach. In 4-H, for instance, there was need for a program for the development of the youth in the state, rather than merely for the farm skills of farm youth. If Extension was in a rut, the specialists agreed that they might be the "ruttiest" in the Service, and resolved to plan broad cooperative programs for the most people possible. Director Aasheim called for a continuing assessment of the direction which Extension should take.

The inquiry continued as the Montana Extension Service studied the SCOPE Report with representatives from the Federal Extension Office. The result in 1960 was a "Montana Scope Report," which said in part that the Montana Extension program would continue to be concerned with:[2]

> Greater ability in acquiring higher income and levels of living on a continuing basis.
>
> Greater ability in maintaining more efficient farms and better homes.
>
> Increased competency and willingness by both adults and youth to assume leadership and citizenship responsibilities.
>
> Increased ability and willingness to undertake organized group action which will contribute effectively to improving their welfare.

This report commented on changing directions which actually had been in progress for a considerable time:

> In the future, the amount of time spent with folks in town will increase because the increasing complexity of relationships centered around agriculture makes it necessary to include all people in the educational program.

Agricultural production, which has long been an area of much emphasis by the Extension Service, is justified only if there is a consuming public.

The original assignment of the Extension Service to agriculture, home economics and related subjects has necessarily become interpreted more broadly. *Extension has been engaged in much service work in the past which has often been questioned as to whether or not it could be classified as educational effort. In the future, there will be less of doing things for people and more emphasis upon teaching and motivating them to do things for themselves and the assistance will have greater diversity of scope.*

The Montana Scope Report is reproduced in its entirety in the Appendix.

Organization Change

Out of the deliberations which developed around the Montana Scope Report came the decision to change the organization of the Montana Extension Service from the traditional three program areas of agriculture, home economics and 4-H to five divisions of emphasis:

1. Agricultural production and conservation
2. Marketing and utilization of agricultural products
3. Home economics—food, clothing, housing, furnishings, health and safety
4. Human relations, communication, citizenship and leadership development, and 4-H
5. Public affairs and community development

This was done in order to insure emphasis upon the less traditional areas.

A program leader was named for each area, and the state was divided into three Supervisory Districts with a man and a woman supervisor named to each district.

The reorganization represented more of a change in emphasis than in content. The trend nationwide was to give greater prominence to programs relating to human relations, communication, leadership development, public affairs and community development. However, many Montanans wished to continue to stress the importance of the economic basis of the good life; namely, the efficient production of quality agricultural products. M. A. Bell's firm statement in the February 1951 *Montana Farmer-Stockman* had wide approval:

> The progressive development of an economically successful system of agriculture under this northern plains and intermountain climate and its survival over a long period of years evidently will require constant attention to the practical application of well-adapted facts and contributions derived from carefully planned animal, plant, soil fertility and conservation research and dissemination of that knowledge through effective programs of extension and education.

In 1965 Director Aasheim summarized trends which were forming as a result of the SCOPE Report of 1958 and the Montana version of 1960. He noted a movement toward:

Broadening Extension programs to better serve the needs of Montana people.

Helping communities to analyze resources and development possibilities.

Developing feasibility studies where experience was limited in such matters as feed manufacturing plants, cattle feeding, hog production, and packing plants.

Greater emphasis upon farm management.

Assisting the federal government in training candidates for such new programs as that for VISTA workers for the Indian Reservations and also for HEAD START workers for the Northern Cheyenne Reservation.

Organization Structure
Montana Cooperative Extension Service
(Implemented July 1, 1967)

Extension Director

- **Information 2½**
- **Finance Officer 1**

- **Program Coordinator 1** — Youth, Family and Adult Education Programs
  - Youth and 4-H 3
  - Family Life 1
  - Home Economics 2
  - Clothing 1
  - *Rural Civil Defense 2
  - *Civil Defense Adult Ed. 2
  - Rural Development Consultants 2 (One serving in an eight-county area)
- **Program Coordinator 1** — Agriculture and Economic Development Programs
  - Agri. Chemicals 1
  - Livestock 1
  - Agri. Marketing 1½
  - Farm Management 2
  - Consumer Education ½
  - Horticulture 1
  - Dairy & Swine 1
  - Range Management 1
  - Agronomist 1
  - Soil Management 1
  - **Wildlife 1
  - Weed Control 1
  - Plant Pathologist ½
  - Sheep Production 1/10
  - Entomologist ⅓
  - Animal Geneticist ⅝
  - Economist 1
- **District Supervisor 3**
  - Co. Extension Agents 96 (Serving 52 counties and 6 Indian reservation, 68 men, 28 women)

    12 of these agents are assigned to Indian reservations and are funded entirely from federal funds—Bureau of Indian Affairs.

NOTE: Numbers indicate number of people in each position.

* Rural civil defense and civil defense adult education specialists are funded entirely from special federal funds.

** The Extension wildlife specialist is financed by the State Fish & Game Department.

Developing women's programs outside of the Home Demonstration Clubs resulting in the first Woman's Week at the University in the summer of 1965 with 300 present. The program offered such classes as 1) Communication in Marriage, 2) Art in Everyday Life, 3) Green Thumb, 4) Families Around the World, 5) Chemistry in the Home, 6) Home Management, 7) Children's Behavior, and 8) Women in Government.

The reorganization of the Service was carried one step further when on July 1, 1967, the former five programs were grouped under two coordinators—one for human resources development concerned rather directly with people, and a second for agricultural and economic development involving production and economic factors. This reorganization, see Table I, encouraged emphasis on the *problems of people,* as well as on the agricultural program of long standing. The human resources area encompassed the program areas of 4-H, community development and home economics.

The Changing County Agent

Following World War II, the expanding body of scientific knowledge began to change the function of the county agent and the home demonstration agent rapidly. One result was the adoption of the term "home agent," since detailed demonstration was less needed than in the earlier days when new utensils, tools, materials, stoves and other devices were being publicized. The work of the agents was supplemented in a helpful way on the farm by the assistance of representatives, for example, of seed, feed and machinery companies, and of utility, home furnishings and food manufacturing companies in the home.

The SCOPE reports, and the reorganization of subjects and administration, helped to implement the recommendations of the 1959 conference; the agent should know where to obtain desired information or help, rather than being expected to render skilled service immediately in many areas. With the continued expansion of programs and the larger number of women's and youth organizations, the county personnel were called on more and more as coordinators of group programs and trainers of leaders. In 1962 the agents asked the state office for increased assistance in programs which would give them training in group dynamics, resource development and program planning.

The advantage of this trend in which the agent took less responsibility for making specific recommendations was illustrated in Gallatin County in 1968. As a result of treating alfalfa with chloridane for weevil control in 1967, excessive amounts of heptachlor epoxide appeared in the milk from several large dairy herds. Thirty milk producers were restrained from selling milk, and some 1,500 cattle were quarantined. The specialists and agents had worked with the best authorities in the state, but had not made specific recommendations, and were, therefore, not subject to censure. They helped the dairymen to adjust their operations to meet the emergency, and they assisted producers to make application for indemnity payments. In order to be most useful, the agent, said the director, should have the best possible training and keep abreast of new developments in every way available, but instead of posing as an authority on every question, he should "discuss the evidence and alternatives but let the farmer make the decision on the basis of information from Extension and elsewhere."

Opposing pressures were regularly placed upon the Extension Service. On the one hand it was urged to expand its role, increase the diversity of its clientele, and expand its programs. On the other hand, if this expansion departed in any marked way from the longtime program, abrasive criticism arose almost immediately. This conflict often created problems in securing adquate financial support from all of the three major fund resources—county, state and federal—particularly from the county and the state.

Shortly after Carl McIntosh became president of the University in 1970, he asked all departments to prepare a statement of goals and indicate priorities.

After much deliberation, the Extension Service prepared a statement titled "Goals and Related Activities of the Montana Cooperative Extension Service." It stated that Extension's major goal was to contribute to human satisfaction by improving the quality of life. The role of Extension was to help people obtain what they wanted most for their community through the development of human resources, together with the development of natural and economic resources. This statement is included in the Appendix.

This was one of the earliest statements of goals emphasizing such a broad objective and urging the use of all available resources. Montana leaders called attention to the increasingly complex conditions of the times and the multiplicity of skillful public and private agencies anxious to give assistance. In such a situation the agents were faced with the need for higher speed facilities for informing themselves and incorporating available opportunities into the Extension program. These might extend from sophisticated theories of social psychology to the most basic practices of agronomy. The agents must have special abilities "to listen, relate to, and gain the confidence of their people," says the statement.

The "Goals" summary observes that Extension personnel have always been concerned with tangible results, giving them highest priority. But an important goal is also to develop skills in attaining the results; the processes useful to a dynamic society. The statement continues, "The Extension process, like the democratic process, is part of the goals system and deserves far more attention." Understanding and teaching the process of perceiving, analyzing and solving shifting situations is an integral part of the changing job of the county agent.

In line with the emphasis on closer ties with research and resident instruction and the use of resources available to Extension through these arms of the University, the 1973 Homecoming activities were dedicated to the Extension Service. Many honors were given to Extension workers, and they in turn prepared a number of floats which interpreted their aims. The event gave the Extension people the feeling that they were appreciated as a major component of Montana State University.

Vital Issues and Great Decisions

In line with Director Aasheim's interest in public affairs and public responsibilities, the Extension Service applied for and received a grant of $40,000 in May 1959. This was one of only six such grants made to the Extension Services in the nation, and was made by the Fund for Adult Education, which had been incorporated in 1951 by the Ford Foundation. The Montana Extension Service set up a broad base for the use of the grant and proceeded to conduct a series of discussions titled "Vital Issues."

> In a democracy, no central authority can determine the benchmarks much less a universal policy line. We must depend on the capacity of leaders to comprehend the issues at large and their relation to fundamental values and the questions at hand, but we must not overlook the need for an informed general public. Thus, our conception of education in public affairs and for public responsibility is:
>
> 1. To identify and clarify the fundamental issues
> 2. To provide the framework and perspective essential for assessment of policy matters
> 3. To supply this education in a form suitable for leaders presently exercising public responsibility and adapted to various levels of competence, interest and responsibility
> 4. To further provide this education in appropriate form to those who may be led to assume public responsibility.

The pilot project consisted of sending Ernest R. Ahrendes to serve as a consultant to County Agent Don Hunter in Roosevelt County. Additional comment will be made later of the notable accomplishments there. J. Glenn Jones was sent to Lewis and Clark County where results were encouraging, but not as immediately productive as those in Roosevelt County.

Interest in the public discussion of a number of current important issues spread through the state, and in the spring of 1960 the Extension Service prepared materials incorporating fact sheets on the following subjects:

1. Education—What Goals for Montana?
2. How Can Montana Grow?
3. Welfare in Montana—What Needs? Whose Responsibility?
4. What Do We Expect from Government?
5. How Should Montana Finance Public Service?
6. What Can Montana Afford?

These issues were somewhat obvious and they were already being widely and vigorously discussed. Desires were voiced in public meetings for additional money for education and public services in the state in the midst of the 1960 political campaign. Meanwhile, the Montana Taxpayers' Association, through its publication, the *Montana Taxpayer,* maintained that the fact sheets were unrealistic in suggesting additional government services without answering the question of where the money was coming from. The Republicans held that these discussions, financed in part by public funds and directed by state employees, supported the Democrats.

In December 1960 the public discussions continued but took a somewhat different trend with kits of materials, titled "Great Decisions," distributed to self-administered citizen discussion groups. The materials were prepared by the Foreign Policy Association, a well-known organization generally recognized as being reputable and impartial, albeit direct in balancing issues. The 1960 series was used in a large number of communities and was well received. A wider demand arose for the 1961 series which consisted of the following topics:

1. Deadlock over Germany
2. Soviet Challenge and World Leadership
3. France and Western Unity
4. Japan—The Future of an Asian Ally
5. The U.N. in Explosive Africa
6. The Americans in Jeopardy
7. Arms and Survival
8. Blueprints for the World Economy

A storm of protest broke over the state almost as soon as the program was announced. The John Birch Society was in the height of its activity, and it maintained that the historic policies and activities of the United States were unpatriotically criticized in the discussion materials, and that a tax-supported agency should not have any part in their dissemination. Additional groups joined in, maintaining that education in public affairs was not a proper function of the Extension Service and supported the contention that the "Great Decisions" materials were "insidious" and "subversive." The dissension was fueled to a certain extent by the tensions still prevailing from the political campaign of 1960 in which liberal versus conservative schizophrenia played a colorful part.

The Vital Issues project was not renewed. While it had received noisy criticism, many who were associated with it appraised the materials favorably. Some 800 kits of the Vital Issues information were used in Roosevelt County alone, and opinion was overwhelming that while the points of view were forcefully presented, they were fair, informative and of value to the citizen. The end result was that the people in Montana communities had a constructive experience in organizing and in discussing public issues. Director Aasheim commented, however, that perhaps the time was somewhat "inopportune" for an impartial consideration of the problems involved.

The Budget Cut of 1961

No sooner had the Extension Service engaged in its quest for a more effective organization and a program of broader scope and importance than it encountered a severe setback which caused continued re-evaluation. The political campaign of 1960 resulted in the election by a narrow margin of a Republican governor, lieutenant governor and House of Representatives, while the Senate had a majority of Democrats. The Republicans had campaigned on the issue of economy, and a special session was required to resolve the hard-fought financial questions. Allocations to the University units were based upon projected enrollment estimates, and appropriations for the University of Montana and Eastern Montana College were increased while those for Montana State were reduced by 2.8 percent, and other units by much more. The Experiment Station, however, received a cut of 34 percent in state funds, and the Extension Service was cut 51 percent. The actual state funds appropriated for Extension were $510,455 for the biennium in 1961-63 in comparison with $1,047,773 for the 1959-61 biennium. Federal funds were increased, however, and in many cases county funds were also increased, so the final result was a 20 percent loss in total operating funds for the Extension Service and 27 percent for the Experiment Station, but the shock of the severe action of the state Legislature continued to be felt.

A natural question of the Extension Service was why their cut had been so drastic. In an administrative letter in March 1961, just after the special session had adjourned, Director Aasheim commented as follows:

> I know we all wonder what caused this sort of action to be taken. As we look at the whole picture there is no question but that the current Legislature was very economy minded and we were victims, as were many other institutions and state agencies. I'm sure many feel that we took more than our share of the cutback and I feel we did, too, which raises the question then as to why? I can't answer why but here are some of the things I heard said about us which seemed to justify, in some people's minds, that Extension could stand a cut:
>
> 1. Too many people in a county. At most, the state should contribute to the salary of only two agents in a county. (General impression received, a man and a woman).
> 2. If counties want more of this kind of service than indicated above, they maybe should pick up the cost.
> 3. Extension gets money from many sources and all-in-all, it's quite a lot of money (Indian funds, Smith-Lever, A.M.A., Fish and Game, counties, state, and grants).
> 4. In some cases they questioned the competency of individuals.
> 5. They were questioning some of the old programs.
> 6. They questioned some of the new programs, such as "Vital Issues."
> 7. Farmers are reduced in numbers and we don't need as big an Extension Service as we have had.
> 8. Too many people on the state staff.
> 9. Should Extension be working with town people?

The Director then suggested that these points indicated a lack of information on the part of many people; a condition which the Extension staff could correct by additional care in public relations.

Changing Organization, Programs and Finance, 1967

The Extension Service adjusted itself to the difficult situation of 1961 with the same resilience with which it had met frequent emergencies in the past. In his annual report at the end of the year, Director Aasheim wrote of the changes which were being made:

> 1. We are adjusting to a new organizational structure.
> 2. We are attempting to adjust to a broader scope of activities.
> 3. We are operating under a new national administration with new programs to service.
> 4. We had a severe reduction in our operating budget due to a cut in state appropriations.
> 5. As a result of the reduction in available funds we have had a reduction in staff.
>
> But we did make progress toward our objectives of increasing involvement of people in program planning activities and in our efforts to achieve more unity as an organization.

The cut in Extension funds continued to be a live issue as the legislative sessions of 1963 and 1965 made modest adjustments upward. In 1965 the Board of Regents appointed three of its members as a committee to meet with the Extension director and other University personnel including the president and the dean of agriculture, to look at Extension's program and organization. Dr. C. Brice Ratchford, then director of Extension in Missouri, and Dr. William Bennett, director of Extension in Utah, were asked to come to Montana to visit with this group. After extensive discussion, the director was asked to prepare a statement for the Board of Regents which would include some of the plans which the Service was considering.

This summary was titled "A Brief Statement Relating to Some Proposed Changes in Program Development, Organization and Financing of the Montana Cooperative Extension Service," dated December 1966. It summarized proposed changes in (a) methods of determining programs, (b) organizational structure, and (c) pay schedules for county Extension agents. A major method of selecting programs was to be by "groups of local people working together in defining problems and opportunities." Extension personnel would help in organizing these groups, and then lend its facilities in working with the people on the problems which had been identified.

A proposed reorganizational structure would consolidate program areas from five to two, which has been noted. One would be designed to deal with people directly: youth, family and community development. The second would consider agricultural and economic development. This concise organization was highly innovative at the time, and it immediately attracted wide attention and influenced reorganization in other states.

Director Aasheim received a National Merit Award for implementing this organization change. The citation read: "For developing and implementing the reorganization of the Montana Cooperative Extension Service with a savings of $25,700.00." An outline of the proposed reorganization has been included under "Organization Change."

The report pointed out the changes which specialization and commercialization were making in agriculture, and the changing role of Extension services. The place of community action and management and the role of the individual in a time of enlarging horizons of interest were noted. The statement also stressed the increasing opportunity for improving the quality of living through programs with youth, the family, and community programs of adult education.

A proposed revision of financing pointed largely toward the county, where county contributions to agents' salaries would be based on the average salary of certain elected officials of each county. State and federal funds would then provide the additional funds for the salary totals which would be determined by Montana State University in line with that of the university faculty.

This entire statement implemented the trend that "increased emphasis would be placed on the participation of local people in determining problems and programs in a systematic manner."

The Regents approved the statement and the reorganization was set in motion on July 1, 1967. By November 1967, after a few months of watching the ideas in motion, Director Aasheim restated certain essentials in his annual report to President Leon H. Johnson:

> Past experience has shown that the most effective method of determining the problems and opportunities of people is to involve them in groups made up of individuals in their community who represent a cross section of interest including those from industry in both town and country.
>
> At present there are cross-sectional groups in existence or in the process of being formed in 16 counties. . . One Extension employee is working intensively in an eight-county area near Billings, assisting with the development of such groups and assisting them after they have been formed. Five of the counties in this eight-county area have cross-sectional groups now.
>
> In the counties where there is not a cross-sectional group in existence or in the process of being formed, the agents are working with established groups and organizations. During the last six months county agents have worked with 10 Chambers of Commerce, two industrial committees, six county and city governments, five planning commissions, two multi-county organizations, and 17 other small special interest or commodity groups. Agents are currently working with four agricultural marketing organizations and are helping seven recreation or tourist business organizations.

Citing the approach which might be made to a specific agricultural industry, the director continued:

> In the field of livestock production there are many choices available to Extension personnel when developing the educational program for the county or state. Where are the greatest needs and opportunities of the

stockgrower? Do they relate to range management, animal nutrition, livestock feeding, animal breeding, marketing, enterprise management, transportation, or is it some other matter that needs consideration? By discussing these alternatives with people from the industry and others, a better job can be done of developing the education program.

Dr. Leon H. Johnson followed Dr. R. R. Renne as president of Montana State University, 1964-1969. President Johnson was initially highly supportive of Extension's expanded program, but he soon received pressure from influential Montanans which caused him to suggest that perhaps Extension should return to its traditional programs in agriculture, home economics and 4-H.

He proposed this policy to a meeting of the University Advisory Council and asked for comments. He received no support for his proposal, and instead a number of members voiced approval of the Extension program. For example, a rancher in eastern Montana replied, "I think the work in the field of human relations and communication is some of the most worthwhile work Extension has done. It should be continued." A farmer and state senator from North Central Montana asked, "Who will help us with objective information relative to water policy and development?"

The Chairman of the Council, a stockgrower from western Montana, was even more supportive: "The most important contribution that the Extension Service has made to Montana is the development of leadership, not the skills of homemaking and agricultural practice. It would be a mistake to limit its role as you are suggesting."

President Johnson was impressed by this response, and did not raise the question again.

Funding and Staffing

The reappraisal which took place in the early 1960s led to the issuance of informative materials from time to time for educational use among the citizens of the state. The "cooperative" aspects of Extension relating to financing as well as programming were emphasized. (Reference to financial detail has been held in this account to only a few illustrative instances.) Comments were made about the slowness with which Montana and its various counties met their responsibilities in making the work of both the Experiment Station and the Extension Service fully effective. During the periods of crisis such as in wars, drought, depression, attacks upon plagues and disease, and experimentation with new ideas, the additional funds from the federal government made it possible for the agencies to prove their worth.

The pattern of finance by the federal government is somewhat complex because of the effort to treat each state with the greatest fairness possible. First, 4 percent of the total appropriation by the Congress is withheld for the Federal Extension Service. Then, 20 percent of the remainder is

Sources of Funds Provided to Support the Montana Cooperative Extension Service (1914 — 1982)

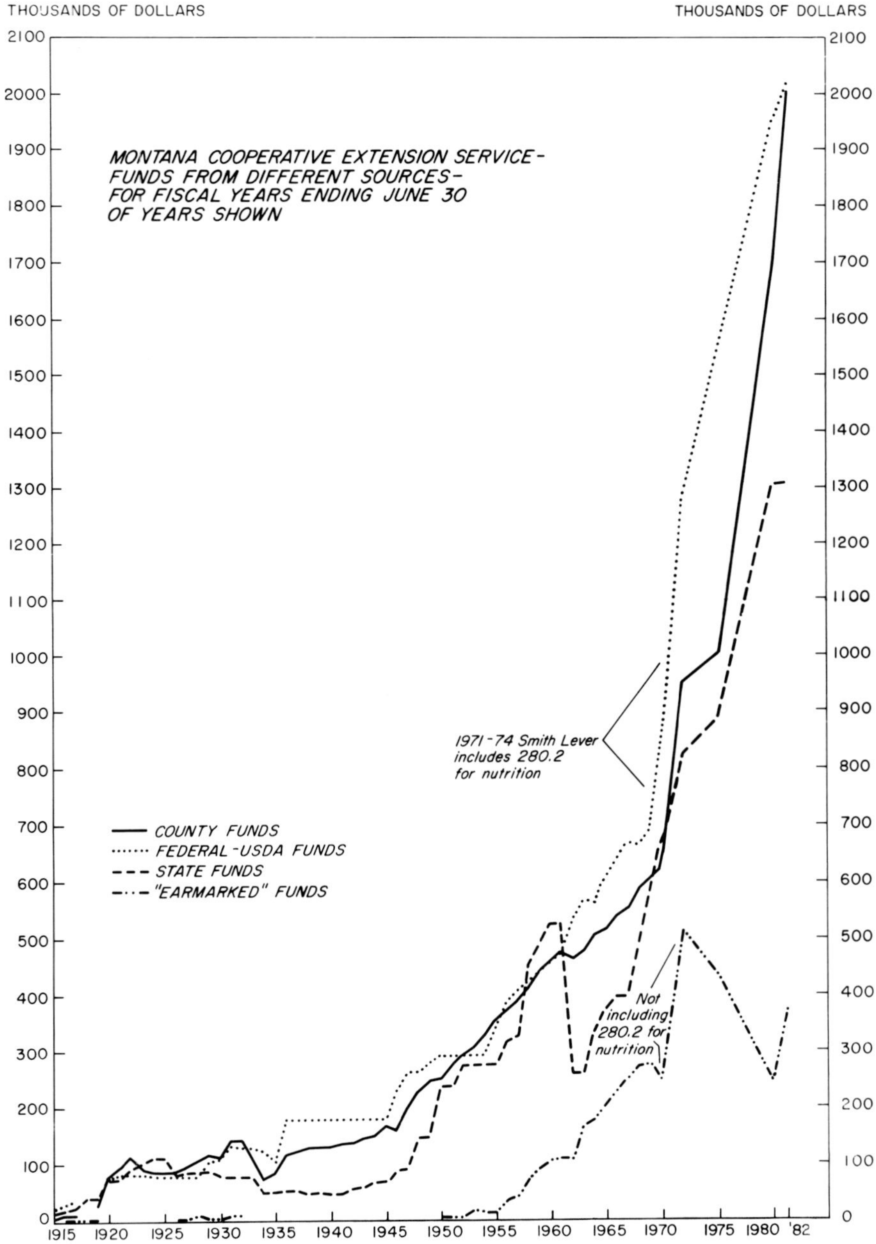

distributed equally among the states. An additional 40 percent is distributed on the basis of the total farm population at the last census, and the remaining 40 percent is apportioned on the basis of numbers of rural non-farm people.

State or county funds are required for matching except for grants for special programs, of which there are many. For the year 1969-1970, for instance, Montana received $860,349 in federal funds, but only $574,064, or 66 percent, required matching. At that time the state supplied $552,309 and the counties $580,757. The counties pay about 40 percent of the agents' salaries, as well as providing office, secretarial and travel funds. The actual portion of an agent's salary paid by a county is based on a schedule established in 1966 by Robert F. Rasmusson.

"Earmarked" funds for special purposes make up an appreciable part of the total Extension budget. From the federal government in 1969, some $140,800 was allocated for special nutrition education for low income families, and matching funds were not required. The U. S. Bureau of Indian Affairs has found it advantageous to finance Extension agents for work among Indian families, and in 1969-1970 the Bureau made $140,900 available for this purpose. The Montana Fish and Game Department provided funds for a wildlife specialist in Extension. Occasional special programs are funded, such as educational work in the use of fertilizers, financed by a special tax on fertilizer sold in the state, to use only one example. The accompanying chart shows the trends in the financing of the Montana Extension Service from 1914 through 1982.

Specialized Areas: Agricultural & Economic Development

Donald J. Luebbe took over the assignment on July 1, 1967, as program coordinator for agricultural and economic development. Since this position represented chiefly a change of title for work which he had been doing since 1960, he was able to tighten administrative reins and continue the work in the major areas. Projects which were active in the new division included:

Farm Management
Marketing & Utilization of Agricultural Products
Agricultural Outlook & Public Affairs
Wildlife Management
Crop Production
Beef Production
Sheep and Wool Production
Dairy and Swine Production
Range Management
Weed Control
Horticulture

Plant Pathology
Agricultural Pesticides & Chemicals

A brief summary of developments in a number of projects reveals trends in rural Montana, and the direction which Extension has been able to give to them since 1958:

Farm Management was one of the first projects to be given more emphasis. For years, farmers resisted taking the time to keep records, but by 1968 Merle E. Quenemoen, Robert F. Bucher and LeRoy D. Luft were all working overtime as Extension economists to keep up with the demand for help. The larger farms soon began to consider the use of computers. Twenty-one county agents and 19 vocational agricultural instructors took special training in computer science in 1968, and in 1969 specialists assisted 22 county agents in holding farm management schools. A decline of 10 percent in the net income per farm in 1968 increased the interest in farm management and careful record keeping.

The *Marketing and Utilization* specialists, Charles H. Rust and James H. Cothern, had a number of feasibility studies under way by 1968 which guided development of commercial enterprises. Included were such projects as a safflower oilseed plant in northeastern Montana, a feed pelleting plant in the southeastern area, a potato processing operation in western

Farm machines became larger and larger in efforts to reduce cost of production.

Montana, and a hay marketing association in Teton County. A *Cattle Feeders' Guide* was issued in 1970.

Assistance was given to the Montana Wheat Research and Marketing Committee in collecting and shipping grain to the West Coast. Railroads agreed to provide rapid transit all-wheat trains with low rates and economical loading and unloading arrangements. Traveling seminars of Montana wheat growers to Pacific Northwest marketing centers were productive. Exchanges between our agricultural leaders and those of Japan were arranged. Robert H. Bastrup began a 10-year period as manager of this program, and a small fee collected on marketed wheat financed its operation. A savings of $1,000,000 was made in the first year, with more expected in the future.[5]

The *Agricultural Outlook and Public Affairs* specialist, John Bower, with part-time assistance from James H. Cothern, attempted to keep the people of rural Montana informed of the pressing problems of the time, one of which in 1968 concerned the comparative merits of the sales tax, the income tax, and the property tax as a means of financing public services.

One of several recurring issues of the period was that of evaluating Montana's water resources, and this led to the enactment of a law in 1969 under which Water Conservancy Districts might be formed. James Cothern coordinated the plans of the State Water Resources Board with those of the Great Plains Agricultural Council's Water Committee.

The position of *Wildlife Management* specialist was authorized in 1955. Eldon Smith held this position for years and worked directly with the Montana Fish and Game Department, from which the majority of funds came. He was able to do much to increase the understanding between the wildlife management groups intent upon the necessary stable soil, adequate water, and good plant cover for fish and wildlife, and the agricultural interests which often felt they were threatened with losses of substantial amounts by such effort.

Crop Production has been a major concern since the inception of the Extension Service. Agronomists Arthur F. Shaw and Manford J. Jackson, who worked together for many years, found the need to devote a great deal of time to obtaining pure seed. Increasing emphasis upon wheats which would produce a high quality flour for the modern baker was only one of numerous demands. The close cooperation of the Montana Seedgrowers Association has been of great value in this work.

These specialists have also been heavily involved in the increasing production of barley for livestock feed, with much of it going to rapidly growing swine and cattle feeding demands. Close attention has been given to the production of superior malting barley; the newer specialty crops such as safflower and sanfoin also received encouragement. A 50-year crop standardization program required considerable time from Shaw.

Jackson worked with agricultural chemical projects which received special federal funding. He has also given attention to hay and pasture

Mike Linsenbigler (left) and Rudy Cicon team up to measure the water depth in a saline seep discharge area.
(below) Areas in eastern Montana where a saline seep is most prevalent.

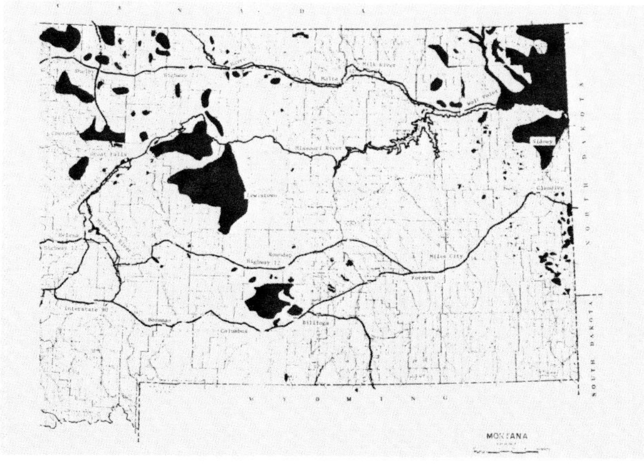

crops. An effort to revive the former highly successful alfalfa seed production led to the promotion of leaf cutter and alkali bees. Where they were used the seed production increased three-fold. The steadily increasing incidence of weeds and weed seed in small grain production has led Jackson to devote most of his time to the eradication of weeds in recent years.

Arthur Shaw was a member of a small national team to visit Russia over a period of years in a Spring Wheat Production Study sponsored by the Foreign Agricultural Service of the USDA. He retired in 1980 and his place was taken by Don Baldridge, formerly of the Southern Montana Experiment Station at Huntley. Considerable attention is presently being given to such approaches as minimum till, flexible cropping and alternative crops.

Soils Management specialists spent several productive years emphasizing fertilizer field trial programs. Increased production values were estimated in the millions of dollars over the years 1965-1970. Fertilizer companies increased their contributions to the program from $2,300 in 1962 to $18,000 in 1968. Experiments with combinations of fertilizer, water, and crop sequence on irrigated lands greatly increased the yields and quality of malting barley; i.e., from 38 bushels per acre in 1960 to 63 bushels in 1968, to use only one example.

A continuing difficulty is "alkali seep," the increasing appearance of alkali in areas once highly productive. Experimentation with summer fallow versus continuous cropping, and types of crop sequence, as well as other approaches have slowed the plague in some instances, but it is still a menace.

Beef Production was a primary concern of the Farmers' Institutes when in 1911 Dr. W. J. Hartman was employed to assist with the state livestock industry. In 1914 he continued as the first livestock specialist. By 1968 Montana beef cattle numbered three million head, and the income from beef cattle amounted to about 40 percent of the total agricultural income.

Since the production of beef for across-the-counter sale is a major industry, the livestock specialists have worked with the Montana Stockgrowers' Association and the Montana Beef Performance Association to produce the most desirable type of distribution of fat and lean demanded by the changing tastes of the consumers. This work was hastened by the greatly increased practice of fattening cattle in Montana during and following World War II. This was partially brought about by the growth of population in the Pacific Northwest. By 1968 some 120,000 cattle and calves were on feed; numerous feed lots sprang up and animal nutrition became immediately important. The Baby Beef projects of 4-H youth were valuable in the studies which were made.

The specialists obtained a great deal of support for their work from the Montana Beef Performance Association. This had been formed in 1956, largely through the efforts of the livestock specialist, Paul Orcutt. A program of careful record keeping of the best practices and certification is summarized in Orcutt's publication, *Cows, Calves and Computers, The First*

Cattle on feed in a commercial feedlot.

Twenty Years of the Montana Beef Performance Association, 1977.

Dairy Production was a favorite program of Extension in the days of diversification because of its nutrition and family cash income possibilities. Insufficient attention was given to high performance, however, until F. M. Hillman began to form cow testing and evaluation associations and to urge the improvement of breeding animals in 1914. J. Oscar Tretsven was a particularly effective specialist in the long span from 1922 until 1955.

Dairying has remained a minor industry, averaging only about 3.5 percent of the total agricultural income. A large number of small dairies operated in the western valleys of Montana in the first half of the century. The number of marketing plants decreased by almost 50 percent between 1960 and 1970, but the increased production per cow has enabled the remaining dairies and marketing plants to prosper. The Dairy Herd Improvement Association has been an active and progressive aid to Extension work. Areas of research have included nutrition, housing and milking equipment, artificial insemination, disease control and marketing techniques. Earl J. Peace assumed the dual position of dairy and swine specialist on July 1, 1959, and continued until July 1, 1969, when he became full-time swine specialist. Dairy production then received less Extension attention and was handled by the Director of the Dairy Herd Improvement Association.

Sheep and Wool Production programs were directed for many years by James Drummond who was in charge of the Montana State Wool Laboratory for the Experiment Station, and spent only about 10 percent of his time with Extension work. The members of the active and closely-knit Montana Wool Growers' Association kept in close touch with the research work done in the laboratory, along with the experimental flocks over the state, some of which were maintained by the association. Montana maintained productive research and kept abreast of the improvements in the quality of the meat and wool, in addition to working on the local problems of shortage of labor, costs of labor, predator control and relations with sportsmen. Nevertheless, the number of sheep in the state continued to decline. The lamb crop in 1968 hit a low of 826,000 head and went still lower in 1972 to 709,000—the smallest number since records were first kept in 1924. The number of sheep sheared in 1971 totaled 902,000 and decreased to 853,000 in 1972.

The Wool Growers retained their enthusiasm, and as the retirement of James Drummond approached, continued requests brought the appointment of the first full-time specialist, Rodney Cott, on February 25, 1980. He continued the work in the field of production and also in problems of marketing. Numbers of sheep continued to decline since prices also declined. This is a frustrating problem tied to national and international relations where high production nations such as Argentina subsidize their exports in order to encourage their growers.

Range Management as a science owes a great deal to Montana. The well-known independence of cattlemen and sheepmen was intensified in Montana because of the immensity of their holdings, but this provided a flexibility which encouraged their associations to assess their problems and cooperate in finding solutions.

Mont Saunderson came to Montana State in the mid-1920s with an accounting background in agricultural production management. He was given the assignment of making an economic and business study of the range livestock industries in Montana, both sheep and cattle ranching.

Saunderson recalled in 1973 some of the incidents related to his study.[6] Director of the Experiment Station F. B. Linfield was concerned with the "mining" of the range which was taking place in Montana and the nation, which he felt was a management problem. He suggested that Saunderson investigate land laws and policies, land tenure, long-term credits, and methods whereby operators could acquire stable ranching units where they could develop the range rather than destroy it.

Director Taylor of the Extension Service urged Saunderson to scrutinize land tenure. He cited examples where ranch operators, in order to obtain even a temporary control over the rangelands they needed for an effective operating unit, must:

> Deal with as many as 10 different kinds of land ownership, many of them absentee. Among these would be public domain lands, state grant lands,

railroad grant lands, homestead lands, lands foreclosed by banks, county tax deed lands, and lands held for investment by out-of-state corporations.

M. L. Wilson had gone directly to the point of view of management:

> Early in the work, M. L. commented to the effect that one of the disciplines of science badly needed by western stock ranching was missing. This, he said, was the science of range management, which should soon "come of age," and do for western stock ranching what the science of agronomy and soils was doing for the farm. He commented, however, that this development probably was some time in the future. . . since largely out of necessity, the western stock ranching operations had been speculative, due to the problems of land tenure, the vagaries of weather and the ups and downs of the livestock markets.

Saunderson carried on numerous studies, formulated ranch accounting systems which were eagerly accepted by range operators, and worked toward the development of the highly useful "animal unit" measure of land use. He added, "Later, I was to use my influence in getting the words 'and Range Management' added to the name on the door of the head office of the Animal Husbandry Department at MSU." This apparently was the beginning of the formal use of the term range management.

Director Taylor also asked Saunderson to work with Nic W. Monte, secretary of the newly formed Mizpah-Pumpkin Creek Grazing District. This, as has been noted, was a landmark in cooperative grazing districts which led directly into the national Taylor Grazing District pattern.

A range management specialist was not appointed, however, until Ray G. Johnson took the post in 1950. His single-year term was followed by a nine-year productive period by Karl G. Parker.

The importance of Montana range lands can be seen at a glance when compared with other land uses:

Land Use	Approximate Acreage	Land Use (%)
Crops, hay, water, urban and other use	15,970,779	17.15
Forested	15,500,000	16.65
Rangeland (federal)	19,107,620	20.52
Rangeland (state)	4,224,146	4.54
Rangeland (private control)	38,297,172	41.14
	93,099,717	100.00

Dr. Ned Jeffries became range management specialist in 1968. He continued to urge the use of the best practices, since up to 70 percent of the private rangelands showed the need of improvement. Some of the ever ending programs included control of overgrazing and the accompanying luxuriant crop of noxious weeds, reseeding, experiments with new grasses, interspersed stock watering installations, and irrigation wherever this was possible.

A *Swine Production* specialist was the culmination of the slow but steady growth of experimental work which had produced the Hamprace hog. Earl

J. Peace divided his time between dairy and swine programs from July 1, 1959, until July 1, 1969, when he became full-time swine specialist. He worked closely with the statewide Montana Pork Producer's Council, serving as secretary, and he was also editor of the quarterly publication, *The Montana Pork Producer*. In 1955 some 112,000 hogs were produced in the state; by 1969 this number was 266,000, and it doubled again to 435,000 by 1971. Close relationships with the Veterinary Research Laboratory and other animal departments helped in the control of disease. The increase in barley production and experimentation in feeding was of great benefit. Rapidly changing methods in confinement housing were highly productive in the cooler climate of Montana. Carcass tests in regional schools, in combination with beef and sheep, improved marketing practices. Probably more productive changes have taken place in swine and pork production than in any other animal unit.

Weed Control was recognized as a major problem immediately upon the organization of the Experiment Station in 1893. The Station botanist at that time listed the noxious varieties of weeds and issued repeated warnings. Not until the introduction of 2,4-D and other herbicides in the 1940s, however, was Extension able to give real assistance. In 1946 Don Luebbe, who was irrigation specialist, was asked to direct weed control as well. He did so until 1947, and then Harold Dusenberry took over for a year until a specialist was appointed. Leslie Sonders provided continuity from 1961 until 1975, and in more recent years Manford J. Jackson has been giving a great deal of attention to the problem of weeds in small grain production.

The enactment of a suitable weed control law in 1969 was a big step forward. The Montana Weed Control Association was formed then and by 1970 Montana had 31 organized weed districts, many of them with experienced personnel carrying on a year-round program. The plant pathologist in the Experiment Station has continuously been of great assistance. Elwood Morris served in this position from 1909 until 1956, when his associate, Dr. M. M. Afanasiev, succeeded him, 1956-1968; then Dr. Robert C. Carlstrom took the position.

The state has suffered from a number of severe weed invasions. Cheetgrass at one time threatened thousands of acres. Leafy spurge, first discovered in 1930, was infesting some 80,000 acres in two-thirds of the counties by 1970. Dalmatian toadflax was first reported in 1950, and within little more than a decade was declared a menace in 31 counties. Spotted knapweed was third on the list in 1968. The most productive areas—the well watered valleys—were also most highly infected. The range and forest lands each had many and entirely different species. Weed seed in small grains lowered their value greatly, particularly in seed grains. The scientists were assisted to a considerable degree by summer fallow practices, and of course by herbicides, but weeds continued to prosper.

Horticulture was the most active department of the early Experiment Station. Director Emery, a professional horticulturist, made extensive orchard plantings in several parts of the state; apples and cherries, particularly, flourished. The first specialist, Edward E. Isaacs, was not appointed until 1926, but by 1969 he was giving 50 percent of his time to the 5,100 acres of seed potatoes under production that year, and the Montana Potato Growers Association had replaced the early, politically powerful Montana Horticulture Society.

Growing seed potatoes continued to be one of the highly profitable occupations, and by 1968, the annual seminar attracted 175 growers. Montana seed was favored for a long period, in part because of the dramatic work in the late 1930s by Dr. Vincent Iverson of the Experiment Station and Dr. Harry C. Kelley, Department of Physics, in controlling disease, followed by the Botany and Microbiology Department in developing a virus-free potato. The production of potatoes for the grocery market in Montana has lost considerable ground to the lower altitude areas of Idaho, Utah and Washington, but is still profitable.

The once thriving apple production activity has also been lost to the above mentioned neighboring states, but the cherry production around Flathead Lake still prospers in a limited area. Experimentation in yards and gardens with improved fruits and vegetables causes the county agent and state offices to be flooded each summer with requests for help.

The horticulturist is also in charge of the large program of shelter belt planning and care, which continues to grow, and experimentation also continues involving kinds of trees and shrubs as well as methods of planting and care.

This brief summary of the work in these specialized areas of the Extension Service is a record of outstanding accomplishment and a reminder that in each area equally challenging opportunities lie ahead. The dollars and cents return, only one measurement in any one of these productive fields, has been more than enough to bear the cost of the entire Service for the half century.

Human Resource Development

The purpose of the Human Resource Division of Extension is to use the economic base of the life of the people, which has just been described, and to build upon this base educational opportunities that will enable Montana people to reach their full potential in every way as individuals, as families and as communities. This program is designed to assist the people in evaluating present conditions, determining needs, planning progressive improvements, and carrying out these plans cooperatively and wisely.

In the 1967 reorganization, Mrs. Vivienne B. Kintz was named coordinator of human resource development. The basic long-time activities remained in full force and the new developments were grouped with them as effectively as possible.

In the reorganization plan, increased emphasis was given to program development. Four area home economists, who were generalists rather than specialists, planned and helped direct programs throughout the state. County programs based on identified needs brought increased interest in communications, study of institutions, community health, nutrition, consumer education and similar areas of interest.

Decision Making and Development of Leadership, two closely related programs, were initiated to complement the human resource approach. These programs, using many of the Dale Carnegie techniques, carried women from their technical programs in home making into areas of planning and directing community development. The decision making groups set up patterns for the examination of problems, the establishment of priorities and for developing action organizations. In the Development of Leadership programs the techniques of self-examination, management of meetings, presentation of objectives and the creation of self-confidence were considered. The expanding role of women was also discussed, including such aspects as interest in self-development, better family relations, legislation, improved education and community improvement. The effective help of the people completing these programs made a striking difference in Extension work.

Home Economics

An earlier chapter on the homemaker noted the establishment of the work by Kathrine Jensen in 1914, and the important emergency work of the World War I period. At the close of the war in 1918, Mary Ann Graber was appointed specialist in nutrition; in 1919 Georgia Roosevelt became clothing specialist; and in 1921 Gladys Branegan was named home management specialist. Changes in personnel and programs were also noted.

By 1968 the home economics work was divided into three areas for supervisory purposes, each supervisor having responsibility for both youth and adult programs. Mrs. Hazel Hinman was in charge of foods and nutrition, Mrs. Dorothy Gallup of clothing, and Mrs. Carol Tarnavsky of furnishings and equipment.

The *Clothing* project always attracted the largest clientele. The homemakers clubs conducted clothing projects in 54 counties in 1968, and the 4-H clothing program included all 56 counties with 6,445 members enrolled in 8,299 projects. Traditional projects were continued and clothing construction and remodeling was requested in 43 counties opposed to requests in 15 counties for the next most popular choice, buying and the care of fabrics and finishes. Consumer information and pattern alteration and repair were also popular. Changing styles and the new fabrics increased the interest of all age groups, including the boys and men in the household, with regard to the psychological impact of clothing. This led in 1969 to a popular six-lesson course on "The Socio-Psychological Aspect of Clothing." Minority groups—Indians, migrant workers and low-income groups—particularly in relation to children's school clothes, were given special attention.

Foods and nutrition projects were always standard, and traditionally enrollment in these subjects was second highest. Thirty-five counties carried out 56 adult programs in 1968 and 4-H projects were carried on in 55 counties with an enrollment of 5,944 members.

Emphasis was placed upon the 4-H program because of its size, and consumer education received full attention. The many new foods, packaged and prepared foods, and costs, provided interesting sessions. Attention was also given to the food-nutrition needs of lower income groups, and the reception here was a hearty one.

In the overall field of foods there was still a steady need and demand at all economic levels for the early-day emphasis upon the basic preparation of meats and vegetables, good breads, and upon variety in meals. There was also a recurrent demand for such subjects as the principles of baking, low calorie meals, nutrition and physical health, and for the always feared problem of food poisoning. Signs of a more affluent time appeared in such subjects as gourmet cookery, holiday cookery, food garnishes, foreign foods and outdoor cookery.

The Expanded Food and Nutrition program (EFNP) beginning in 1969 came as a part of a national program to improve nutrition in low income families. Tariff money was allocated to the Department of Agriculture to be used in this nutrition education program. An initial $10,000,000 was followed in the next two years by two increases of $20,000,000 each, creating a $50,000,000 national fund for better nutrition. The $274,000 for Montana was the largest increase in allocation of federal money the Extension Service had ever received.

The allocation was made on the basis of the percentage of people in Montana families whose income was $3,000 or less. A State Food and Nutrition Committee worked closely with county agents in determining the number and location of these families, and in developing county committees.

Eleven locations were designated in the counties of Missoula, Silver Bow, Lewis and Clark, Cascade, Hill, Fergus, Yellowstone, Richland, and Roosevelt, and on the Blackfeet, Rocky Boy, Northern Cheyenne, Fort Belknap and Crow reservations.

Paraprofessionals were trained in food buying, food preparation, nutrition, and human relations. They then worked with the designated member families, one-on-one in the home or in small centers in small groups. By June 30, 1973, there were 31 paraprofessionals working throughout the state. Four of these were Indian, one was Black, and two were Spanish American. A total of 1,332 families were involved.

In the summer of 1970 additional federal funds were provided for a 4-H youth phase of this program. It was instituted in Missoula, Silver Bow, Lewis and Clark, Cascade, Fergus and Yellowstone Counties, and professional direction was given. By mid-1973 some 6,551 youth from low income families were included in this program. These activities included both rural and city families.

The development of the paraprofessionals, who were recruited from underprivileged familes where possible, was spectacular, and many officers of local government who followed the work carefully reported that the entire program was particularly successful.

Furnishings and Equipment continued to provide a basic area of interest. The consumer approach was made to new materials, fashions, methods and costs, and comprised groups in 21 counties. 4-H-centered programs were highly popular in this area with projects in 55 counties, many of these involving the do-it-yourself approach. Twenty-six counties set up projects using older standard programs designed for a consideration of overall home living. In this area, also, considerable attention was given to the low income groups. On many Indian reservations, families wished to improve their furnishings in the new or improved homes which were being constructed. Upholstery was particularly popular, and many Indian families proved to be especially adept in this craft.

Human Development and Family Relations

When the position of family life specialist was created in 1949, Miss Amy Wold set up a series of highly stimulating annual Family Life Conferences. They focused attention upon the possibilities of creating a better climate in Montana for improved child care and human relations generally. Additional approaches were emphasized when Dean K. Hoffman became specialist in August 1962. Quirks of the individual and the family which were foremost in the projects on Decision Making and the Development of Leadership were capitalized upon to give the work a real-life touch. Family Life Conferences and Woman's Week considered a wide range of factors in stable and unstable families. Again, the problems of the low income group and Indian families were noted and assistance given where possible. The Montana Council on Family Relations, for which the specialist was secretary, gave considerable aid to the movement to introduce the study of family life and sex education into the schools. Marriage preparation materials were sifted for use at various age level discussion groups. The needs of senior citizens was another area which was investigated.

Communications Training

National programs in communications training had met with wide approval, especially in relation to leadership development conferences. This program developed just when several disturbing incidents of misunderstanding had taken place in Montana because of lack of communications. Workshops were first held for all Extension staff members. Techniques were developed for group discussion to insure, first, a clear presentation of the subject; then an equally well-organized discussion; then evaluation; and finally a majority decision was sought, with the hope of obtaining acceptance.

The choice of difficult real-life questions gave rise to the need for devising methods of proposing alternatives, with the elimination of certain aspects until wide acceptance was reached. The specialists were surprised to find that a lack of communication was so readily recognized as a problem. A presentation of certain communication techniques to a state Churchman's Workshop in Bozeman in 1967, for instance, brought immediate approval. The ministers became so intrigued with the possibilities for use in their churches that they set up three conferences for 1968, for three days each, entirely devoted to the suggested methods.

The University administration, the Experiment Station and the Extension Service tabulated a number of incidents where a lack of clear communication had caused disturbed relations with groups in the state. A recent rather severe disagreement between the Animal Science Department and the Montana Stockgrowers concerning the beef performance testing program was cited as a clear example. Additional training was given to new members of the staff and students in preparation, and the English Department was asked to give special assistance on the subject of communication skills.

Resource and Community Development: A New Direction

The expanding demands of the rural people for assistance in broader areas than agricultural production, and the demands of people in towns for different types of assistance, led to planning for the larger community, then the county and then the multi-county region. Instead of subjects relating only to the farm and to the farm home, the 1952 Rural Progress conferences when compared with a similar series 25 years earlier, as has been noted, showed a striking change of emphasis by suggesting such topics as schools, taxes, vocational training, libraries, citizenship, communication, leadership training, health and safety, and religious and cultural values.

This broadening horizon was influenced by The Montana Study, directed by the University System under a grant from the Rockefeller Foundation, 1944-1947. Study groups were set up in a number of communities, notably at Lonepine, Darby, Libby, Conrad and Lewistown, where resources were inventoried, trends were evaluated and action groups formed. The entire community became involved in a completely democratic way; careful records were kept, and numerous projects were carried on to implement the purpose of the program: "To study ways to stabilize and enrich the small community and the family in Montana and America, in the belief that this is one of the most critical problems facing Western civilization."[7]

The Bitterroot Valley project became one of the first of the broadened Extension projects. The people of the valley had a long history of cooperation through their enormous summer picnics hosted by the county cooperative dairy associations. The Ravalli County Improvement Association was organized in the 1930s, and the Montana Study had formed study groups in Hamilton, Darby, Stevensville and Victor. In 1957 the Extension Service set up a Montana Rural Development Committee and offered aid to any county which would formulate a promising plan. Ravalli County organized a county committee, and with the assistance of long-time County Agent William Thomas, presented the first project. This included the traditional aims of increasing farm income and adjusting farm operations, then proposing small industries associated with agriculture, mineral resources and forestry.

Ravalli County expanded its program when it secured assistance from the Resource Conservation and Development Project (RC&D), established in 1962 and administered by the Soil Conservation Service and the Extension Service. A reorganization into the Bitterroot Valley Resource Conservation and Development Project was made in 1965. It is unique in that it is the only state project to incorporate an entire river valley, that of the Bitterroot in Ravalli and Missoula counties. By 1968 the project had received 175 proposals for specific enterprises, with some 5,000 people directly involved. Some of the proposals undertaken included weed control, stabilization of river banks, the Ricketts Memorial Museum, the Senior Center, community irrigation projects, opening a stone quarry, and a plastic fabrication plant.

A Roosevelt County Planning and Improvement Council formed in 1960 made significant progress with relatively little support. Director Aasheim secured a grant from the Fund for Adult Education in 1959 which enabled him to send Ernest Ahrendes to Roosevelt County for a time to assist County Agent Don Hunter, who had an effective county advisory council in operation. Some accomplishments included:[8]

Preparation of an overall economic development plan for the county

Organization of three development corporations to promote industrial growth

Implementation of a program which placed 12,000 acres of Missouri River bottom land under irrigation

Initiation of a fire fighting training program for the Fort Peck Reservation Indians.

Development of a county-wide nurses aide training program

Promoted local arts and crafts, initiated a museum and sales center, and developed a tourist program

Initiated a broad program of outdoor recreation

Provided additional specialized educational services for young people at a lower cost by joint hiring and use of teachers among neighboring high schools

Increased agricultural income by tripling hog production, increasing cattle feeding, encouraging new crops—1,000 acres of safflower and making adjustments in farm and ranch management

The major printed materials used to initiate the discussions from which the above project ideas developed were "Vital Issues," previously mentioned. A faction violently opposed these materials, but an overwhelming majority supported them. A publication titled *Teamwork Equals Progress* included comments favoring the program:

> We believe the work of the County Planning and Improvement Council has been very beneficial not only to the Fort Peck Tribes, but to the community in coordinating activities of all county, state and federal agencies. . .and creating public understanding of the community problems and recommending solutions, generating support for action programs.
>
> William Youpee, Chairman, Fort Peck Tribal Executive Board, Poplar

> The program should grow in importance as people realize the necessity for continuous adult education.
>
> John Simard, Farmer, Bainville

> The efforts and accomplishments of this pilot program during the past two years will reflect an era of broader understanding and cooperation on the part of farm and non-farm groups for years to come. We feel that Roosevelt County has benefited immensely by this program.
>
> Board of County Commissioners, Roosevelt County

This type of reaction came as a result of the wholehearted support of the program by Don Hunter and Ernest Ahrendes, who had the courage to venture into the area of public affairs and involved the people with a positive attitude.

Many Extension agents and state staff members were not supportive of such programs and were not displeased when some people chastized the director for venturing into these controversial areas.

The South Central Montana Development Federation is an enlargement of the Community and Rural Areas Development Project which Hunter and Ahrendes had worked with in Roosevelt County in 1965.[9] On July 1, 1966, Don Hunter went to Billings to help coordinate a multi-county development. This was made possible by a special federal appropriation, earmarked for community development. The program was making satisfactory growth when Hunter died suddenly of a heart attack in May 1968. Interested agents, however, continued the work.

The eight counties which made up the federation—Yellowstone, Musselshell, Sweet Grass, Golden Valley, Wheatland, Carbon, Stillwater and Big Horn—came together slowly, but by February 11, 1971, all had set up development councils. The chief unit was a 15-member executive committee which represented a cross section of interests and areas in each county. Subcommittees of the federation and the counties included those on agriculture, industry, roads, education, youth and tourism. Wheatland

County's committee worked closely with the Harlowton Chamber of Commerce. Big Horn County and the Crow and Cheyenne reservations formed an *Economic Development Association (EDA) District* which received encouraging funding. Stillwater and Carbon counties also formed Resource Conservation and Development Project committees.

Some important projects which were initiated rather early included a 4-H Junior Leadership Camp, the Billings Mayor's Human Relations Committee, a Bureau of Indian Affairs Community Development Seminar, a Yellowstone County Fair Board plan to upgrade its physical plant, and inter-county projects such as a roads study for Carbon County. Ten district weed control advisory boards were formed. An alkali study group comprising six counties assisted in a statewide intensive study of newly formed alkaline areas. Feasibility studies made from time to time included such subjects as poultry feeding and marketing, beef feeding, seed producing and marketing, and the increased use of fertilizer.

The Economic Development Association of Eastern Montana, consisting of 18 counties east of Yellowstone County, comprised still another evolving multi-county enterprise. These counties included Phillips, Valley, Daniels, Sheridan, Roosevelt, Richland, McCone, Garfield, Prairie, Dawson, Wibaux, Fallon, Carter, Powder River, Custer, Rosebud and Treasure. The formal adoption of the constitution and bylaws took place in Circle in September 1970.

The region had a background of common concern and cooperation in a study and action project on health facilities, but many other problems continued to plague the area. It suffered, for instance, a 14,000 loss in population during the 1960s. Difficulties such as drought, storm damage, insect and rodent infestation called for improved communication for regional action, together with state and federal assistance, as did remedial projects such as reclamation, range improvement, highway development and marketing.[10]

The bylaws state the purpose of the organization as: (a) to provide the leadership required for developing and implementing plans for the orderly development, improvement, conservation and efficient use of the human and natural resources; to take the necessary action to insure improvement of the economic opportunities for the people in eastern Montana; (b) to secure federal, state, local governmental, private individual and private corporation assistance; and (c) to promote mutual cooperation between local, state, federal and private agencies and individuals in the furtherance of the above named purpose.

The association formed slowly and perfected an organization close to the grass-roots concerns. Officers then worked closely with the Extension Service and the State Department of Planning and Economic Development to assist in forming development groups and comprehensive plans of work for each county. Next, a plan for the area was formulated.

James DeBree, an Extension staff member who had experience in community development work, was named to assist the association from an office in Sidney. He was paid jointly by the Extension Service and the Department of Planning. When Planning's funds were reduced in 1971, federal community development funds were made available to the extent of providing an assistant for DeBree.

The 1971-72 report outlined programs for six major committees: agriculture, education, finance, transportation, recreation and tourism, and water development and conservation. Important specific problems under consideration included saline seep research, large stock feedlots and marketing points, weather modification, safflower research, and environmental aspects of coal mining and power development. State and federal agency representatives have been involved in these studies, as well as in those of broad civic concern such as health planning, law enforcement, manpower training, education and housing.

Rural Areas Development—RAD

In 1961 the United States Department of Agriculture continued the rural development work through a program renamed Rural Areas Development, RAD, as it is commonly called. Working through a state committee, RAD has given a great deal of atttention to Montana problems and opportunities.[11]

The first meeting was held at Montana State College in August 1961, and the state bylaws and organization were completed in February 1962. In each state the Extension Service was instructed to assume leadership in the program and this was guaranteed in Montana by Director Aasheim serving many years, even after retirement, as its chairman. The stated objective of RAD is: To help develop the human and natural resources of the rural areas of the state to their fullest potential by developing processes among the people for continually solving their problems.

The organization of RAD included an executive committee of eight members who were representatives of important agriculturally related groups, including the presidents of the Montana Farm Bureau, the Montana Farmers' Union, and the Montana Grange. Working subcommittees were then established for various projects or areas of development, and counties were urged to set up a county committee with which the state committees could cooperate. Subcommittees reporting in 1973 included:

1. Range Management
2. Recreation
3. Swine
4. Weeds
5. Water
6. Soil Fertility
7. Forestry
8. Manpower
9. Community Development

The quarterly meetings and the activities of RAD have served as a clearing house in evaluating resources, planning major projects, testing the climate of opinion, and making contact with action agencies of the federal, state and local governments. The federal government continued to establish agencies which were able to assist in implementing the planning work which was RAD's special function.

In 1963 the United States Department of Agriculture ordered all of its agencies to cooperate closely at the state level by forming Technical Action Panels (TAP). Frequent inter-agency meetings and Extension's representation on the state RAD committee insured immediate cooperation. In 1969 the Technical Action Panels were reorganized and renamed, becoming the USDA Committees for Rural Development. The assignment and purpose were not changed.

The Montana organization for the state committee was completed in January 1970. Director Aasheim found himself chairman of the new USDA Committee as well as of RAD. Following 1967 and the establishment of the Montana State Department of Planning & Economic Development, this important action agency was added to the battery operating under RAD to accelerate the community development effort.

Through the close cooperation of county committees of lay people and the county USDA Committees for Rural Development, each county went to work developing situation statements and development plans. State RAD and USDA committees could then work directly with the counties. The exact influence of these committees is difficult to assess. One result, however, apparent from the beginning, was the greater harmony achieved among the farm organizations. This need had been continuously expressed since the days of Director Cooley.

The work of the several RAD subcommittees was highly productive. The Manpower Subcommittee sought to obtain better facilities for vocational technical training. A result was greatly increased state funding for vocational technical programs. A film, "More Power for the Job," was produced by the University Film & TV Center for the Extension Service.

The Roosevelt County program, previously referred to, continued its plans and in a February 1966 RAD meeting representatives reported the following important achievements for the county and the county seat of Wolf Point: (1) a new hospital, (2) a new retirement home, (3) a new library, and (4) a new city office building. In addition, a private business, the CB&F Corporation, had received $350,000 from the Office of Economic Opportunity for a bulk feed plant in Culbertson.

The RAD subcommittee on soil fertility promoted successfully a phenomenal increase in the use of fertilizers. The carefully recorded results of substantial profits encouraged the 1971 Legislature, after two previous urgent requests, to impose a small tax on fertilizers sold, and assign this income to the Extension Service and the Experiment Station to continue the promotion of fertilizer use.

The subcommittee on swine widely publicized the new methods of housing, handling and feeding hogs. The projection increased to such an extent that the Legislature funded a full-time swine specialist beginning July 1, 1969. Earl Peace, who had been dividing his time between dairy and swine, took the new post.

With the several community programs in mind, a prominent Montana newspaper editor commented on October 21, 1973:[12]

> Something is happening in Montana that bodes well for those who come under its persuasive influence. That something is somewhere between a revolution and evolution...
>
> The "something" is the Community Resource Development Program of the Montana Cooperative Extension Service...
>
> If ever there was an Operation Bootstrap, with the right kind of help, CRD is it.

Many, many proposals have come to the RAD committees through the years for consideration and implementation. The broad scope of these activities has continued to cause confusion as to the appropriate involvement for Extension. Director Aasheim reminded his staff in an administrative letter of March 21, 1966, when the questions first began to come, of the original broad concept of the Extension Service by referring to the report of the House Committee on Agriculture when submitting the Smith-Lever Bill of 1914. This report read in part as follows:

> The theory of this bill is to extend this system...to the entire country by providing for at least one trained demonstrator or itinerant teacher for each agricultural county who in the very nature of things must give leadership and direction along all lines of rural activities...social, economic and financial.
>
> He is to assume leadership in every movement, whatever it may be, the aim of which is better farming, better living, more happiness, more education and better citizenship.

While social and civic problems received attention, the basic thrust of the RAD committees remained economic in nature, and in 1970 "A Montana Situation Statement with Special Emphasis upon Agriculture" was prepared and published by the Extension Service.

This was followed by the 1971-1972 Plan of Work by the USDA Committee for Rural Development which outlined the following areas:

1. Farm Programs
2. Farmstead Windbreaks
3. Feasibility of Feeding, Packing and Processing Plants
4. Housing
5. Land Use Alternatives
6. Planning and Zoning
7. Pollution
8. Reclamation of Strip Mines
9. Rural Credit Needs
10. Soil, Land and Water
11. Saline Seep Problems
12. Timber Resource Use
13. Transportation
14. Weed Identification and Control
15. Wilderness and Wildlife

Civil Defense

The Rural Civil Defense program began in June 1961 as an education program exclusively, under the direction of the Federal Department of Health, Education and Welfare, using Extension Service personnel. Additional funds were made available in 1963 and again in 1966. In 1968 support for the program declined or shifted to some forms of safety and natural disaster relief.

Specialist Roy Linn, 1966-1974, directed an educational program which meshed state and federal activities. Construction and use of fallout shelters received considerable attention. School administrators and teachers were widely used to incorporate civil defense concepts into the school curriculum and train community instructors in the practice of personal and family survival.

Another phase of the project was the training of selected groups to become radiological monitors. By 1968 some 438 monitors had been licensed and assigned to a station. Trainees included staff members of the Forest Service, Soil Conservation Service, Agricultural Research Service, Consumer Marketing Service and the Federal Aviation Administration.

The specialist and his staff also conducted workshops for Extension agents, USDA personnel, county defense boards, citizens groups and the Montana Civil Air Patrol. A civil defense training exercise was conducted each year at the week-long 4-H Conservation Camp in the form of an educational game which simulated the conditions of a community hit by disaster.

The Montana State College mobile campus as it appeared in 1962.

The Mobile Campus

In 1960 the Alumni Association gave Montana State College a large house trailer to be used by the Cooperative Extension Service for exhibits and educational activities. The fund reduction in 1961 prevented its use in that year, but in 1962 a campus-wide committee using the theme of the 100th Anniversary of the Land-Grant College provided aid. The "mobile campus" was sent over the state where the exhibits were viewed by 58,000 people in 74 towns in 55 counties, and was considered to be a new and important technique for informing the public.

In 1963 the Extension Service used the mobile campus to introduce the ideas of civil defense to the state. A set of well-planned exhibits showed:

1. A huge mold of the atom—the "Tiny Giant"
2. Information concerning the atom
3. Ways of harnessing the atom
4. The two sides of the "Tiny Giant"—peaceful and constructive
5. The story of Montana State University—teaching, research and Extension

In 1963 the trailer traveled some 8,000 miles and was seen by about 65,000 people in all counties in Montana, including showings at eight fairs. The preparation of exhibits and the transportation and staffing of the trailer proved to be costly and time consuming. Although the two trips had been judged successful, the trailer was sold to the University of Arizona in 1965.

MEMIS

The computer age overtook the Federal Extension Service in 1969, and in accord with its directives to the states, the Montana Extension Management Information System (MEMIS) became effective July 1. Don Luebbe, program coordinator, was given chief responsibility for organization and contact, together with Robert Bucher, Extension economist, who had used computer techniques in connection with his Farm Management and Accounting projects.

The adoption of the information system called for a tight schedule of program planning for as much as five years ahead. It demanded a specific plan of work of line items for each year, beginning with 1970. These were to serve as the basis for planning and reporting the time spent on each project, and were designed to make the analysis and comparison of results more direct and specific. The depth of reflection on such factors as the human element and cause and effect relations were uncertain.

In his plan of work for the fiscal year 1972 Aashiem stated:

> The management information system has been in operation since July 1, 1969. It is well supported by the staff and is generally accepted. We are beginning to use it more effectively in management.

This plan of work is an aggregation of the plans of work prepared by each planning unit or discipline. The staff was asked to plan 70 percent of their time for fiscal year 1971-1972. The totals returned and included in this report are 72 percent of staff time.

Data from the machine printout was summarized by Extension elements and is shown in the following table for fiscal year 1972:

Line Item No.	Extension Element	Planned Man-Months	Percent of Total
01	Improving Farm Income	228.4	19.3
03	Marketing	15.0	1.3
06	International Programs	0.0	0.0
09	Food and Nutrition	50.0	4.2
10	Safety and Emergency Preparedness	4.7	0.4
11	4-H Youth Development	229.8	19.4
12	Improved Family Living	113.8	9.6
16	Community Development	220.2	18.6
18	Forestry Production and Marketing	1.7	0.1
51	Soil and Water Conservation	18.8	1.6
52	Recreation, Wildlife and Natural Beauty	17.4	1.5
53	Resource Protection	11.3	1.0
90	Program Leadership and Administration Support	272.1	23.0
	Total Planned Months	**1,183.2**	**100.0**

Only 72 percent of the time was planned in order to allow for desirable adjustments in time spent on certain line items because of unforeseen developments which could occur during the year. Approximately 10 percent of the figure shown for program leadership and administration support is charged to staff leave time, and 6 percent to staff training.

The annual report for fiscal year 1974 showed a distribution of time for the various projects similar to that for 1972, with the exception of much less time planned for community development, and proportionally more for family living.

Leadership Education

One of the most encouraging developments in the entire history of the Montana Cooperative Extension Service began July 1, 1971, when the Kellogg Foundation of Battle Creek, Mich., made a grant of approximately $600,000 to the Montana Extension Service. The grant supported a five-year program of leadership training for Montana citizens. It took the name of the Kellogg Extension Education Program, familiarly known by its participants as KEEP.[13] This program revolved around the training of 600 Montana men and women with the intent that they should assume leadership roles in their generation. The proposed outline of formal subject matter included local, national and international problems, in addition to leadership techniques. Intensive training sessions were held for a three-week period each year for three years, for 90 carefully chosen persons in three groups, 30 persons in each group. To support this primary group of 90 in their communities and sections of the state, an additional 510 persons were given, in groups of 40, broadly based, intensive training in leadership and communications for one week each year for the three-year period.

In addition, the 90 principal trainees participated in a study-travel seminar each year. The first year the tour was restricted to Montana; the second year important points in the United States were visited; and in the third year a foreign nation of special interest was visited by a selected number of the trainees. The group which visited the People's Republic of China in 1978 was one of the first to be admitted under the new treaty arrangement with the United States.

The Kellogg Foundation provided additional funds which eventually brought the total grant to approximately one million dollars. More than 1,000 Montanans were involved in the program. M. E. (Gene) Quenemoen directed the entire program and remained on the Extension staff as economist. An academic policy board was named to advise, and this RAD committee served in a consultant capacity and provided a check on reaction over the state.

Each year a two-day seminar was held between as many of the trainees as possible together with 50 active leaders of the state in positions of responsiblity—"the power structure." These meetings provided opportunities for mutual acquaintance and exchange of information and ideas. Informal relations with county agents and state staff furnished additional contacts for the trainees, and kept the entire Extension Service personnel aware of the program. The trainees were encouraged to work in their home counties with Extension personnel and community leaders on problems of importance to test and develop their leadership skills.

Following the close of the leadership program, its members fulfilled the expectations of the Kellogg Foundation and the Extension Service by accepting responsibility for community and state activities. Finding that they could accomplish more by keeping in contact, a Montana Leadership Development Association was formed in August 1979. This has supported an executive director, sponsored a quarterly publication, held an annual four-day "Leadership Week," and promoted a number of forums and workshops each year. Individual members have been elected to the Legislature, to county boards, and have carried many other leadership positions. Governor Schwinden praised the association and its members in

January 1982 as "a new breed of talented and articulate leaders who will guide the state on its future course."[14]

A People and a Spirit

M. L. Wilson remarked, in the comparatively quiet year of 1929, "We should do all we can to make certain that we are facing the future and not the past." He participated fully in the dedicated effort to meet the problems of the following years of depression, drought, war, and unequalled change sketched in this book, which left little time to look to the past.

During the past decade or two, additional changes have been taking place in the nation, in which Montana has given progressive leadership. These have been slow, but revolutionary shifts giving emphasis to people in their social relationships—what the late University president Leon H. Johnson, in speaking to the Extension staff, described as "the truly human problem of man, involving his community, his home and his family."

The Extension Service has continued to work on the multitude of programs which for many years have been closely related to this emphasis upon people, such as health, food, clothing, home planning, community relations and recreation, as well as on the programs of a more economic nature such as increased production, more profitable marketing and better management practices.

Another aspect of this shift of emphasis has been to expand its clientele from rural and small town residents exclusively, to the people in the cities, people wherever they are.

Accompanying this broader concern has been a replacement of the long-time concept of an "Agricultural Extension Service," centered in the college agricultural divisions and little known by other departments of the institution, to a "University Extension" effort involving all of the resources of the University staff and facilities.

This broader concern has been caused in part by an awareness that in the midst of the steady acceleration of productivity and increased affluence by the majority of the people in both rural and urban America, there has been another and enlarging segment of "The People Left Behind"—a large number of people who need assistance in finding new directions.

A study by representatives from the Extension Service and the National Association of Land-Grant Colleges and Universities was released in 1968 under the title "A People and a Spirit." It raised questions of priority, and urged enlargement of programs which Montana already had under way such as working with disadvantaged youth in the cities, retraining and resettlement projects, and working with minority groups.

With these additional areas of need in mind, the Montana Extension Service, in 1972, reviewed its trends and issued a statement of "Goals and Related Activities," one of the first of its kind in a national reassessment. This declaration keeps in mind the original Smith-Lever charter statement,

Cover pages of some very significant publications which appeared during Director Aasheim's tenure as Extension director.

which has been frequently mentioned, that the purpose of the Extension Service was "better farming, better living, more happiness, more education, and better citizenship" for the "entire country." The Goals summary recognizes that "a significant portion of Montana's citizens still perceive the Extension Service to be an agricultural advisory service," and it affirms that Extension proposes to continue to fulfill this function. However, the report continues, now that the urgency for food and fiber and manpower has eased, the requests of the people reveal that they now foresee "a substantially broader purpose." Some suggested "new goals to fit the times," including, among others, the following:

1. Continuing to develop better communities in order to assure more productive and enjoyable lives for the people.
2. To help in providing a sense of purpose to many segments of society, particularly with youth and the aged, and removal of many of the behavior tragedies which beset the nation.
3. To enlarge activities which will improve the nation's ecology as related to its air, water and land pollution, and minimize the loss of its forests and wildlife.
4. To assist in family and community generated recreational facilities.
5. To enlarge "life-long learning" opportunities from vocational train-

ing for youth to retraining for displaced adults, and for broadening technological and cultural interests in all ages.

6. Strengthening the forces which provide for constructive consideration and preservation of national and world cooperation.

As an educational arm of the government, first to rural America, and now increasingly to urban America as well, the Cooperative Extension Service, now rapidly becoming the University Extension Service, is unique. It represents the world's first and greatest attempt at mass adult education. The astonishing success of the movement has been one of its chief difficulties, since its immediate objectives have been so frequently achieved as to make its long-time goals a changing target. Its very success has raised a frequent question in recent decades as to its continued usefulness.

Certainly the Extension Service of the United States and of Montana has made a contribution of the first magnitude in increasing agricultural production and in improving the quality of living in rural America. It has made another contribution which, in the ongoing unfolding of a government by the people, may be judged of equal or even of greater value. This is the Extension "method"; that of consulting with the people and identifying leadership, and then leaders and people together transforming the life of the community, all in the best democratic tradition.

"There Is Much Yet To Be Done."

Dr. Roland R. Renne, who appointed Aasheim director of the Montana Extension Service in 1958, was president of Montana State University from 1943 until 1964. Renne was a man of great social conscience and he supported Aasheim's efforts to expand the scope of the Extension programs. In 1974 Dr. Renne was selected to give the M. L. Wilson Distinguished Lecture, which he titled "The Land-Grant Universities in Today's World: An Appraisal." In this lecture he devoted considerable time to the Extension Service. His comments, pointed toward future needs, were particularly timely and excerpts are included here: "A review of changes that have occurred since the Extension Service was established will point up some of the current conditions in which the land-grant institutions are now operating:

1. We have changed from an agrarian to an urbanized society.
2. Land-grant universities, in which agriculture was originally dominant, are now complex institutions. Now, agriculture is a considerably smaller part of the total academic scene.
3. Goals of society have gradually changed, including a shift from preoccupation with the materialistic 'standard of living' to a concept of 'quality of life.'
4. As agriculture has progressed, there has been a specializa-

tion and fragmentation:

A. In the structures of colleges and departments.
B. In the fabric of research.
C. In the business of farming.
D. In the maze of farm organizations and institutions that serve agriculture.

"Under these conditions, colleges of agriculture have narrowed their scope of concerns to nearly an exclusive preoccupation with agricultural production and closely related activities. There is less concern for problems of the family, of health care services, of social institutions and programs, and of education. This is, indeed, a major departure from the approach M. L. Wilson emphasized in his efforts to help farm families survive. M. L.'s approach was much more relevant to the solution of the major problems encountered by Montana's rural families.

"The serious aspect of all of this is that while colleges of agriculture have dropped these issues from their agendas, the universities of which they are a part have not taken up these responsibilities. For example, most colleges of education do not give major attention to rural schooling and educational opportunities, just as most colleges of medicine do not concern themselves in a comprehensive way with rural health care services.

"Who will fill the void and take up responsibility for providing needed important services to rural people if the colleges of agriculture and the land-grant universities do not? Dr. Russell G. Mawby, president of the Kellogg Foundation, has drawn up what he calls three 'dangerous generalizations' based on his host of impressions, contacts and experiences gathered from through the country. These are:

1. 'Most colleges of agriculture are not really concerned with rural community development.' Mawby bases this conclusion on the fact that he has found little evidence that either colleges of agriculture or the universities of which they are a part, have given a sense of priority and urgency to rural problems.

2. 'Most colleges of agriculture are not equipped to deal adequately with the broad range of issues including health care delivery, education, business and industry, political institutions, social services. While faculty members with broad responsibilities and training in community development can provide leadership, the knowledge resources of the larger university must be brought to bear.'

3. 'The inadequacy of resources within colleges of agriculture is compounded by the organizational rigidities of most universities. The typical extension service, which may have a community development component, simply does not have available to it the intellectual and technical resources of the university at large. This circumstance is further compounded by an 'insular mentality' of

many colleges of agriculture which seem to set them aside from the mainstream of academic life.'

"One thing is certain, that the involvement of more of the whole university, and not just the college of agriculture, will be essential for maximum success of major rural community improvement programs. A wider range of disciplines than those available in the college of agriculture, covering the political, social and economic sciences is needed. The tripartite program of teaching, research and public services must involve more than just the college of agriculture.

"Colleges of agriculture have increasingly concentrated on the science and technology of agricultural production, but they need to become more concerned with the problems of rural people and rural communities because the viability of the agricultural sector in our economy is only one component of rural life. Of course, it is not expected that it is the role of the colleges of agriculture alone to serve these needs fully, but their leadership may be essential, particularly if there is no other focal point within the university for the rural voice to be expressed.

"Concerning the future, there is evidence that leaders of the land-grant universities are aware of the existence and significance of current problems and needs of rural communities in their states. But, the approach and organizational structure that will be most effective in solving the rural problems is not clear. The administrative structure for providing instruction, research and extension services by the Universities varies widely among the 52 institutions. It is apparent, however, that there have been major changes recently in the administrative structure of many of the institutions, in efforts to achieve maximum performance and efficiency in carrying out their programs, many of which have been recently established and cover a wide range of subjects.

"Nearly half (25) of the 52 institutions now have a vice president for research who has university-wide responsibility for the general administration of research. Of these 25, 14 are located among the 22 separated state universities, and the other 11 are in the single university states. Thus, nearly half of all of the universities have centralized general supervision for all research in the institution in one office: Montana State is one of these.

"Twenty-one of the 52 universities have a vice president for administration, and 19 have a vice president for academic affairs. Montana State is included in each of these.

"Ten of the 52 have a vice president for Extension, and Montana State is not one of these. Thus, Extension coordination at the university level has lagged behind such coordination for research and instruction. Three of these vice presidents for Extension are in western states: Oklahoma, Utah and Wyoming. (Montana adopted this plan in 1975.)

"In conclusion, as we look ahead to the end of this century, we can only speculate on what new programs and services will be added to the land-grant university 'mix'; but we can predict with considerable confidence

Dr. Carl J. Hoffman, vice president of Extension and Continuing Education, 1975—. Dr. Hoffman came to Montana after 20 years of Extension work in Colorado. His bachelor's and master's degrees are from Colorado State University and his Ph.D. from the University of Wisconsin. In addition to directing a vigorous program in Extension, he has received national recognition for his work in Great Plains development, dairy production and Extension education.

that, as in recent years, there will be many changes in subject matter, courses and departments, in research programs, both basic and applied, and in Extension services and methods. And, no doubt, significant changes in organizational and administrative structures will be required. There is much yet to be done, and it is hoped that our land-grant universities will make the best of these challenging opportunities in the years ahead.

"The central idea of the land-grant institutions since their creation has been to stay close to the people and their needs, and at the same time to work with society for its continued improvement. If this concept prevails, we need not worry about the American people's willingness to meet the needs of a changing world in the coming years."

When Director Aasheim tendered his resignation, effective July 1, 1974, Dean J. A. Asleson sent a letter to all Extension personnel asking for suggestions for the future of the Service. The director replied with a strong recommendation that the new director have the title and the responsibility of vice president of the University for Extension.

A committee appointed by President Carl McIntosh to consider this change, recommended that it be made. Therefore, when Dr. Carl Hoffman succeeded Director Torlief Aasheim on February 1, 1975, his position was "Vice President for Extension."

REFERENCES

1. Resolutions at annual American Farm Bureau Federation Convention, December 11, 1958.
2. *The Cooperative Extension Service Today—A Statement of Scope and Responsibility.* Federal Extension Service, April 1958. "Objectives and Responsibilities and Scope of the Montana Extension Service." Montana Extension Service Circular, August 1960.
3. Goals and Related Activities of the Montana Cooperative Extension Service, October 1, 1973. 14 pp.
4. These summaries come largely from the 1968 *Annual Report* of the Montana Extension Service; from the Plan of Work, 1965-1970; and from *EMIS Narrative Progress Report, July 1, 1969-June 30, 1970.*
5. *EMIS* (Extension Management Information System), *op. cit.*
6. Saunderson's paper is in the Montana State University Archives.
7. Poston, R.E., *Small Town Renaissance, A Story of the Montana Study.* Harpers, 1950; James O. DeBree, "Community Development Efforts in Montana," Montana Extension Service Mimeographic Circular, 30 pp. August 1969, contains a good summary of the Montana Study in context.
8. Roosevelt County Council Report. Montana State University Archives.
9. DeBree, James O., *op. cit.;* William R. Lassey, Anne Williams, Howard Huffman, "Multi-County Areas as a Tool for Development in Montana," *Research Report #2,* Department of Sociology Center for Planning and Development, Agricultural Experiment Station, December 1970. pp. 42-54; Williams, Ellis E. "Leaders Views About Community Development in South Central Montana," Montana Extension Service Circular, August 1970; Interview with Doyle Stocks, July 27, 1971.
10. DeBree, James O., *op. cit.;* Lassey, et al., *op. cit.;* Kraenzel, Carl F. and MacDonald, Frances H., "Cost of Space Facts for Health Services in Eastern Montana," and "Health Needs as Reported by Residents of Fifteen Eastern Montana Counties," Montana State University Circulars, Oct. 24, 1966, prepared for the State Board of Institutions.

    The health study started in the early 1960s was directed by sociologist Dr. Carl F. Kraenzel, and former health specialist Frances H. MacDonald, both of Montana State University. This work has been continued by various agencies.

11. Minutes of the RAD Executive Committee, in the files of the chairman, give a running account of the discussion and action of the quarterly and annual meetings. Also see, Torlief S. Aasheim, "Organization of Rural Development Efforts in Montana," address given before a USDA Committee for Rural Development Workshop in Las Vegas, May 11-12, 1971. 6 pp.; and DeBree *op. cit.*, pp. 14-19.

12. Bowler, Duane W., "Operation Bootstrap," The *Billings Gazette*, October 21, 1973.

13. "A Proposal to the Kellogg Foundation for an Educational Program in Leadership Development for Leaders and Potential Leaders in Montana," Montana Extension Service, March 1971; "Five-year Grant Totals $600,000 for Rural Leadership Education," *Bozeman Daily Chronicle*, July 13, 1971; Reports in office of Extension Economist; *The Montana Leader*, the newsletter of the Montana Leadership Development Association, quarterly, since 1979, Bozeman.

14. The Newsletter of the Montana Leadership Development Association, January 1982.

APPENDIX A
Montana Extension Service Personnel— State Staff—1914-1982

DIRECTORS

Fred S. Cooley* . . . . . 7/1/14—7/1/24
Alfred Atkinson** . . . . . 9/1/24—6/30/25
John C. Taylor . . . . . 7/1/25—12/31/45
Robert B. Tootell . . . . . 1/1/46—6/30/53
Newell E. Beers . . . . . 11/1/53—8/31/58
Torlief S. Aasheim . . . . . 9/1/58—7/1/74
Robert Bucher (acting) . . . . . 7/1/74—2/1/75
Carl Hoffman . . . . . 2/1/75—

* Cooley was superintendent of Farmers' Institutes as well as director of Extension until 6/30/23.

** Atkinson was officially director of Extension, but Taylor was actually in charge as acting assistant director from 7/1/24 to 6/30/25.

ASSISTANT DIRECTORS

John C. Taylor . . . . . 7/1/24—6/30/25
Newell E. Beers . . . . . 8/1/46—10/31/53
Burl Winchester . . . . . 10/14/53—7/1/61
Danny Cheatham . . . . . 8/11/82—

ADMINISTRATIVE ASSISTANTS

Thomas Booth Holker . . . . . 9/1/46—10/1/49
Helen McCarthy
  Secretary to Director . . . . . 10/1/46—12/31/55
  Extension Finance officer . . . . . 1/1/56—6/30/78
Clyde Carrol . . . . . 7/26/78—

MANAGEMENT INFORMATION OFFICER

Robert Bucker . . . . . 1/1/74—7/1/74

PERSONNEL AND TRAINING LEADERS

Harry A. Cosgriffe . . . . . 9/1/56—11/1/60
J. Glenn Jones . . . . . 7/1/61—8/31/63
John C. Bower . . . . . 7/1/67—12/30/73
Allen Nelson . . . . . 1/1/74—

PUBLICITY, PUBLICATIONS & INFORMATON

Extension Editors, Directors of Publications, and Program Leaders

Ray B. Bowden, Extension Editor . . . . . 1925—1930
John Dexter, Director of Publications . . . . . 5/11/30—6/30/35
  (Also assistant editor from about 1927?)
Louis G. True, Director of Publications . . . . . 4/1/35—6/30/64
  (Acting editor—4/1/30—4/1/35)
Burl Winchester, Program Leader,
  Information . . . . . 7/1/63—11/1/65
Sam Rosen
  Assistant Editor . . . . . 9/1/64—12/1/65
  Program Leader, Information . . . . . 12/1/65—7/1/67
  Extension Editor . . . . . 7/1/67—12/30/82

Assistant Editors and Specialists

Paul T. DeVore, Assistant Editor . . . . . 10/20/34—9/15/35
Herbert M. White, Assistant Editor . . . . . 9/1/35—3/20/65
Mary Louise Dunne, Assistant in Publications . . . . . 7/1/48—6/30/51
F. William Howell, Information Specialist . . . . . 2/15/64—3/24/66
E. Raymond Ozmund, Information Specialist . . . . . 4/18/66—1/13/67
William R. Beasley, Information Specialist . . . . . 4/15/67—
Marilyn Wessel, Assistant Information Specialist . . . . . 1/1/73—7/1/74

PROGRAM LEADERS

Donald J. Luebbe, Agricultural Production and Conservation . . . . . 7/1/60—6/30/67
John Bower, Public Affairs (Also Extension Economist) . . . . . 7/1/61—6/30/67
Louise J. Shunk, Acting Program Leader for Home Economics . . . . . 4/1/61—12/31/63
Vivienne B. Kintz, Home Economics . . . . . 1/1/64—6/30/67
Burl Winchester
Human Resources . . . . . 7/1/61—6/30/63
Information . . . . . 7/1/63—1/31/65
Paul J. Moore, Human Resources . . . . . 7/1/63—9/17/66
E. Dean Vaughan, Marketing and Utilization (Also Extension Marketing Specialist) . . . . . 7/1/61—1/17/63
Charles H. Rust (followed Vaughan) . . . . . 1/1/63—6/30/67

PROGRAM COORDINATORS:

Vivienne B. Kintz, Human Resource Development . . . . . 7/1/67—6/12/76
Donald J. Luebbe, Agricultural and Economic Development . . . . . 7/1/67—7/31/70
Charles Rust, Agricultural and Economic Development . . . . . 8/23/70—
Marian Moline, Human Resource Development . . . . . 6/12/76—7/1/77
Martha Johnson, Human Resource Development . . . . . 7/1/70—7/1/80
Andrea Pagenkopf, Human Resource Development . . . . . 7/1/80—7/1/81
Ruth Conone, Human Resource Development . . . . . 7/1/81—

COUNTY AGENT LEADERS AND SUPERVISORS

County Agent Leaders

M. L. Wilson . . . . . 9/1/14—12/31/21
J. C. Taylor . . . . . 4/1/21—6/30/24
E. H. Lott . . . . . 7/1/25—8/31/25
Fred Bennion . . . . . 8/1/26—6/30/30
R. E. Bodley . . . . . 4/16/34—6/30/43
T. Booth Holker (Acting) . . . . . 4/1/44—8/31/46

Assistant County Agent Leaders

J. C. Taylor . . . . . 1/1/17—3/31/21
E. H. Lott . . . . . 7/1/22—6/30/25
T. Booth Holker . . . . . 5/1/39—3/31/44
George W. Gustafson . . . . . 1/1/46—4/30/47

County Agent Supervisors

George W. Gustafson . . . . . 5/1/47—6/30/53

William C. Stucky, Junior County Agent
Supervisor . . . . . 7/1/48—6/30/55
Harry A. Cosgriffe . . . . . 10/1/51—10/30/56
J. Glenn Jones . . . . . 7/1/54—6/30/59
Robert F. Rasmussen . . . . . 9/16/55—6/30/72
(District Supervisor since 7/15/65)
Donald J. Luebbe . . . . . 4/1/59—6/30/60
(Program Leader, 7/1/60—6/30/67
and Coordinator from 7/1/67—see above)
Allen P. Nelson . . . . . 9/1/66—12/31/73
Arthur Jacobs . . . . . 7/15/65—6/30/74
Rex Campbell . . . . . 7/1/58—9/1/66
Lloyd Pickett (Indians) . . . . . 1/15/70—8/1/75
Doyle Stocks . . . . . 8/15/70—
Thadeus Wojcichowski . . . . . 1/15/70—6/30/78
Dale Bergland . . . . . 7/1/72—
Dana Schrupp . . . . . 7/1/74—6/30/78
Jim Johannes . . . . . 8/1/79—
Jim DeBree . . . . . 8/1/78—7/1/81
William Peterson . . . . . 11/13/82—

HOME DEMONSTRATION LEADERS AND SUPERVISORS

State Home Demonstration Leaders

Katherine Jensen . . . . . 7/1/14—8/1/16
Bess Rowe . . . . . 9/15/16—10/15/20
Anna N. Turley . . . . . 1/1/21—9/1/23
Blanche L. Lee . . . . . 7/1/23—6/1/36
Miriam Hawkins (Acting) . . . . . 8/15/28—3/13/29
S. Frances Smith . . . . . 9/1/36—9/30/46
Doris Ingram Anderson (Acting) . . . . . 7/1/37—2/28/38
Bessie E. McClelland . . . . . 10/1/46—3/31/61
(Program Leaders and Coordinators from 4/1/61, see above)

Assistant State Leaders

Mignon Quaw Lott . . . . . 4/1/18—9/30/21
Lucile Reynolds . . . . . 10/1/19—6/20/20
Blanche L. Lee . . . . . 7/1/20—6/30/23
Bessie E. McClelland . . . . . 7/16/46—9/30/46
Marjorie Downs . . . . . 10/1/46—5/31/50
Martha Dunlap . . . . . 2/15/51—1/31/60

Supervisors and District Agents

Louise Shunk, Supervisor . . . . . 7/1/56—3/31/61
Acting Program Leader . . . . . 4/1/61—12/31/63
District Agent . . . . . 1/1/64—6/11/65
Mae Farris True
Home Economics Supervisor . . . . . 4/1/60—6/30/60
District Supervisor . . . . . 7/1/60—6/30/65
Extension Home Economist . . . . . 7/1/65—6/30/67
Vivienne B. Kintz
District Supervisor . . . . . 1/1/61—6/30/61
District Extension Agent . . . . . 7/1/61—12/31/63
Dianne M. Undem
Extension Home Economist . . . . . 7/1/65—6/30/67
Carol Tarnavsky
Area Home Economist . . . . . 7/1/67—7/1/76

Hazel U. Hinman
  Area Home Economist . . . . 9/1/67—6/30/69

Dorothy Gallup . . . . 7/1/68—7/1/76
Marian Moline . . . . 8/1/69—6/12/76
Jane Brunsvold . . . . 6/9/69—6/30/70
Allene Pierce . . . . 8/8/70—4/1/76

BOYS' AND GIRLS' CLUB LEADERS

State Leaders

Augusta D. Evans . . . . 7/1/14—5/31/16
Martin J. Abbey . . . . 9/1/16—8/31/19
Charles E. Potter . . . . 1/1/20—3/31/35
R. E. Cameron . . . . 4/1/35—9/30/46
T. W. Thompson . . . . 10/1/46—6/30/49
Paul J. Moore . . . . 12/1/49—9/17/66
Clifford W. Vaughan
  Acting State Leader . . . . 7/1/61—6/30/62
  State 4-H and other Youth Specialist . . . . 7/1/62—7/1/74
Vivienne B. Kintz
  Coordinator for Human Resource
  Development from 7/1/67—(See above)

Assistant State Leaders

Mina Ogilvie . . . . 3/1/17—7/31/20
Charles E. Potter . . . . 4/1/17—12/31/19
E. E. Hupp, "Emergency" Assistant Leader . . . . 4/1/18—8/31/19
Emily Linhoff . . . . 8/1/20—8/31/23
Doris Ingram Anderson . . . . 7/1/23—10/15/26
  and 5/1/28—9/15/28
Vangel M. Russell . . . . 10/15/26—4/30/28
  and 9/1/28—4/30/29
Myrtle G. Euren . . . . 7/15/29—6/30/34
Pauline Bunting . . . . 7/10/34—12/31/45
Geraldine Fenn . . . . 7/20/46—3/31/67
T. W. Thompson . . . . 4/15/46—9/30/46
Dana O. Schrupp . . . . 10/1/48—6/30/49
Janice Fenske (Peace) . . . . 8/1/50—6/30/51
Margaret F. Kohl . . . . 1/18/46—6/30/52

Associate State Leaders

Margaret F. Kohl . . . . 7/1/52—6/30/53
Esther R. Rugland . . . . 7/1/53—12/7/57
Clifford W. Vaughn . . . . 3/16/56—6/30/61
Margaret A. Kuhl . . . . 4/14/58—1/18/61
Mary Jane Stangl . . . . 2/1/61—6/30/61
  Home Economics 4-H Specialist . . . . 7/1/61—6/30/64
James Sargent . . . . 7/1/67—
Harold Strobel . . . . 7/1/73—
Andrew Hubble . . . . 9/1/73—11/30/74
Harold Gaustad . . . . 6/18/73—6/30/80
Betty McCoy . . . . 4/1/78—

EXTENSION SCHOOLS AND MEETINGS:

J. R. Campbell, Assistant Superintendent of Farmers' Institutes and Leader of Extension Schools . . . . 9/1/20—8/31/23
Samuel J. Hampton, Lecturer . . . . 9/1/20—9/1/22
Mignon Quaw Lott
Recreation Specialist . . . . 9/1/20—9/30/21
Rural Specialist . . . . 10/1/21—8/31/22

(F. S. Cooley served as Superintendent of Farmers' Institutes from 1907 until they were terminated in 1923.)

PUBLIC AFFAIRS, PROGRAM CONSULTANTS

Ernest R. Ahrendes
Roosevelt County . . . . 7/1/59—6/30/62
State . . . . 7/1/62—9/30/64
J. Glenn Jones . . . . 7/1/59—6/30/61
John Bower, Program Leader . . . . 7/1/61—6/30/67
James DeBree . . . . 7/1/67—7/1/70

COMMUNITY AND RURAL AREAS DEVELOPMENT

Agents-at-Large and Associate Agents

Theodore W. Witzel, Lake County . . . . 2/1/58—8/8/60
Sidney J. Tietema, Ravalli County . . . . 2/1/58—10/19/59
and Lake County . . . . 10/20/59—8/31/62
Robert J. Hanson, Ravalli County . . . . 11/1/59—6/30/61

Program Consultants

Carlton A. Infanger, Missoula Area . . . . 12/16/59—6/30/61
Ernest R. Ahrendes, State . . . . 8/20/65—6/30/66
Donald L. Hunter, Community and Resource Development Specialist, Billings Area . . . . 7/1/66—5/17/68
Rex F. Campbell, Community Development Consultant . . . . 9/1/66—12/31/73
James O. DeBree, Municipal and Community Assistant . . . . 7/1/67—11/1/70
Ellis E. Williams, Billings Area, replaced Hunter . . . . 10/1/68—8/30/69
Darryld Kautzmann . . . . 9/1/69—6/16/70

Extension Specialists

AGRICULTURAL ECONOMICS

Farm Management "Demonstrators" and Extension Economists

H. B. Cannon . . . . 7/1/14—9/30/15
F. Josiah Chase . . . . 10/15/15—8/31/17
Arthur J. Copeland . . . . 10/1/18—6/20/21
M. L. Wilson . . . . 1/1/22—6/30/24
Virgil D. Gilman . . . . 8/1/24—9/30/27 and 4/1/29—6/30/32
Ben F. Alvord . . . . 11/1/27—12/31/28
Paul Carpenter . . . . 9/10/28—8/31/34
Carl H. Peterson . . . . 7/1/32—6/30/33
Austin Goth . . . . 8/1/34—6/30/38

Clyde Howard . . . . 10/1/35—6/30/38
Willard Lamphere . . . . 10/16/35—3/4/39
John C. Bower . . . . 7/1/38—6/21/41
and 5/1/55—7/1/67
H. Ralph Stucky . . . . 1/1/46—11/10/54
Merle E. Quenemoen . . . . 7/1/55—9/30/65
and 8/1/68—7/1/72
Warren L. Trock . . . . 7/1/57—8/31/62
Sidney J. Tietema . . . . 9/1/62—4/7/67
LeRoy D. Luft . . . . 7/1/66—7/1/81
Robert F. Bucher . . . . 8/15/66—12/31/73
Gerald Schaeffer . . . . 1/1/76—5/25/77
Brad Garnick . . . . 8/15/77—11/30/80
Duane Griffith . . . . 9/1/77—
Keith Suta . . . . 7/1/78—2/29/80
Ernest Ahrendes 1/16/46-10/15/46
John Wicks . . . . 7/1/70—9/30/73

Land Economists

Robert B. Tootell . . . . 7/1/31—3/31/34
Horace G. Bolster . . . . 9/1/34—10/16/45
Carl A. Peterson, Assistant . . . . 1/16/39—2/31/41
Ernest R. Ahrendes . . . . 1/16/46—3/31/47
Verne House . . . . 7/1/71—

Marketing Specialists

Rev. R. W. Beers . . . . 12/10/16—10/30/20
J. M. Brandon, Assistant . . . . 7/1/20—9/1/20
John F. Ware . . . . 10/1/20—10/31/21
C. C. Conser . . . . 7/1/30—11/1/33
Peter L. Slagsvold, ... 8/1/33—12/31/33
also Extension Economist . . . . 5/1/34—2/28/39
and 10/1/38—2/28/39
E. W. Bird, Dairy Manufacturing . . . . 7/13/36—9/13/36
Bruce L. Brooks . . . . 5/22/50—7/31/56
E. Dean Vaughan . . . . 1/1/57—1/17/63
Owen S. Wirak . . . . 1/1/61—6/30/61
Ramon W. Sammons . . . . 1/1/64—7/15/66
Charles H. Rust . . . . 1/1/63—8/31/69
James H. Cothern . . . . 12/18/67—6/30/72
Roger Brownson . . . . 10/1/69—7/1/71
Joe Kropf . . . . 11/8/72—9/17/74
James Cornelius . . . . 8/16/76—7/1/78
Henry Bahn . . . . 6/8/81—

Agricultural Engineering

Armin J. Hill, Rural Electrification . . . . 7/1/38—8/31/41
Eric B. Wilson . . . . 5/1/46—6/30/61
Weldon R. Seaton, Assistant . . . . 2/15/50—2/28/50
Gerald Westeson . . . . 7/1/76—
Roy Linn . . . . 7/1/74—
Larry King . . . . 10/15/81—

Extension Agronomists

F. E. Fuller . . . . 5/18/17—2/29/20
A. J. Ogaard . . . . 3/1/20—8/31/28
Virgil D. Gilman, Assistant . . . . 6/1/22—7/31/24
F. E. Price, Assistant . . . . 6/27/22—10/27/22

Laurence D. Kurtz .... 4/1/28—9/30/29
Waldo Kidder .... 9/10/30—8/31/31
Sam L. Sloan .... 8/1/31—11/30/35
Ralph D. Mercer .... 3/1/36—12/30/60
Arthur F. Shaw .... 7/1/52—7/1/80
Manford J. Jackson .... 7/1/64—
(Agricultural Chemicals - half time)
Howard Bowman .... 10/1/70—
Grant Jackson .... 3/1/77—5/31/79
Don Baldridge .... 1/8/81—

Dairy Specialists

F. M. Hillman .... 10/1/14—3/31/17
W. E. Thomson .... 11/1/17—8/31/20
J. Oscar Tretsven .... 8/15/22—2/28/55
Arthur O. Jacobs, Jr. .... 4/1/56—6/30/59
Earl J. Peace .... 7/1/59—7/1/69
(Dairy and Swine Specialist)
Buelon (Pete) Moss .... 7/1/70—7/1/72
John Bryant .... 10/1/77—

Horticulture Specialists

Edward E. Isaacs .... 7/1/26—6/30/53
Homer N. Metcalf .... 3/1/47—6/30/79
Leonard A. Yeager, Assistant .... 5/16/49—6/30/52
Horticulturist .... 7/1/52—6/30/57
Orville W. McCarver .... 9/16/57—
Robert A. Wearne, Area Horticulturist .... 11/11/57—6/30/61

Irrigation Specialists

Blaine Ferguson .... 7/1/21—11/30/22
G. H. Bingham .... 12/1/35—2/23/46
Donald J. Luebbe, (Irrigation and Weed
Specialist) .... 7/1/47—10/31/47
Harold L. Dusenberry .... 7/1/48—4/1/52
and 4/12/54—9/14/56
Stanley W. Howard .... 5/1/52—5/1/54
and 12/1/56—6/30/61

Livestock Specialists

W. J. Hartman .... 7/1/14—5/31/18
E. H. Riley, Field Agent .... 9/1/18—8/31/20
G. L. Cleland .... 10/1/20—10/14/20
R. B. Millin .... 1/24/21—8/31/28
R. L. Waddell .... 9/1/24—8/31/28
D. E. Richards .... 9/20/28—2/29/32
I. M. C. Anderson .... 7/1/32—6/30/38
Howard Lewis .... 10/6/37—1/31/40
E. Paul Orcutt .... 2/1/40—10/15/44
and 1/1/48—6/30/61
F. A. Ralston .... 6/1/45—6/30/51
Harold L. Dusenberry .... 5/1/47—12/31/47
N. A. Jacobsen .... 8/1/51—5/31/71
Burl Winchester, Animal Geneticist .... 11/1/65—6/30/78
Roger Brownson .... 7/1/71—

Conservation Specialists

Martin P. Hansmeier .... 1/1/37—2/15/50
and 2/15/52—6/14/52

Burnell W. Brink, Acting . . . . 4/1/50—1/15/52
Rex F. Campbell . . . . 9/1/52—6/30/58
Bernard L. Brown . . . . 4/14/58—6/30/60
Charles M. Smith . . . . 2/6/61—12/5/75
Roger Wilson . . . . 2/16/70—11/24/79
Neil Christensen . . . . 9/15/71—10/31/78
Paul Kresge . . . . 7/12/76—
James Morgan . . . . 12/1/74—4/30/78
Ray Garlak . . . . 7/25/78—
Bill Schafer . . . . 7/10/80—
James Bawden . . . . 10/1/80—

Weed Specialists

Donald J. Luebbe* . . . . 6/1/46—10/31/47
Harold L. Dusenberry* . . . . 1/1/48—6/30/48
Robert L. Warden . . . . 7/1/48—6/30/51
Parker E. Heikes . . . . 11/1/55—9/10/61
Leslie W. Sonders . . . . 12/1/61—6/30/75
Mike Jackson . . . . 9/1/70—
*Combined with Irrigation Extension work 7/1/47—6/30/48

Rural Civil Defense Specialists

Richard W. Stedfield . . . . 3/1/63—12/20/65
Dorothy E. Gallup . . . . 7/1/64—6/30/68
Clifford C. Rylander* . . . . 4/1/66—7/1/70
Roy Linn* . . . . 8/1/66—7/1/74
Carol Tarnavsky* . . . . 8/1/66—6/30/67
Harold Strobel* . . . . 5/1/67—9/30/68
*Rural Civil Defense and Adult Education Specialists

Poultry Specialists

Ralph L. Smith . . . . 3/1/19—12/31/21
E. E. Weaver . . . . 5/1/22—7/31/22
Harriette E. Cushman . . . . 7/22/22—12/31/54
Edmund Guenther-half time . . . . 10/1/55—9/30/56
Glen Hart . . . . 11/1/56—12/31/58

Range Management Specialists

Ray G. Johnson . . . . 1/1/50—11/16/51
Karl G. Parker . . . . 1/23/52—3/20/61
Donald E. Ryerson . . . . 12/1/61—5/31/67
Ned W. Jefferies, Range Specialist . . . . 2/1/68—7/10/70
Carl Wambolt . . . . 10/1/61—6/30/80
John Lacey . . . . 8/1/81—

Wildlife Specialist

Eldon Smith . . . . 2/23/55—6/30/77

Sheep and Wool Specialists

James Drummond . . . . 11/1/61—7/1/72
Rodney Kott . . . . 2/25/80—

Swine Specialists

Earl Peace . . . . 7/1/69—3/30/79
Wayne Gipp . . . . 6/1/80—

Entomologists

Gary Jensen . . . . 9/15/75—
William McCollum . . . . 7/1/79—
Robert Madsen . . . . 5/1/79—4/30/81
Scott Nissen . . . . 7/15/80—

Plant Pathologists

M. M. Afanasiev .... 7/1/55—6/30/68
Lorry Clafin .... 5/1/73—12/31/74
Ed Burns .... 7/1/73—6/15/78
Dallas Batchelor .... 3/24/75—7/1/77
Robert Carlstrom .... 7/1/68—7/1/71
Mike Sun .... 3/1/78—
John Riesselman .... 6/1/79—
Robert Davidson .... 9/1/75—6/30/82

Community Development Specialists

Jack Barringer .... 10/1/70—10/1/73
Michael Moyer .... 7/1/70—10/1/73
James DeBree .... 11/1/70—7/1/76
Herb Pepion .... 12/1/70—3/31/73
Cal Oraw .... 2/15/72—7/1/76
John Burton .... 4/1/79—7/31/81
Dave Sharpe .... 7/1/82—

Farm Forestry Specialists

John P. Drummond .... 12/1/46—4/30/49
O. M. Patten .... 7/1/49—6/30/53
Richard T. Marks .... 7/1/57—11/30/65

HOME ECONOMICS

Clothing Specialists

Georgia A. Roosevelt .... 7/15/19—9/19/20
Ruth Patchin .... 9/1/20—8/31/22
Inez LaBossier .... 9/1/22—8/31/26
Edith I. Mott .... 9/1/26—7/31/29
Hazel Kidder .... 9/1/29—4/15/32
Josephine Pollock .... 9/1/32—5/15/37
(See also Family Life)
Mary Omen .... 7/1/37—6/30/38
Martha Hensley .... 4/1/38—12/31/45
Lora Hilyard .... 2/1/46—6/30/62
Harriet Toan .... 8/1/62—4/20/68
Dorothy Gallup .... 7/1/68—7/1/78
Pamela Cutright .... 10/16/78—7/1/80
Barbara White .... 1/15/82—

Nutrition Specialists

Mary Ann Graber .... 9/1/18—8/31/22
Martha Hunter .... 9/1/22—7/31/24
Suzanne Thompson .... 9/1/24—8/31/26
Jessie McQueen .... 9/1/26—7/31/29
S. Frances Smith .... 8/1/29—1/31/36
Bessie E. McClelland .... 9/15/36—7/16/46
Mary E. Loughead .... 1/1/47—6/30/61
Dianne Undem .... 11/1/64—7/1/67
Carol Tarnavsky .... 7/1/67—5/15/77
(See also Rural Defense and Adult Education)
Andrea Pagenkopf .... 10/1/71—7/1/80
and 7/1/81—
Susan Butler (E.F.N.E.P.) .... 8/1/77—2/6/80
Anne Shovic .... 10/31/81—6/30/82

Home Management Specialists

Gladys Branegan . . . . 1/1/21—7/31/21
Lottie Milan . . . . 1/1/22—8/31/23
Gertrude Hoffman . . . . 9/1/23—8/15/25
Miriam E. Hawkins . . . . 8/1/25—8/15/28
Edith A. Sloan . . . . 9/4/28—6/26/29
Nora L. Hulbush . . . . 9/1/29—7/31/33
Oona Stautz . . . . 9/1/33—9/30/36
Margaret Tuller . . . . 9/15/36—6/15/47
Ermina J. Fisher . . . . 8/6/46—10/18/52
(Acting: 8/6/46—6/30/49)
Vilma Golde, Assistant . . . . 1/1/49—6/30/50
Mae Farris (True)
Home Management Specialist . . . . 7/1/51—2/28/59
Specialist in Housing, etc. . . . . 3/1/59—3/31/60
Supervisor, etc. (see above) . . . . 4/1/60—6/30/67
Equipment & Furnishings Specialist . . . . 7/1/67—4/30/68
Eleanor F. Wilson . . . . 8/17/53—6/30/56
Agnes Sunnell . . . . 9/1/58—4/17/60
Alberta B. Johnston . . . . 8/1/60—10/31/61
Family Economics Specialist . . . . 11/1/61—1/20/63
Allene Pierce . . . . 9/1/70—4/1/76
Sandra Dellinger . . . . 10/16/78—10/31/80
Marsha Goetting . . . . 4/1/77—
Carol Jo Thompson (Env.) . . . . 1/1/78—

Family Life Specialists

Josephine Pollock, Child and Parent
Education Specialist; Part-time . . . . 12/1/33—11/30/37
Amy Wold . . . . 10/1/49—6/30/62
Dean K. Hoffman
Family Life Specialist . . . . 8/1/62—6/30/63
Human Development Specialist . . . . 7/1/63—6/30/72
Robert Neils . . . . 1/1/73—6/30/75
Robert Lind . . . . 8/15/78—

Rural Health Specialists

Frances McDonald . . . . 9/1/46—8/31/56
Monte Gagliardi . . . . 6/1/80—

SOURCE OF EXTENSION FUNDS

Year	County Funds	Federal USDA—Funds 3b & 3c	State Funds	Earmarked Funds	Totals
71-72	795,860.00	1,079,864.00	740,040.00	303,640.00	2,919,404.00
72-73	829,760.00	1,234,290.00	750,300.00	421,012.00	3,234,362.00
73-74	950,057.00	1,277,943.00	822,535.00	514.257.00	3.564.792.00
74-75	1,071,394.00	1,506,387.00	886,983.00	432,082.00	3,896,846.00
75-76	1,532,467.00	1,962,412.00	1,377,838.00	484,158.00	5,356,875.00
76-77	1,304,294.00	1,647,306.00	1,441,907.00	294,023.00	4,687,530.00
77-78	1,352,616.00	1,755,428.00	1,302,215.00	249,611.00	4,659,870.00
78-79	1,489,347.00	1,891,704.00	1,206,665.00	225,146.00	4,812,862.00
79-80	1,701,137.00	1,949,054.00	1,304,620.00	245,344.00	5,200,155.00
80-81	1,857,352.00	2,165,714.00	1,371,874.00	225,286.00	5,650,226.00
81-82	2,005,940.00	2,155,158.00	1,838,163.00	369,480.00	6,368,741.00
Total	14,890,224.00	18,625,260.00	13,043,140.00	3,794,039.00	50,352,663.00

APPENDIX B
A Chronological Summary of Some Landmarks Contributing to the Growth of the Montana Extension Service, 1862-1982

1862 The Morrill Land Grant College Act, July 2
1862 The Homestead Act, May 20
1864 Montana becomes a Territory, May 26
1866 Nelson Story makes first cattle drive to Montana up Texas Trail
1873 First Montana chapter of the Grange organized, 1938 reorganization more permanent
1875 *Rocky Mountain Husbandman* begins publication by Sutherlin brothers
1883 Montana Wool Growers Association forms
1883 Northern Pacific Railroad completed across Montana, September 9
1884 Montana Farmers' Institute board established. Not organized until 1901.
1885 Montana Stockgrowers Association organizes
1886 Seaman A. Knapp sets up, in Louisiana, the first extension-type cooperative experiment.
1886—1887 Hard winter in Montana turns absentee ownership of cattle ranches into a home-owned industry
1887 Hatch Act established funds for an Experiment Station at each Land Grant College
1887 Association of American Agricultural Colleges and Experiment Stations first convention, October
1887 Great Northern Railroad reaches Great Falls
1889 Montana becomes a state, November 8
1892 Farmers' Alliance forms state organization
1893 Agricultural College of the State of Montana created, February 16
1893 Montana Agricultural Experiment Station established. S. M. Emery first director
1893 State Bureau of Agriculture, Labor and Industry created
1893 First Farmers' Institutes held
1893 Augustus M. Ryon named first president of the College
1893 Montana wins many prizes at Columbian World's Fair in Chicago
1894 James Reid succeeds Ryon to become second president of the College
1894 Montana Experiment Station Bulletins 1, 2, 3, 4 published
1897 Montana Horticultural Society formed
1899 State Board of Horticulture created by Legislature
1900 Samuel Fortier becomes second director of the Experiment Station
1901 Board of Administration for Farmers' Institutes created
1902 Montana Agricultural Association formed
1902 Frederick B. Linfield comes as professor of agriculture and director of the Farmers' Institutes
1903 Montana State Fair at Helena authorized by law

1903 F. B. Linfield becomes third director of the Experiment Station. Continues until 1936
1903 Montana Registered Cattle Breeders' Association formed
1904 Alfred Atkinson comes as agronomist
1904 James M. Hamilton becomes the third president of Montana State
1905 Experimental "dry farms" established with railroad support
1906 F. B. Linfield outlines need for "agricultural expert" in each county.
1906 First position comparable to a county agricultural agent, held by C.W. Stallings in Smith County, Tex.
1907 First Montana chapter of the American Society of Equity established
1907 Fred S. Cooley becomes superintendent of Farmers' Institutes
1907 Beginnings of Western Montana Branch Experiment Station in Corvallis
1907 Central Montana Branch Experiment Station, Moccasin, established
1908 Northern Pacific special agricultural train featuring the dairy industry
1909 Fourth Dry Farming Congress & International Dry Farming Exposition in Billings, in October
1910 Huntley Branch Experiment Station established
1910 Northern Pacific "Better Farming Special"
1910 Boys' and Girls' Corn Clubs formed in eastern Montana provide background for 4-H organization
1910 James J. Hill makes a $1,000 gift for prizes to the Corn Club contestants
1910 Farmers' picnics begin
1911 Dr. W. J. Hartman becomes livestock specialist under Farmers' Institutes
1911 M. L. Wilson joins Farmers' Institute and Experiment Station staff
1912 Montana Seed Growers Association formed
1912 Northern Pacific, Great Northern and Milwaukee railroads all run agricultural special trains
1912 First Country Life Convention in Great Falls, following White House Country Life Conference
1913 Department of Agricultural Extension created at Montana State College
1913 First Farmers' Week in January at Montana State
1913 M. L. Wilson goes to Custer and Dawson counties as first county agricultural agent, May 1
1913 Carl H. Peterson goes to Fergus County as second county agent, May 7
1913 Northern Montana Branch Experiment Station organized near Havre
1913 *The Montana Farmer* begins publication, July 1 in Great Falls
1913 Director Linfield becomes dean of agriculture
1914 John (Jack) C. Taylor becomes third Montana county agent, in Custer County

1914 Smith-Lever Act, May 8, creates the Extension Service
1914 Fred S. Cooley becomes director of the Montana Extension Service July 1
1914 Kathrine Jensen becomes first home economics state leader July 1
1915 First regional 4-H Exposition and Conference held in Wibaux, in November
1915 The C.B. & Q. Railroad runs a special agricultural train
1916 Montana Extension Service begins to organize Farm Bureau chapters
1916 Montana Farmers' Union organizes, April 3
1917 U.S. declares war on Germany, World War I, April 6
1917 Montana Extension Service cooperates with State Defense Council in World War I
1917 Clara Aust Bush serves during summer in Missoula County as first Montana home demonstration agent
1917 Federal Smith-Hughes funds create Future Farmers of America and Future Homemakers in the high schools
1917 Montana Dairymen's Association formed, January, at Montana State College
1918 First 4-H summer camp held in Blaine County
1918 Mignon Quaw writes and directs the Extension Service play, "Long, Long Trail"
1919 Legislative statute authorizes relationship between Extension Service and the Farm Bureau; statute repealed in 1949
1919 Alfred Atkinson becomes fourth president of the College
1920 A. J. Ogaard becomes Extension agronomist and secretary of the Montana Grain Growers
1921 Position of Extension economist established April 1, with M. L. Wilson as economist
1921 "Summer Fallow Club" organized by William Reed on the Big Flat near Turner
1921 Leonard Geiseker begins Land Classification study
1921 State Department of Agriculture, Industry and Labor established
1922 Dry Land Farm Success Study begun in the Triangle Area
1922 Bureau of Agricultural Economics established in USDA
1922 Annual "Outlook" survey and report begins in USDA; Montana reports begin in 1923
1922 State Board of Equalization established; adopts Robert Clarkson's tax evaluation plan
1922 Women's Camps begin, continue until 1946
1922 Harriette E. Cushman becomes poultry specialist, July 1
1922 First Poultry (turkey) Cooperative Marketing Association in nation formed in Pondera County
1923 Extension Service *Bulletin No. 66*, "Dry Farming in the North-Central Triangle," published
1923 Blanche Lee becomes state home Extension leader, July 1
1924 Fairway Farms experiment begins
1924 John (Jack) C. Taylor becomes second director of the Extension Service

1924 Retiring director, F. S. Cooley, writes *A History of the Montana Extension Service, The First Twenty-five Years*
1924 U.S. Range Livestock Experiment Station established at Miles City
1925 Purnell Act provides funds for departments of Agricultural Economics in Land Grant colleges
1925 Department of Agricultural Economics established at College
1926 Extension Service *Bulletin No. 81*, "Basic Facts About Montana's Agriculture," published
1927 Extension Service *Bulletin No. 84*, "An Agricultural Program for Montana," based on *Bulletin No. 81*, published
1927 First County Home Demonstration Councils formed in Valley, Carbon and Rosebud counties
1927 Northern Pacific special agricultural train, "Better Livestock, Better Feed"
1928 Mizpah-Pumpkin Creek Cooperative Grazing District established
1929 Federal Farm Board established to assist farmers' cooperatives and to stabilize prices
1929 Extension personnel granted leave for graduate study
1930 First annual 4-H Convention at Montana State College
1930 4-H County Councils formed
1930 Northern Pacific "Low Cost Farming Special" advertises results of Fairway Farms experiment
1931 State Drought Relief Committee appointed
1931 Robert B. Tootell becomes first Extension land economist in the nation
1932 M. L. Wilson assists in planning FDR's "New Deal" agricultural policy
1932 Home economics "Live at Home" and "Buy Wisely" programs begin
1933 Experiment Station *Bulletin No. 278*, "Farm Organization as Affected by Mechanization," summarizes technological findings of the Fairway Farms experiment
1933 State Grazing District Law passed
1933 M. L. Wilson goes to Washington to work with the USDA in the "New Deal" agricultural program
1933 The Agricultural Adjustment Act (AAA) signed May 12
1933 Wheat Allotment program begins with M. L. Wilson in charge
1933 Emergency Livestock Purchase program in operation
1933 The "Malta Plan" of resettlement developed by Henry Lantz in Phillips County
1934 Taylor Grazing Act passed
1934 Division of Subsistence Homesteads established in USDA with M. L. Wilson in charge
1934 Montana Water Conservation Board established in a special Drought Relief session of the legislature
1934 Grasshopper control project in operation
1935 Bankhead-Jones Act passed to assist land purchase program
1936 Clyde McKee becomes the fourth director of the Agricultural Experiment Station

1936 Johnson-O'Malley Act passed giving aid to work with Indians. Montana enters program in 1955
1936 Frances Smith becomes state home economics leader
1937 A. L. Strand becomes fifth president of Montana State College
1938 Mount Weather Agreement
1938 Martin Hansmeier named first Extension conservationist
1938 Regional Junior Fat Stock Show established in Billings
1942 Roland R. Renne becomes sixth president of Montana State
1942 Rural Fire Control program begins
1942 Second World War begins, November 9
1944 Agricultural Advisory Council created at Montana State College
1944 Veterans' Resettlement project initiated
1944 Montana Study instituted, ended in 1947
1945 Bankhead-Flannagan Act supports post-war services
1946 Outlook Reports take form of "Looking Ahead with Montana Farmers," discontinued in 1959
1946 Robert B. Tootell becomes third director of Extension, January 1
1946 Bessie E. McClelland becomes state home demonstration leader, September 30
1946 Frances McDonald becomes Extension health specialist
1947 Northwestern Branch Experiment Station established at Creston
1947 Eastern Branch Experiment Station established at Sidney
1947 International Fellowship Youth Exchange (IFYE) begins, Montana enters in 1948
1949 State Home Demonstration Council formed
1949 Family life specialist added
1950 Range management specialist added
1950 Civil Defense program begins
1951 State Department of Agriculture established
1952 Rural Progress Conferences Extension *Bulletin No. 274*, "What the People Said at the Montana Rural Progress Conferences"
1953 Newell E. Beers becomes fourth director of Extension, July 1
1953 Maurice Kelso becomes the fifth director of the Agricultural Experiment Station
1955 Wildlife Management specialist position established
1956 Montana Beef Production Association formed
1957 State Rural Development Committee formed
1958 Torlief S. Aasheim becomes fifth director of Extension, September 1
1958 Roy E. Huffman becomes sixth director of the Agricultural Experiment Station
1958 "The Scope and Responsibility of the Cooperative Extension Service" (SCOPE) Report published
1958 Grant of $40,000 from Ford Foundation, "Fund for Adult Education," used for work in public affairs and public responsibilities
1959 Extension personnel receive rank and tenure equal to resident faculty
1960 Montana SCOPE Report completed
1960 "The Mobile Campus" project launched; discontinued in 1965

1960 "Vital Issues" program initiated
1960 Montana Extension Service changes its structure from three program areas to five
1961 Severe budget cut for the biennium
1962 J. A. Asleson becomes seventh director of the Agricultural Experiment Station
1962 State Animal Health Laboratory built at Montana State College
1962 Resources, Conservation and Development (RC&D) project launched
1962 Montana Extension Service emblem (Flaming E) adopted
1962 "Cooperative" included in the title of the Extension Service
1963 Technical Action panels (TAB) formed
1964 Leon H. Johnson becomes seventh president of the College
1965 First Woman's Week at College June 20-25
1965 Montana State College becomes Montana State University, July 1
1966 Report on "Proposed Changes in Program Development," submitted to the Board of Regents
1966 Don Hunter employed as area agent in Community and Resource Development
1966 Symposium at Montana State University honoring M. L. Wilson
1966 South Central Montana Development Federation formed
1967 Montana Extension Service changes structure from five program areas to two
1968 New policy instituted for eighth quarter leave
1968 USDA publication, "A People and a Spirit" suggests new directions
1968 Council on Indian Affairs formed
1969 Expanded Food and Nutrition Program (EFNEP) provided funds for work in urban areas
1969 Montana 4-H Foundation created
1969 Montana Extension Management Information System (MEMIS) adopted
1970 Carl W. McIntosh named eighth president of the University
1970 Economic Development Association of Eastern Montana is formed
1971 Kellogg Foundation grant of $600,000 funds leadership training program, "Kellogg Extension Education Program" (KEEP)
1972 Statement of "Goals and Related Activities" prepared
1974 Director Aasheim retires July 1; recommends that new director be named vice president for greater unification of University work
1974 Touch-tone telephone link computer service for farm-ranch decision making instituted
1974 Saline Seep program started
1974 Civil Defense program shifts to a Safety program
1974 Community Pride, a 4-H community betterment program, established
1974 "Mulligan Stew," a 4-H television program, established
1975 Carl Hoffman becomes vice president and sixth director of the Extension Service
1976 Agricultural Engineer employed; initiated irrigation program
1977 Indian reservations terminate Extension programs: Rocky Boy's, 1977; Crow, 1978; Blackfeet, 1980

1977 Farm-ranch Recreation Guide developed
1977 Estate Planning correspondence course initiated
1977 Energy program initiated; first project, no-till in sugar beets
1977 William Tietz becomes ninth president of Montana State University
1978 AGNET initiated
1978 Full-time dairy specialist expanded dairy work
1978 Money management correspondence course initiated
1979 Leaf cutter bee alfalfa pollination project successful
1980 Promotion of filing water claims under new law
1980 Full-time sheep specialist employed
1980 Cooperation in PCB feed poisoning investigation
1980 Full-time human health specialist, expanded program
1980 Statewide Extension Advisory Council organized
1980 James R. Welsh became eighth director of Agricultural Experiment Station
1981 Assistance with alternate sources of energy programs—gasohol, bio-fuels, etc.
1981 Water Policy Education (WET) initiated
1981 Youth program, "After Graduation—What Next?" conducted
1982 All counties set up county-wide Advisory Councils
1982 Integrated spotted knapweed control program in western Montana initiated
1982 New project developed on Organization and Leadership Skills
1982 Government and Public Policy Processes project begun
1982 Outside Review Committee made recommendations on Extension organization
1982 Assistant director position re-established
1982 Program area for 4-H re-established

APPENDIX C
The Montana Association of County Agricultural Agents

By Robert F. Rasmusson

The first meeting held to consider the formation of a county agent organization was on January 8, 1920, in Bozeman. County Agent A. D. Anderson was elected temporary chairman and C. M. Yerrington was elected temporary secretary. Each of the 22 agents attending the meeting was given one minute to express his views in regard to forming an organization of county Extension agents. Two committees were appointed, one to confer with the Extension director and the county agent leaders regarding plans for forming an organization and the other to draw up a constitution and bylaws for the organization. Attending the first meeting were: A. D. Anderson, E. W. Anderson, Paul Banker, H. S. Brossard, J. Chase, Robert Clarkson, W. R. Gordon, F. M. Hillman, H. W. Jones, H.N. Kauffman, G. E. Lewis, J. W. Manning, F. E. MacSpadden, D. W. Mendenhall, C. A. Michels, R. B. McKee, W. E. Pollington, R. Spring, Wm. Stapleton, Murray Stubbins, M. A. Thorfinson and C. M. Yerrington.

The second preliminary meeting was held on the next day, January 9, 1920. At this meeting the constitution and bylaws committee reported. Their report was adopted and the constitution was approved. The name of the organization included in the new constitution was The Montana Association of County Agents. The constitution stated that "The object of this association shall be to develop a greater spirit of fellowship between county agents in assisting them to solve problems that are common to all and with the cooperation of the State Extension Service, secure greater efficiency in putting into effect the things that the Farm Bureau board may direct."

Membership in the organization as stated in the constitution was based on the payment of membership dues which included a membership fee of $1.00 and annual dues of $1.00. Probably due to the fact that the association did not get 100 percent cooperation in the payment of dues, it was short of funds during the early years and a special assessment of $1.00 per member was made in 1924.

The first constitution included the county agent leader and the assistant county agent leader as honorary members in an advisory capacity. The constitution specified that the annual meeting of the association was to be held in connection with the annual Extension Conference.

An election of officers was held in connection with the January 9th meeting. The officers elected were: A. D. Anderson, president; F. E. MacSpadden, vice president; and C. M. Yerrington, secretary-treasurer. One director was to be selected from each supervisory district. Directors elected were: M. A. Thorfinson, Northern District; R. Spring, Western District; and G. E. Lewis, Yellowstone District. The minutes of this meeting indicated that it was moved and seconded that the meeting be adjourned. Mention is made that the motion did not carry, but no further additions were made to the minutes.

The association was a little slow in becoming an active organization. Little was done the first year after the organization meeting. No time was provided for a meeting of the association at the annual Extension Conference in 1921. A similar situation existed in 1922 through 1925. At the 1925 annual meeting of the association the members decided that the association had better do something or disband.

In 1926 The Montana Association of County Agents joined the National Association of County Agricultural Agents and dues for the Montana association were increased to $2.00 per year. It was in 1926 that *The Montana County Agent* was conceived and publication by the agents was started on a monthly basis. It was changed to a semi-monthly publication in 1927. This publication served an important purpose by providing an opportunity for communication and an exchange of ideas between agents. It was a tool which helped to maintain the morale of the staff.

The minutes covering the 1927 annual meeting of the association refer to the organization as The Montana County Agents' Association." This name was used frequently as the years passed, but there is no record of changing the name of the association until 1960, when the name was officially changed to The Montana Association of County Agricultural Agents. This change was made to bring the Montana association name in line with the National Association of County Agricultural Agents.

In 1930 Director Taylor reported that county agents could take one month leave with pay plus their regular two weeks annual leave, making a total of six weeks leave for study. Salary for this leave was to be paid out of state and federal funds. Also, at the 1930 association meeting, both the president of the Farm Bureau and the state president of the Farmers' Union spoke to the association members. Both assured them of the cooperation of their organizations. The state president of the Farm Bureau said that the Farmers' Union was a good thing and that all farmers' organizations are good.

The association secretary's report for 1944 included a rather complete suggested schedule for classification of counties for use in establishing starting salaries for county Extension agents. This schedule was revised and updated in the 1945 report of the association secretary. The schedules were developed by the Salary and Leave Committee but were never used by the Montana Cooperative Extension Service.

The constitution of the Montana County Agents' Association was revised and distributed to members in 1936. Dues were $2.00 per year at that time. They have steadily advanced over the years until in 1982 it cost $50.00 to belong.

The purpose of The Montana Association of County Agricultural Agents at the time it was organized, and as it has remained over the years, was to promote better fellowship among the Extension family and to assist in the development of a better Extension Service in the state.

In recent years the association has been quite active. It served an important function as a connecting link between the male Extension agents, the Montana Association of Extension Home Economists, and the members of the state Extension staff. Many of the committees included in the two state County Agent Associations have had the same purpose and they frequently have held joint meetings where their committee activities and plans were coordinated.

The purposes and activities of the Montana Association of County Agricultural Agents probably will change as the Montana Cooperative Extension Association develops. Some form of working arrangement will need to be implemented so that good communication will take place between the two organizations.

APPENDIX D
The Montana Association of Extension Home Economists

By Robert F. Rasmusson

The organization of the Montana Home Demonstration Agents (the first title of county Extension agents specializing in home economics) was completed during the annual Extension conference held at Montana State College in January 1929. The organization meeting was held in the Fireplace Room of Herrick Hall. Officers elected were Frances Smith, president, and Fanny Gorton, secretary-treasurer. The name selected at that time was the Montana Home Demonstration Agents' Association. More recently, when the title of women agents was changed to county Extension agents, the name of their organization was changed to the Montana Association of Extension Home Economists.

Home demonstration agents of record in January 1929, all of whom are presumed to have been present at the organization meeting, were: Mrs. Doris J. Anderson, Yellowstone County; Alberta Coleman, agent-in-training; Julia Corkill, Rosebud County; Mary Ebersole, Fergus County; Mrs. Florence Elliott, Valley County; Fanny Gorton, Richland County; Vesta Hanson, three-county district—Daniels, Sheridan and Roosevelt; Frances Smith, Flathead County; and Neva Woods, Lewis and Clark County.

The desirability of organizing as a professional group had been under consideration for some time. The agricultural agents had been organized since 1920 and the home demonstration agents became anxious to form a parallel group. Miss Blanche Lee, state home demonstration leader at that time, was most cooperative and encouraged the agents to form an organization.

Accordingly, at the organization meeting a Constitution Committee was appointed. The constitution was completed and mailed to the members during February or March of 1929. It provided, among other things, for the assessment of $1.00 annual dues per member.

Since home demonstration agents at that time were located so far apart, the need for keeping in touch with one another was expressed by all agents present at that first meeting. It was decided that a monthly newsletter would be a morale builder for the then isolated home demonstration agents. The group voted unanimously to make the newsletter idea their first project. The name "Shu Tan Ka" was selected for the newsletter. "Shu Tan Ka" is an Indian term meaning "to council together." The word later was worked into a cover design for the newsletter. The first issue of the *Shu Tan Ka News* was published by Mrs. Doris Anderson on March 21, 1929. Agents took turns in publishing the monthly paper and their names were listed alphabetically to determine the order of publication. The distinctive name of the newsletter became the inspiration for introducing ritualism into the association's annual banquet.

All professional women employees and former professional employees of the Extension Service were eligible to contribute items to the *Shu Tan Ka News*. Contributions were to include Extension items of common interest or personal items of interest to other home demonstration agents The newsletter was published more or less regularly until 1970. At that time the currently employed Extension members decided to discontinue the publication because it no longer was needed to fulfill its original purpose.

During the years following the organization of the Montana Home Demonstration Association, a social affiliate also developed under the name of the Shu Tan Ka Council. A ritual was developed and used in connection with the initiation of new council members. This ceremony usually took place at the time of the annual banquet held by the agents and, as a result, the banquet was called the Shu Tan Ka banquet. All women Extension agents and former Montana women agents were eligible for membership in the Shu Tan Ka Council. Originally, women state staff members who had not served as county Extension agents in Montana were invited to the banquet and the Shu Tan Ka ceremony but were not eligible to become members of the council. The eligibility requirements for membership have since been relaxed.

The Montana Association of Extension Home Economists is still an active organization involving the women county Extension agents. The Montana association is a member of the National Association of Extension Home Economists. The purpose of the Montana Association of Extension Home Economists is to provide the women county Extension agents with an opportunity to discuss their special problems, exchange ideas, and promote a closer fellowship among the county Extension women agents. In addition, the association provides an opportunity for better communication and understanding between the women county Extension agents, the men county Extension agents, and the state Extension staff members. The members of the association cooperate closely with the Montana Association of County Agricultural Agents and their common committees usually have joint meetings where coordinated action may be planned. The annual meeting of the association is held during the annual Extension conference.

APPENDIX E
Montana Cooperative Extension Association

By Robert F. Rasmusson

The idea of creating a Montana Extension Association that would include all county Extension agents, district personnel and state staff members was introduced by Extension Director Aasheim in 1967. This idea was received with mixed reactions by both state and field staff members. It was discussed at least annually each year during annual Extension conferences.

The question of forming an overall Extension organization was considered by both county agent associations at one time or other during 1967 and the years following. At a joint meeting of a representative number of men and women agents held in 1967, the group seemed in accord regarding the formation of an overall Extension organization and they even went so far as to suggest that the name of the new organization be The Extension Council of Montana.

The Montana Association of County Agricultural Agents (MACAA), at a later meeting, passed a resolution recommending the formation of an organization of this type. Their recommendation suggested that all men and women Extension agents, district personnel and state staff members, with the exception of administrative personnel, be eligible for membership in the organization. The resolution included a plan for implementing the recommendation. The Montana Association of Extension Home Economists (MAEHE) was presented with a copy of the MACAA resolution but their members failed to act on it during their annual meeting. Some concern was expressed by some of the MAEHE members as to the effect an overall Extension organization would have on their present association. The resolution was later referred to at the 1970 "Overall Extension Organization Committee."

Finally, in 1969, a preliminary committee was appointed by Director Aasheim to explore the possibilities of the reorganization of professional Extension workers' associations in Montana. The committee included Allen Nelson, chairman; Vivienne Kintz, Eugene Quenemoen, William Gibson and Lois Stephens.

At their first meeting, held August 19, 1969, the committee decided to make a situation statement covering the present organization of staff members of the Montana Cooperative Extension Service. They outlined some of the possible courses of action for the committee to take. They studied various materials available covering Extension organizations. One of the resource materials used by the committee was a report prepared by the Extension subcommittee of the National Committee on Organization and Policy. The title of this 1965 report is "Evaluation of Purposes and Effectiveness of State and Local Agent Associations."

As a result of the work done by the preliminary Committee on Reorganization of the Professional Extension Workers' Associations, and because of the interest shown by the Montana Extension agents in a service-wide Extension organization, Director Aasheim appointed an Overall Extension Organization Committee in 1970. Included on this committee were:

Allen Nelson, chairman; Phillip Wilson, Arthur Shaw, Allene Pierce, William Gibson, Eugene Quenemoen and Lois Stephens. The first meeting of this committee was held on December 18, 1970; special attention was given to reviewing the organizational structures of overall Extension organizations in Wisconsin, Oregon, Ohio and Indiana. The resolution passed by the MACAA was also reviewed by the committee.

Other meetings of this Overall Extension Organization Committee were held. State staff members on the committee reported that the agricultural specialist staff was reluctant to make a commitment on the question. According to the state staff committee members, a proposal was made by the agricultural specialists which included a recommendation that the two county agent associations form one organization which would operate over a trial period. A decision as to whether specialists would participate in the new organization could be made after the trial period was completed. Purposes and possible goals were suggested by the new committee and attempts were made to diagram the organization of an Overall Extension Association. Much frustration existed and many members of the committee apparently were of the opinion that the chances for an organization of this type to succeed were poor. As an alternative, three other approaches to reorganization without forming an overall Extension association were studied. No specific recommendations were made by this committee.

Another Overall Extension Organization Committee was appointed in 1971. The membership of this committee included William Richter, chairman; Stanley Howard, Lois Lohr, Phillip Wilson, Allen Nelson, Eugene Quenemoen, Lois Stephens, Vivienne Kintz and Marian Moline. This committee continued the discussion regarding the question of forming an overall Extension organization. The committee members took every opportunity to discuss the question of Extension staff organization with both state and field staff members. It became evident that a more favorable attitude by both state and field staff members toward the idea of creating a new Extension organization was beginning to develop.

At a meeting of the committee held on September 23, 1971, it was decided that a proposal for forming an overall Extension organization would be laid before the entire professional staff of the Montana Cooperative Extension Service. The committee completed work on a tentative constitution and bylaws. They made preliminary plans for developing a new organization; they proposed that the name for such an organization be the Montana Cooperative Extension Association. The tentative constitution was reviewed and revised by the committee at meetings held during the year, and on February 15, 1972, the tentative constitution was reproduced for distribution to the Extension staff. A plan for submitting the issue of organizing a new association to the staff was completed.

During the last part of February 1972, a ballot was mailed to all state and field staff members of the Montana Cooperative Extension Service. Included with the ballot was a copy of the suggested constitution and bylaws prepared by the committee. The statement on the ballot was "I favor the formation of an Overall Extension Organization as proposed by the suggested constitution and bylaws dated February 15, 1972. Yes_______ No_______." The judges for the balloting were Beverly Wallace, Mike

Jackson and Gene Quenemoen. The results of the balloting tabulated on March 3, 1972, follow:

Staff Members	**Yes**	**No**	**Undecided**
Men Agents	31	13	——
Women Agents	11	9	——
State Staff	24	4	1
Total	**66**	**26**	**1**

Plans were made by the committee to complete the organization of the Montana Cooperative Extension Association during the 1972 annual Extension conference. This included the election of a board of directors and adoption of the constitution and bylaws. At a meeting held on March 29, 1972, the Extension Organizaiton Committee selected Eugene Quenemoen as temporary chairman for the fall organizational meeting.

On October 2, 1972, during the annual Extension conference, the first meeting of the Montana Cooperative Extension Association was held. Members of the board of directors elected at this meeting were: Eugene Quenemoen, Merle Lyda, Larry Brownell, Betty McCoy, Dorothy Hoffman, Marian Moline, Lois Stephens, Gregg Carlson, Carolyn Dufner, Ellis Williams, Darryld Kautzmann and Nancy Horsfall. The board of directors elected Eugene Quenemoen chairman and appointed Charles Egan, secretary, and Patricia Sias, treasurer. The final draft of the constitution and bylaws as adopted by the new association was published on October 6, 1972.

Following the organization of the Montana Cooperative Extension Association, the first meeting of the board of directors was held in Lewistown June 1, 1973. All directors were present. This was remarkable since all staff members took annual leave to attend association meetings and they paid their own mileage, board and lodging.

The second annual meeting of the Montana Cooperative Extension Association was held in October 1973 in connnection with the annual Extension Conference. Association committees also held meetings at this time. Membership of the association's standing committees, as provided in the constitution, was largely determined by the preference indicated by the members. Seven new members were elected to the board of directors at the 1973 annual meeting of the association. The board of directors for 1973-1974 were: Larry Brownell, Merle Lyda, Patricia Geroge, Dorothy Hoffman, Eugene Quenemoen, Marian Moline, Lois Stephens, Gregg Carlson, Dale Bergland, Bruce Eckert, Robert Johnson and Kathy Greenwold. The board elected Larry Brownell chairman and appointed Charles Egan, secretary, and Patricia Sias, treasurer. The All Extension Banquet, sponsored by the association during the 1973 Extension Conference, was the first major undertaking of the association.

The constitution for the Montana Cooperative Extension Associaton includes the following objectives:

1. To create and maintain a climate for effective communications and coor-

dination of efforts for all aspects of Extension work in Montana.

2. To strive for and promote high standards of professional ethics and foster good public relations.
3. To secure through collective efforts benefits which cannot be achieved individually.

Membership in the new association includes any member of the professional Extension staff whose dues to the association are paid. Membership is voluntary and retirees of the Montana Cooperative Extension Service are eligible for continued membership. Functions of the association and obligations of individual members to the association are intended to be conducted separately from the operation of the Montana Cooperative Extension Service.

A board of 12 directors is the governing body of the association. Each director serves a term of three years, and four directors are elected each year. One director is elected from each of the program coordinator areas and one from the state staff. The remaining seven directors are elected at large. The director of the Montana Cooperative Extension Service is to be an ex officio member of the board. In addition to the chairman, the officers of the board are vice-chairman, secretary, and treasurer. The chairman and vice-chairman are elected by the board from the membership of the board. The secretary and treasurer are appointed by the board but need not be board members; in which case they do not have the right to vote as a director of the association.

Dues for the Montana Cooperative Extension Association are determined by the budget committee, annually, but are to be approved by the membership at the annual meeting.

Provisions were made in the bylaws for seven committees. Included in the seven are two committees representing the Montana Association of Extension Home Economists and the Montana Association of County Agricultural Agents. These committees include the officers and directors of the respective associations, and problems or questions developed through these committees are to be routed to the Montana Cooperative Extension Association through other or regular special committees of the association.

APPENDIX F
Epsilon Sigma Phi

By J. C. Taylor, Director Emeritus
Montana Extension Service

Epsilon Sigma Phi is primarily an honor society. It is based on tenure and quality of service. To be elected a member is in itself an honor.

The official birthday of Epsilon Sigma Phi is January 10. It was on this day in 1927, at the home of M. L. Wilson, that the first chapter was instituted at Bozeman, Montana. Those present were Fred Bennion, Robert Clarkson, Gertrude Erickson, William Jones, William A. Lloyd, Carl H. Peterson, Charles E. Potter, Madge J. Reece, John C. Taylor, and Milburn L. Wilson. These Extension Service members may be considered the honor society's founders. The pledge and creed were prepared by W. A. Lloyd, but in recent years these have been changed by action of the national board.

The idea of an honor society occurred to William A. Lloyd of the Montana Extension Service, he relates, while he was collecting data for a paper, "Some Factors Affecting Tenure in Extension Work." Lloyd presented this paper at a meeting of the Extension section of the land grant colleges held in Washington, D.C., in November 1926. In the preparation of this paper, correspondence was conducted with some 350 Extension workers including directors, county agents, home demonstration agents, supervisors and specialists, all of whom had been in the service for a period of more than 10 years. The letters from these veteran employees disclosed an idealism which appeared to characterize extension workers generally and, particularly, those who had remained in the service long enough to acquire a professional attitude toward it.

The more important of these characteristics as disclosed by the letters were 1) a confidence in American agriculture, 2) a belief in the fundamental values of American country life, 3) a genuine country mindedness on the part of extension workers, 4) a real love for extension work and a genuine happiness in doing it, 5) a persistent optimism, 6) a belief in the opportunity that extension offers to do something worthwhile for agriculture and the farm home, and 7) a loyalty to the organized Extension Service and a fraternal regard for fellow workers.

The National Association approved of the society at once, and the organization of chapters by state spread rapidly. By the end of the first year, 26 chapters were organized. Within four years, chapters had been formed in 45 states and two territories and the total membership exceeded 2,000.

In 1934 a Committee on Awards, chaired by the writer, made a report proposing that appropriate awards be made to the members upon recommendation of the chapters. The report provides that one-fourth of the chapters shall be privileged to nominate one candidate per year for a Certificate of Recognition. The following is a list of such awards made by Alpha Chapter:

J. C. Taylor	1936	T. B. Holker	1962
Doris Ingram Anderson	1939	Bessie E. McClelland	1966
W. P. Ostby	1944	Charles E. Potter	1969
R. E. Bodley	1947	Theodore S. Fosse	1973
Harriette E. Cushman	1950	Geraldine Fenn	1978
J. O. Tretsven	1954	Robert Rasmusson	1980
Frank M. Harrington	1957		

The Distinguished Service Ruby is the highest honor awarded by the fraternity, and not more than one award is made each year. The following members of Alpha Chapter have received this award:

William A. Lloyd
M. L. Wilson
J. C. Taylor
Madge J. Reece
Harriette E. Cushman

Epsilon Sigma Phi stands for high professional attitude and accomplishment and a fraternal relationship that encourages the best that is in us. The purpose and accomplishment of the fraternity's effort can point to the national legislation concerning retirement of cooperative employees and the progress of state legislation to extend retirement privileges; to progress in obtaining academic rank, sabbatical leave and other privileges for Extension employees at colleges and in counties; to making possible professional improvement to a larger number of Extension workers; to preserve the history of Extension work in this county by sponsoring the preparation and publishing of a state Extension history in practically every state or territory; and the perpetuation of the foundation of Extension work through the Wilson and Knapp Memorial Arches in the national Capitol.

APPENDIX G
The Ancient and Abnormal Order of the Buffalo

By Carl H. Peterson, Second County Agent
in the Montana Extension Service
With an Update by Robert F. Rasmusson

Shortly after the Extension Service was established and the staff came together annually in convention, a need for a formal or informal occasion for good fellowship was recognized. The stress of the war years and the rapid change in staff prevented this from developing.

However, during the Extension conference in January 1922, an event occurred which inadvertently formed the basis for such an occasion. This ceremony, largely humorous but partially serious, became a regular part of the annual banquets of the conferences. At this time the entire staff, as well as wives, husbands and co-workers in federal and state agencies, are initiated into the Ancient and Abnormal Order of the Buffalo. The geniality which prevails in the preparation and carrying out of this ceremony has done much through the years to foster a spirit of good will among all of the members of the Service.

The incident at the 1922 conference which precipitated the first ceremony was the address given by the chancellor of the Greater University system, Dr. Edward C. Elliott. Older members of the Extension Service will recall that the four units of the Greater University of Montana were at that time directed by a chancellor. His principal duties were to create a closer coordination between the various units of the Greater University. At the time Dr. Elliott, later president of Purdue Univesrity, was the chancellor, and one of his numerous duties was to meet the field force of the Agricultural College at the annual conference.

This was in the days before the extended building program had taken place on the campus of the State College; hence, all general assemblies were held in the assembly hall on the top floor of what is now known as the administration building, Montana Hall.

The address of Dr. Elliott was scheduled for three o'clock in the afternoon. As is always true in such a conference, a number of groups were meeting in various places throughout the campus. Among those was a meeting held in the dairy section of the agricultural building, at which Professor Martin was discussing with a group of agents the dairy projects throughout Montana, and he kept them approximately 10 minutes overtime.

The Extension men in those days, in common with the young men of today, were full of energy and hurried up the four flights of stairs in Montana Hall, hoping not to be late to Dr. Elliott's lecture. Those who are familiar with the construction of the building recall that the acoustics are such that the lightest step on the lower flights of stairs is plainly audible to those sitting in the asembly room. Therefore, when the dozen agents rushed up the stairs, it mildly disrupted the lecture of Dr. Elliott, already in progress.

As soon as the men were seated, Dr. Elliott ceased to speak on the subject which he was discussing at the time the agents entered the building, and proceeded to talk to the energetic agents specifically. He stated that the entrance of the agents reminded him of a charging herd of buffalo across the Great Plains of America. He went on to remind those in attendance that in the days of transportation by means of a prairie schooner, these charging herds of buffalo oftentimes cost a two- or three-day delay on the part of the travelers until the buffalo had crossed their path and proceeded into more luxuriant grazing areas.

Dr. Elliott intimated at the close of his special lecture to the charging agents that the buffalo which ranged the plains in the West had very thick skulls—that even a cold chisel or a crowbar would have to be used in order to allow any knowledge to penetrate their thick skull.

It is presumed that Dr. Elliott made these remarks in a jocular frame of mind. However, at the close of the regular session, an indignation meeting was held by the agents in question. They sought to devise some sort of means of informing Dr. Elliott that they did not approve of his remarks to them that afternoon.

During the early Extension conferences, an elaborate program was usually arranged to follow the Extension banquet. The nature of the program frequently centered around some take-off on the personnel of the Extension Service. At the indignation meeting, it was decided that the program for the following Saturday evening when the banquet was to be held would be centered around Chancellor Elliott in such a way that he would never forget the remarks that he had made on that particular afternoon. At this banquet program the ritual which is now being used during the initiation ceremonies of the Ancient and Abnormal Order of the Buffalo was developed.

The evening of the banquet the writer happened upon Dr. Elliott in the lobby of the Bozeman Hotel. The chancellor remarked that he had not intended to offend the agents with his remarks. He was informed, however, that the agents, while they highly respected him, nevertheless did feel offended. Moreover, at the approaching banquet and program, they intended to indicate to the chancellor and to those in attendance what they thought of the remarks. At first, the chancellor was somewhat hesitant about attending the banquet, but the writer assured him that if he could be a good sport, it was best that he attend the banquet—which he promised to do.

Among those taking a leading part in the development of the ritual and designing the costumes were A. J. Ogaard, Blaine Ferguson, Murray Stebbins, Mrs. E. H. Lott (Mignon Quaw), Carl H. Peterson and W. H (Bill) Jones.

At the close of the banquet, the Ancient Order of the Buffalo was instituted and the first group of candidates initiated in due and proper manner. It was not only impressive to those in attendance, but also forever placed the Extension Service in a most enviable position among the other divisions of the institution.

Dr. Elliott was so well pleased with the initiation of this particular order that he desired to have it made national in character. But those in charge felt that it was distinctly a Montana institution and, as such, should remain so.

UPDATE

There has to be a motivating reason for setting up any organization. The founders of the Ancient and Abnormal Order of the Buffalo had a purpose in mind when they responded to the chancellor's message at the 1922 annual Extension conference. They perhaps did not visualize at the time that the organization would still be going strong a half century later, nor that it would be as useful to the Montana Extension Service as it has been. The ritual, though somewhat humorous, is intended to and does teach some rather pointed and timely lessons to the beginning Extension workers.

Membership in the Ancient and Abnormal Order of the Buffalo originally was limited to Extension workers. Later the wives of Extension workers, Experiment Station employees, Montana State University faculty, and other friends of Extension were included in the groups initiated. Persons involved in annual conference programs, though frequently not closely related to Extension work, were invited to participate in the initiation. Persons may not participate in the Buffalo initiation ceremony unless they have been invited by the members of the organization. No person is compelled to join the organization although all new Extension employees are urged to present themselves for initiation. The annual initiation is the only meeting held each year. All members of the organization are invited to attend this meeting. There are no dues.

Groups have been initiated each year since the organization was formed, except in 1929 and in 1942-1944, when the war disrupted normal Extension activities. The largest class of initiates was the first group of 49 persons who took part in the first ceremony, held in 1922. The second largest group was initiated in 1945, following the war period.

The first group of 49 initiates was headed by Chancellor Elliott, but also included were Alfred Atkinson, president of Montana State College; F. S. Cooley, director of the Montana Cooperative Extension Service; M. L. Wilson, county agent leader; J. C. Taylor, assistant county agent leader and later director of the Montana Extension Service; and Charles E. Potter, state 4-H Club leader.

There has been some confusion in regard to the Ancient and Abnormal Order of the Buffalo and the *Montana Buffalo* publication. Both the organization and the publication were started by the Montana County Agents' Association, but there really is no relationship between the two. The *Montana Buffalo* was started by the men county agents and it has continued as an official publication of the Montana County Agents' Association. The fraternity Epsilon Sigma Phi was started about the same as the

Montana Buffalo but there is no relationship between the fraternity and the publication, either.

APPENDIX H
The Montana Buffalo Publication

By Robert F. Rasmusson

The Montana Buffalo is the name of the official publication of the Montana Association of County Agricultural Agents. The publication was started in 1926 under the name of *The Montana County Agent*. During that year it was published once a month. Starting in 1927 it was published semi-monthly, and the cover of the 8½-by-14 inches publication included the outline of a small buffalo with the MCA brand on its ribs.

The name *The Montana Bufffalo* was given to the publication in 1928, and the first issue published under that name was prepared by Sanders County Extension Agent F. J. Keishisnik. It was 8½-by-11 inches and its cover featured a large outline of a buffalo with the MCA brand. In later years the brand on the buffalo was omitted.

The Montana Buffalo, throughout the first 20 years or so of its existence, ran excerpts from letters received from county agents and items on Extension activities; programs and problems were frequently included. The editor added items of interest, poems, jokes and sometimes just nonsense. County agents took turns publishing the *Buffalo*, and it was published on the first and 15th day of each month.

When the Montana Cooperative Extension Association was formed on October 2, 1972, some concern was expressed regarding the publication of the *Montana Buffalo*. With the approval of the Montana Association of County Agricultural Agents, the Public Relations Committee of the MCEA recommended that the *Buffalo* be published by members of the new association. This recommendation was approved, and the first issue sponsored by the MCEA was published by John Maki on November 15, 1972.

The policy established concerning the publication as stated in the November 15, 1972, issue is as follows:

"The primary purpose of the publication is to provide a means for Extension personnel to express their views, beliefs and ideas. The publication is to go only to Extension personnel who have paid their dues to the Montana Cooperative Extension Association. The articles in the publication are confidential and for Extension personnel only. If the publication is to serve as a useful sounding board for ideas, it is important that all association members recognize the confidential nature of the publication and treat it accordingly.

"The editor is expected to send out notices two weeks before his publication date to all MCEA members asking for contributions and suggesting topics for comment. The MCEA secretary is responsible for notifying agents when it is their turn to edit the publication. The secretary will also provide a supply of covers for the publication.

"Publication procedures are the responsibility of each respective county. Assistance in publication may be obtained from the state office if desired."

At the 1973 annual meeting of the Montana Cooperative Extension

Association it was decided to change *The Montana Buffalo* to a monthly publication. A new cover was designed and it was first used on the October 1973 issue. The cover of *The Montana Buffalo* now includes a buffalo bull, cow and calf. It is printed on 8½ inches by 11 inches green paper and bears the statement: "A monthly publication edited by and for the members of MONTANA COOPERATIVE EXTENSION ASSOCIATION."

It appears that members of the Extension staff are making use of the new *Montana Buffalo.* It is providing an opportunity for better communications among all staff members and helps build a feeling of closer unity between the three Extension segments—the men agents, the women agents and the state staff.

APPENDIX I
Silver Buffalo Award

Purpose—The purpose of this award is to encourage and recognize excellence in the field of professional Extension work in Montana counties during the most recently completed fiscal year.

Eligibility—A nominee must:

1. Be a member in good standing of MCEA and MACAA or MEHA.
2. Have worked as a Montana county Extension agent for the entire fiscal year of merit.
3. Longevity of employment with the Montana Cooperative Extension Service is not a criteria for this award.

Basis of Selection—The nominee shall have carried out to the point of excellence a program in one given field of work and/or a totally well-rounded County Extension program.

Nomination—Nomination for this award may be made initially by any county Extension agent(s) by submitting the nominee's name ad project to the local district MCEA director. It is the responsibility of each director to have a nominee from his/her district.

Selection

1. The Selection Committee will be composed of one person from each supervisory district. These committee members shall be named by the MCEA president. The chairman of the MCEA Recognition Committee shall call and chair the meeting, but will not have a vote except to break a tie.
2. The nominee's program of work shall be submitted to the Selection Committee at least one month prior to Annual Conference. It is recommended that the agent's program of work be accompanied by suppporting statements from area program coordinators, agents submitting the name, or by other co-workers.

Award—The recipient of the award will receive a Silver Buffalo plaque which is his/hers to keep. A name plate will be placed on a large Silver Buffalo Plaque which is on display at the Extension Building on the Montana State University campus.

Restrictions

1. Any changes in requirements of this award must be submitted by the Recognition Committee for approval at a general meeting of MCEA.
2. If there is an attempt to transfer this award to another organization or jurisdiction it shall revert to the MACAA.

NOTE: The Silver Buffalo Award was developed by the Montana Association of County Agricultural Agents and the criteria for selecting the award winner was basically the same as in effect now; the winners were to be chosen from Extension agents. Recent award winners are: 1968—Theo Fosse; 1969—Charles E. Egan;

1970—Larry Brownell; 1971—Leonard Harris; 1974—Don Picard; 1975—Anne Rehbein; 1976—Eugene Hoff; 1977—Gregg Carlson; 1978—John Ranney; 1979—Dorothy Hoffman; 1980—Woodrow Ecegren; 1982—Darryld Kautzmann. (Not awarded in 1972, 1973 and 1981.)

APPENDIX J
The Quarter Horse Award
By N. A. Jacobson

On a cold Saturday afternoon in January 1935 the Montana County Agents' Association met for their last session of that year's annual Extension conference, in Room 301 of the agricultural building. After formal business was completed, Willard H. "Doc" Lamphere, agent for Custer and Powder River counties, at Miles City, asked to have the floor. Doc mounted the rostrum carrying a small package. He began to unwrap the package very deliberately while explaining he thought the association should have some sort of award in recognition of any member who might distinguish himself in an unusual way above and beyond the call of duty. Doc referred to the fact that the buffalo was the chief totem of the association, and that he had commissioned a sculptor to produce a plaque displaying such an emblem. Unfortunately, however, the artist selected for this job had never seen a buffalo; therefore, as a model he had to use another animal with which he was more familiar. While still in the unwrapping process, Doc began to enumerate various accomplishments by which an agent might be able to earn this coveted award. At last, he nominated A. C."Pondera" Peterson, Pondera County agent, to be the initial recipient of this award for a variety of achievements attained during the past year under great strain and duress. Pete was then called to the rostrum and the presentation was made.

The ceremony was received with wild acclaim by all those present, and Scotty Cameron moved that the association permanently adopt the award and that it be known as the Lamphere plaque. The motion was duly seconded and carried unanimously, after which Pondera Pete made a very gracious acceptance speech indicating how he proposed to handle this object of art in the future.

During the next year's annual conference in January 1936, the association held a meeting in the sample room of the Baltimore Hotel in downtown Bozeman. At the appropriate time, A. C. Petersen was called to the front of the room to preside over the second presentation of the Lamphere plaque. He called on Director J. C. Taylor along with Ralph Bodley, county agent leader. In the presence of Director Taylor, Pondera Pete then passed the award to Ralph Bodley, having with him as evidence of Bodley's right to the trophy a huge sheaf of papers and other documents, most of which later proved to be copies of Pete's expense account and monthly reports. The award was duly accepted by Bodley but without much comment. The next day, however, it was learned there had been some misunderstanding as to the spirit of the occasion the night before, and ill feeling arose between the donor and the recipient. Pondera Pete, therefore, immediately retrieved the plaque and kept it in his personal possession for an undetermined number of years. The Lamphere plaque, therefore, remained inactive and forgotten about until the association revived the idea in 1952 and it is now known as the Quarter Horse Award. It is awarded to the agent who, in the judgment of the previous year's recipient, made the biggest "boner" during the past year.

The original wood carving was made by Paul Orcutt, then ranching in Powder River County. Paul's brand is seen on the left thigh of the "North end of the horse going South."

Written by N. A. Jacobson, October 21, 1970, who was present at the 1935 session of the Montana county agents, as mentioned above.

NOTE: The original plaque is now stored in the Gallatin County Extension Office. The plaque, which is being used at the present time, was made by using the rear end of a commercially made metal horse. This is mounted on a wood plaque with a metal plate attached for use in recording the names of the recipients and the years the award was made. *R.F.R.*

The Quarter Horse plaque awarded annually to a Montana county agent. The award is of dubious merit.

APPENDIX K
The Oefstos Demonstration

By Robert F. Rasmusson

The use of method demonstrations was the first and the most effective teaching technique used by early Extension workers. This technique is still used and still is one of the most effective teaching methods.

During the period around 1945 to 1955 many county Extension agents arranged demonstrations to show how dynamite could be used to make ditches and pits for stock water. H. M. "Hank" Oefstos, county agent in Phillips county, arranged to have a DuPont representative stage a ditching demonstration on one of the irrigated farms in the Milk River Valley. Hank was a likeable gentleman with a keen sense of humor. He was known for his easy going but efficient way of getting Extension jobs done well.

A crowd of several dozen farmers had gathered to watch the demonstration. Their cars were parked not far from where the demonstration was to be held. The DuPont expert had completed the placement of the explosives, inserted the cap and fuse in the key stick of dynamite and was ready to ignite the fuse. He warned the observers that the explosion would throw mud, soil and water unpredictable distances and that they had better move their cars farther from the area to be blasted. Hank wetted his finger and held it in the wind. Then he told the group that the mud would not reach their cars and even went so far as to say that he would personally wash the mud off, if any fell on the cars. The audience decided that the expert's advice was probably better than Hank's so, with the exception of Hank and the local banker, all the other car owners moved their cars to a point farther from the demonstration.

The fuse was lighted and a tremendous explosion occurred. Mud, masses of Milk River Valley gumbo, pieces of sod and water made an impressive display as it was blown 100 feet or more into the air. As the material fell to the earth, one large piece of gumbo weighing 100 pounds or so fell on the roof of Hank's car. The impact was such that it shattered the back window and flattened the roof in no uncertain manner. Damage of a similar nature was done to the banker's car. Both cars were almost completely covered with mud.

The incident of course received wide publicity in Phillips County. Farmers and others had a lot of fun with Hank as a result of the backfiring of his prediction, but Hank seemed to enjoy the whole affair. He did mention that one of the most difficult phases of this particular demonstration was to explain the incident to his wife.

This demonstration earned the Quarter Horse Award for County Agent Oefstos. Hank did report that the drainage ditch made by the explosion functioned very well.

APPENDIX L
County Agent Ferguson and the "Reds"

By Robert F. Rasmusson

Drought, low farm prices and the depression brought about radical political changes in Sheridan County during the mid-twenties and the early thirties. The organization of a communistic party took place primarily as a result of the promotional efforts of the *Producers News*, a weekly newspaper published in Plentywood with Charles "Red Flag" Taylor as its editor. Mr. Taylor was a personable man who served two terms in the State Legislature, the first in 1922 and the second in 1926.

The communistic party thrived in Sheridan County during the twenties until 1928 when one of the best crops in the history of the county was harvested. But unfavorable crop conditions following 1928 and the crop failures of the thirties starting in 1931 brought about a revival of the "Red" party organization and Red Flag Taylor regained most of the prominance he enjoyed during the twenties.

The Feed and Seed Loan program was developed by the federal government during this second drought period. Applications for loans were reviewed and either approved or rejected by local Feed and Seed Loan committees. It was not uncommon for the county Extension agent to serve as secretary for this county committee. In Sheridan County this arrangement was made, and Mr. E. G. Ferguson, Sheridan County agent, assumed the responsibilities of the committee secretary. The reports on the loan applications were signed by the secretary as well as by the committee chairman.

An application for a feed loan submitted by one of Red Flag Taylor's coworkers had been rejected by the committee for very legitimate reasons. The rejection report was sent to the area Feed and Seed Loan office. By some means or other, a copy of this rejection report was obtained by Red Flag Taylor. The report was reproduced in its entirety in the *Producers News*. Taylor chose to blame the county agent for the rejection of this particular feed loan and conducted a newspaper campaign to have Mr. Ferguson removed from his position as county agent. The campaign produced results and many farmers chose to agree with the recommendations published in the *Producers News*.

Taylor had once strongly supported the county agent. On November 22, 1918, the *Producers News* carried a headline "THEY HAVE CANNED COUNTY AGENT ANDERSON." The story related to the headline included the following statement: "County Agent Anderson is a man of too much honesty of purpose and uprightness of character to be allowed to remain in the office of county agent. It is claimed that he is a hindrance to the complete looting of the farmers by the Loyalty League and that the honorable board of county commissioners. . . have decided to put the skids under him while the chance is good."

But in the January 1, 1932, issue of the *Producers News* the following headline was included: "COUNTY COMMISSIONERS MUST FIRE STOOL PIGEON COUNTY AGENT." Referring to County Agent Ferguson, the accompanying article stated: "Elsewhere in this issue we give the story of the double dealing that has made the county agent's office the den of probably the most vicious enemy of the farmers of Sheridan County."

County Agent Anderson was relieved of his Sheridan County responsibilities for about three months but was reinstated in April 1919, and continued as Sheridan County agent through July 1920. County Agent Ferguson was transferred to Blaine County in April 1932.

Sometime after Taylor's campaign to get rid of the county agent got under way, the county Feed and Seed Loan Committee scheduled another meeting for the purpose of considering applications for loans. The chairman of the committee was Senator Lars Anvick from Reserve.

The meeting was held in the county agent's office, which was located in the basement of one of the bank buildings. The only way to enter or leave this office was by means of an outside stairway.

The meeting was well under way when the sound of people coming down the stairs was heard. Farmers started to file into the office and continued until the room was filled. Probably 40 or 50 people entered the office. It has been said that they had tar, feathers and a pole and were there to forceably take the county agent and escort him out of town. County Agent Ferguson was really worried; the only exit was blocked by 40 or 50 farmers.

Senator Lars Anvick was a diplomat. He spoke slowly with a Norwegian accent, but he spoke well before an audience. In his slow and easy going manner, he arose from his chair and started to talk. It was a friendly approach and he was talking to fellow farmers, many from his home community, and he knew their families and something about their farms.

Lars started with a discussion related to crop conditions, the problems farmers were having and other topics of common concern. He finally started talking about individual farmers in the crowd. He mentioned some of the things that County Agent Ferguson had done for them and how they had benefitted as a result of the activities of the agent. He mentioned working with their children in 4-H Clubs and called the children by their first names. He mentioned the locating of good seed, good breeding animals—beef, dairy and hogs. He gave examples of how they had been helped in their turkey marketing and pointed out that some of them were serving as 4-H Club leaders. This approach began to hit some real sensitive spots.

The door to the office needed to have its hinges oiled and it squeaked when it was opened or closed. Lars could not see the door because of the crowd, but soon the hinges of the door began squeaking. Farmers began to fade towards the back of the crowd with the idea of reaching the door before Lars got around to talking about them. The door squeaked more frequently, and the crowd got smaller. Lars continued to visit with them, and after a half hour or so, his audience had left.

Senator Anvick, the diplomat, had saved the day for the county agent. The *Producers News'* county agent campaign did not end, but farmers in general lost interest in supporting it. "Fergy" as County Agent Ferguson was usually called, mentioned time and again his gratitude for the job Senator Lars Anvick did for him that wintery evening.

APPENDIX M
Objectives, Responsibilities and Scope of the Montana Cooperative Extension Service, 1960

The following report was prepared to implement the national SCOPE statement. Director Aasheim introduced this summary as follows:

"This statement was developed by the staff of the Montana Cooperative Extension Service after consultation with many Montana citizens and staff members from other divisions of Montana State College.

"It is hoped that this statement will serve as a guide to the entire Extension staff in developing an educational program which meets the needs of Montana people and fulfills the responsibilities of the Extension Service."

Original objectives set forth by the Smith-Lever Act stated the general legal framework defining the function of the Cooperative Extension Service. This agreement created the cooperative partnership between the United States Department of Agriculture, the land-grant colleges and the local governing boards.

The Smith-Lever Act states that the legal function of the Cooperative Extension Service is:

> To aid in diffusing among the people of the United States useful and practical information on subjects relating to agriculture and home economics, and to encourage the application of the same.

This legal statement commissions the Cooperative Extension Service as an educational agency. The intent is to apply information to the practical situations in the everyday life of people. In this act, the federal government, the state educational institutions, and the local leaders in charge of local government set out to raise the level of living of people.

Later amendments to the Smith-Lever Act defined the function further as an informal education for persons not attending the land-grant college:

> Cooperative agricultural Extension work shall consist of the giving of instruction and practical demonstrations to persons not attending or resident in said colleges.

These general outlines permitted much freedom for interpretation by various groups and individuals interested in the movement.

The Joint Committee report on Extension programs, policies and goals of the organization of land-grant colleges and universities stated the major objectives in August 1948, as follows:

> The development of people themselves to the end that they, through their own initiative, may effectively identify and solve the various problems directly affecting their welfare.

Through practice over time, the service has supported and will continue to provide an educational program to help people attain:

Greater ability in maintaining more efficient farms and better homes.

Greater ability in acquiring higher incomes and levels of living on a continuing basis.

Increased competency and willingness by both adults and youth to assume leadership and citizenship responsibilities.

Increased ability and willingness to undertake organized group action when such will contribute effectively to improving their welfare.

The Extension Service is the informal educational arm of the United States Department of Agriculture. This assignment states a responsibility for educational service to the other agencies of the United States Department of Agriculture which involves coordination of and leadership in educational activities.

The interdependence which has developed between rural and urban areas makes it evident that Extension does not serve rural areas best by working only with rural people. The Montana Extension Service has always worked with rural and urban people, but the majority of effort has been with people on the farm. In the future, the amount of time spent with folks in town will increase because the increasing complexity of relationships centered around agriculture makes it necessary to include all people in the educational program. Agricultural production, which has long been an area of much emphasis by the Extension Service, is justified only if there is a consuming public.

Educational programs are forced to face *whole* problems and attack them with whomever is concerned in the problem, regardless of age, sex, place of residence or occupation. Educational programs must also create an awareness of alternatives and help bring about an *objective evaluation* of the *alternatives* and then *motivate constructive action.*

The original assignment of the Extension Service to agriculture, home economics and related subjects has necessarily become interpreted more broadly. Extension has been engaged in much service work in the past which has often been questioned as to whether or not it could be classified as educational effort. In the future there will be less of doing things for people and more emphasis on teaching and motivating them to do things themselves and the assistance will have greater diversity of scope.

Broadly speaking, Extension's objective is education for the development and wise use of human, natural and physical resources for the greatest good for the greatest number. Essential to this is a philosophy of developing and using the resources of a community or area to its greatest advantage if the individual, the family, the community, and the nation are to benefit the most. Essential to this objective is also a philosophy of working with other agencies and groups in a spirit of cooperation.

The Montana Extension Service will contribute to this objective in many ways. Some of the important areas of emphasis will be:

1. Agricultural Production and Conservation
2. Marketing and Utilization of Agricultural Products
3. Home Economics—Food, Clothing, Housing and Furnishings, Health and Safety
4. Human Relations, Communications, Citizenship and Leadership Development
5. Public Affairs and Community Development

1. AGRICULTURAL PRODUCTION AND CONSERVATION

Agriculture should strive to produce *efficiently* a *quality product* which the consumer wants.

The trend toward fewer and fewer farmers producing food and fibre for a larger and larger population is an indication that efficiency of production has been increasing. This trend toward fewer farmers will no doubt continue and technology will improve, resulting in greater production efficiency.

Even though supplies of agricultural products are generally in surplus, this does not eliminate the need for efficient production and the need for Montana farmers to compete effectively with farmers from other areas. Surpluses are only such when considered in a nationalistic sense. There are still many hungry people in the world.

Agriculture must also plan so as to be able to provide an adequate food supply for our ever-increasing population. The fact that agricultural surpluses are a current problem presents another opportunity for working with people in bringing about agricultural adjustment which will be for the common good.

A brief history of the nations and national welfare reveals one vital axiom. The welfare of the people, including such concepts as better living, more happiness, more education, can be developed and maintained only within the limits imposed by the quality and quantity of natural resources available. The natural resource base of soil and water imposes its own limits on such renewable resources as forest, grass and wildlife. These resources can be underdeveloped, or in the other extreme, overexploited, to a degree that can result in human competition at lower levels of human existence.

The success of area planning and development, communications and human relations, family living, management, marketing and utilization, public affairs, and youth development depends on a natural resource base. If this base is underdeveloped or deteriorated, the entire structure is weakened, and the first evidence of decline can be a strain on the finer and basic things in family living.

Our natural resources in Montana are not presently in a state of emergency. Neither are they so abundant that we can afford underdevelopment or mismanagement.

Montana is in an enviable position for providing blue ribbon recreational facilities to a growing demand. By virtue of its geographic location, topography, low population density and land ownership pattern, it seems that retaining and developing the natural resources for recreational purposes may well be desirable from the standpoint of future use. This does not imply that such use will be the only use, but rather that over-all management be such that recreational values are considered along with other uses.

Forests in Montana are a great resource on both private and public lands. The public forests are administered through the Forest Service, but small non-industrial private forests will require special consideration because of increased pressures from all major forest interest groups.

Conservation and natural resource development involves the development of our natural resources for the maximum long-time use and yield and the management for the most intensive use consistent with area planning, family goals and capabilities of the resource.

2. MARKETING AND UTILIZATION OF AGRICULTURAL PRODUCTS

For purposes of discussion here, marketing is defined as "all of the things that happen to a commodity from the primary producer to the ultimate consumer." Some additional insight into the original intent of the Cooperative Extension movement may be gained from the following quotation taken from Report No. 110, House of Representatives, 63rd Congress, Second Session, December 8, 1913, regarding the purpose of Extension:

> To teach the farmer the best methods of increasing production is exceedingly important, but not more vitally so than is the importance of teaching him the best and most economical methods of distribution. It is not enough to teach him to grow bigger crops, he must be taught how to get the true value from these bigger crops, else Congress would be put in the attitude of regarding the work of the farmer as a kind of philanthropy. The itinerant teacher or demonstrator will be expected to give as much thought to the economic side of agriculture—the marketing, standardizing and grading of farm products—as he gives to the matter of larger acreage yields. He is to assume leadership in every movement, whatever it may be, the aim of which is better farming, better living, more happiness, more education and better citizenship.

A review of the current economic situation does in some measure support the view that research and education have placed the farmer in something of a philanthropic position. Through mechanization, the application of agricultural technology and mass production techniques, he has produced beyond the consuming capacity of the population. On the other hand, he has removed fear of hunger from the minds of most Americans, reduced tremendously the amount of labor which families must expend for food, has assisted materially in the development of a tremendous array of agricultural associated industries and at the same time finds himself more and more economically disadvantaged as compared to his non-agricultural neighbors. At the same time, he has been working himself off the farm so that now farming represents a small minority.

If the Extension Service is to proceed in the new era, to represent a government service designed to strengthen the agricultural sector of the national economy, it seems clear that Extension must shift some of its production emphasis to the areas of marketing and utilization.

To do this job, close liaison will need to be maintained with commercial firms, government agencies and marketing groups involved in processing, handling and the distributive phases of marketing. Of prime importance will be the informing of producers of quality requirements, demand situations, government program assistance and data pertinent to aiding orderly production planning and the subsequent marketing of commodities.

Marketing education work, of course, cannot be confined to work with producers. Production specialists should incorporate the application of marketing problems of individuals and areas in their subject matter programs. More emphasis must be placed on work with marketing firms to increase their efficiency for the benefit of marketers, producers and consumers.

A well nourished populace requires consumption of foods which are wholesome and of nutritional quality and balance. A well nourished people should be an objective of the educational efforts of the Extension Service. In this respect, it must be kept in mind that even though agriculture is interested in its products being consumed, special interest programs should not be encouraged at the expense of the public's general welfare.

The efficient utilization of all agricultural products is a prime concern of all people and must be included in an educational program along with production and marketing.

3. HOME ECONOMICS—FOOD, CLOTHING, HOUSING AND FURNISHINGS, HEALTH AND SAFETY

The well-being of people is, to a great extent, dependent on food, clothing, shelter and safety, because these things contribute to their comfort, health and very existence.

Montanans are generally well fed, but there is need for emphasis on good eating habits focused particularly on three nutritional problems: overweight, teenage nutrition, and nutrition for the aging.

In the area of clothing and textiles, the major need is to provide the consumer with information which will guide selections from the large variety of new products on the market. Planning the wardrobe and clothing construction techniques are other important areas.

Families need assistance in developing housing and home furnishings plans to give maximum economy consistent with comfort, convenience, utility and safety.

Health and safety are a concern of all people and all Extension workers have a responsibility to educate and motivate people toward safe and heathful living.

The Extension Service should help people and families inventory their resources of time, money, energy, and abilities and to use them effectively in providing food, clothing and shelter. Help should also be given in developing a high level of competence in the homemaking skills and techniques.

4. HUMAN RELATIONS, COMMUNICATIONS, CITIZENSHIP AND LEADERSHIP DEVELOPMENT

The period since World War II has been described as a period of agricultural resource adjustment. It has also been a period of substantial industrial adjustment, a period of cold war and a period of rapidly increasing population. More and more it is becoming clear that agriculture is neither a social nor economic island. Practically none of the major concerns of individuals, communities, counties or nations are handled by the leadership in any one, two or three segments of our society. The problems of agriculture, likewise, are today intricately woven with those of other vocational groupings. If the agricultural minority is to be heard and understood, other leaders with whom agriculture must work, must operate by the same rules and communicate in some common language, else the "best good for all" principle will not likely be approximated. Extension must accept a major responsibility for finding ways of making democratic leadership effective in every important area of public concern.

The growth of farm organizations, cooperatives, labor unions, the agribusiness concept and integration in agriculture are evidence that many individuals' problems and interests also involve others. Human relations, group development and communications workshops and training centers have sprung up all over the United States during the past 15 years. They are patronized largely for the training of industrial management because skills in these areas are related to productive efficiency. If understanding and skills in the humanities, group processes and communications are in fact related to success on the farm—are related to making the farmers' organizations more fruitful—are helpful to the farmer in dealing with those problems which involve other groups, then training of people in these areas deserves a high priority.

From the beginning, Extension has placed leadership development high among its objectives. Help has been given to organize farm groups such as marketing cooperatives, cattlemen's associations, fire districts, weed control districts, home demonstration clubs, advisory boards and so forth, but most of the leadership training has been indirect. Because of the large volume of research now available in the areas of human relations, group process and communications which are recognized as being basic to effective leadership and democratic action, it seems now that Extension, through its own and other resources, should undertake training in these areas directly to achieve these basic objectives:

1. To help the leadership in Montana acquire an understanding of basic human needs and motivations.
2. Understand basic communications processes.
3. Understand and develop skills and some facility in the use of the major communications media.
4. Understand and develop skills in group process.
5. Understand and develop human relations skills.

The *family* has always been a basic unit in the minds of Extension workers. The present situation and trends point to a need to emphasize a human relations program designed to help families adjust to the many influences affecting families, groups and communities. The problems associated with old age perhaps need more emphasis than they have received in the past in order that more intelligent provisions can be made for this stage of life.

In addition to the need for work with the family and other groups in this area, there is obviously the need for work with *individuals* who make up these groups. The Extension Service places much emphasis on *youth development* which has been effectively accomplished through the 4-H program and other methods. The primary aim of this effort is to provide educational opportunities for the physical, mental, social and spiritual growth of young people. These opportunities are designed to help boys and girls develop so that they may lead useful, satisfying lives and be better prepared to accept their full share of adult responsibilities. The informal education provided by the Cooperative Extension Service is especially designed to supplement the training received in the home, church, school, and other youth-serving agencies.

In working with youth, materials and services are available to all boys and girls to the extent that resources are available.

5. PUBLIC AFFAIRS

The aims of educational efforts in public affairs are to create:

1. A more active interest in public issues.
2. A better understanding of the issues and principles involved.
3. A sound basis for critically examining evidence and for logically arriving at intelligent decisions.
4. A greater desire to participate effectively in the solution of public affairs problems.

Public affairs involves all people. All people are concerned with the problems, such as public land policies, subsidies, agricultural programs, schools, health facilities, taxation and public services. Extension has the responsibility to bring these problems into open discussion and to create a feeling of responsible citizenship in regard to public problems and issues.

APPENDIX N
Goals and Related Activities of the Montana Cooperative Extension Service

The following report was prepared by Director Torlief Aasheim and his staff in 1973 to acquaint the State Board of Regents and other state agencies with the broadened structure, the goals and activities of the Montana Cooperative Extension Service.

I. WHAT IS THE COOPERATIVE EXTENSION SERVICE?

A. A Branch of Montana State University

The Extension Service is a branch of Montana State University. Its purpose is to extend useful knowledge to people off campus for their use in achieving satisfaction in their lives and for improving the quality of life in their communities.

B. Authorization

Extension was created by the Smith-Lever Act which was passed by Congress in 1914. The Cooperative Extension Service at Montana State University was organized when state legislation in 1915 accepted the Smith-Lever Act.

C. Background and Philosophy

The Extension Service was organized to help accomplish the national goals of creating a bountiful supply of efficiently produced food and improving the quality of rural communities.

The changing problems of society and the interdependence of the farm and the city have caused Extension, in cooperation with other agencies, to work toward helping all people understand and solve the problems and concerns of modern living.

Some of the dominant features of the Cooperative Extension Service nationally have been its high commitment, along with agricultural production, to 4-H Clubs, family life education, educational programs with women, and broad-spectrum community development.

Extension is people-oriented. Our American society believes that each individual should have the opportunity and the right to participate in all decisions that affect him. This requires that society be a democratic one and that all people have an opportunity to participate. Therefore, in keeping with this principle, Extension involves many people in determining Extension programs.

D. Financing

Extension is funded by federal, state and county governmental units as well as by grants from the private sector. Some federal funds are earmarked for specific purposes to meet current national needs, such as those for civil defense, marketing, and nutrition programs with low income people. Certain state funds are also earmarked for special purposes; for example, recent appropriations for work with fertilizers and soils. County funds are used to support Extension agents in their respective counties. As to funds from the private sector, these are committed for specific purposes.

E. Responsibilities

Sources of funding imply obligations and responsibilities to many segments of society. The memorandum of understanding between Montana State University and the U.S. Department of Agriculture indicates that these two institutions, through the Extension Service, will cooperate in conducting many programs of benefit to Montana people.

Moreover, the Montana Cooperative Extension Service has various educational obligations related to this memorandum of agreement. As a part of the land grant university, the Cooperative Extension Service conducts workshops in many towns, using highly trained instructors and resources, and is in fact the largest off-campus arm of the Montana University System. As a part of the national government, it is the educational arm of the U. S. Department of Agriculture and is responsible for informing people about its many national programs, about other agencies, and about relationships of each to local endeavors. Working as it does in Montana counties through its local county Extension agents, it has obligations to county government and to the local citizens.

Figure 1, Levels of Goals, is a way of illustrating the responsibility the Extension Service has to all levels of society, and to each with goals related to the others:

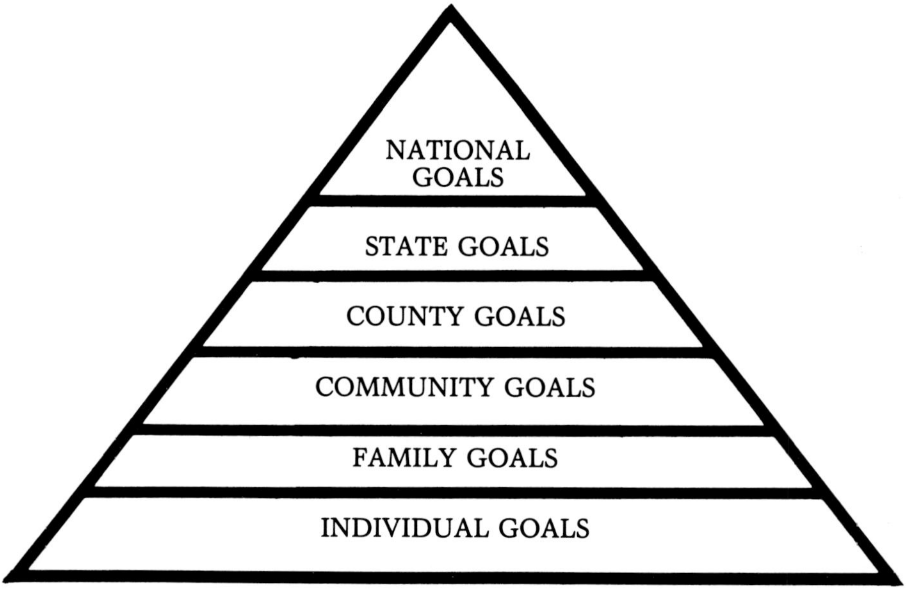

FIGURE 1 — Levels of Goals

F. Organization

The Montana Cooperative Extension Service employs about 100 county Extension field staff and 45 state Extension specialists. (See Appendix N for organizational chart.) All are professionals and members of Montana State University faculty. The efforts of the entire staff are divided into two major program areas. These two program areas are directed toward achieving the same major goal of human satisfaction. However, one is concerned, more specifically, with agriculture and economic development, the other with human resource development. Thus, Extension's goals include programs that are easily measured by monetary means and those that are not. (See Figure III for a way of illustrating how Extension relates to the broad goals of people.)

II. HOW THE EXTENSION PROGRAM IS DEVELOPED

A. Program Development

The broad, overriding effort of the Cooperative Extension Service is to help people so that, through their own initiative, they can identify and solve problems affecting their welfare.

In the process of developing its programs, the Extension Service meets and listens to the people. Field and state staff listen to many leaders throughout the state and note their problems. State organizations and groups also heed what leaders say and transmit these concerns as well to the Extension Service. Leaders from state government also have an important input. They indicate their analyses of problems, suggestions, solutions, and plans of action in specific statements directed to Extension.

National groups and the federal government provide their inputs into state Extension programs through Congress and the USDA.

Listening to and being responsible to many units of society results in being responsive to many masters. For 60 years, however, Extension has found this advantageous in planning complete programs because of the checks and balances involved. Many facts are always considered.

A general statement follows about the goals and activities of the two program areas: 1) Agriculture and Economic Development and 2) Human Resource Development.

Program Areas*

I. AGRICULTURAL PRODUCTION AND ECONOMIC DEVELOPMENT

A. Production and Economic Development

THE MAJOR GOAL:

To develop and sustain agriculture and other business and industry in such a way as to support communities and provide efficient production of goods and services for the consuming public while maintaining a reasonable profit for the producer.

1. Educational Programs—for example:
   a. Beef Production—To insure the efficiency of beef production and strengthen the economic position of Montana ranchers, farmers and livestock feeders and, thus, their contribution to community development.
   b. Soil and Water Resources
      (1) To improve the management of soil and water resources for the production of adequate food and fiber of high quality at a cost the consumer can afford and in a manner consistent with the protection of the environment and a fair return to the producer.
      (2) To develop an understanding of the alternative uses of soil and water resources and respective consequences of such uses.
   c. Crop Production—To secure maximum return from crops through use of improved varieties and updated cultural practices.
   d. Weed Control—To assist all weed control districts to comply with the Montana noxious weed law in order to reduce economic losses caused by weeds and yet preserve the environment.
2. Other education efforts are directed:
   a. To secure maximum return from agricultural products through improved marketing systems.
   b. To strengthen agriculture through improved farm policy.
   c. To secure maximum return from farm operations through use of improved business methods.
3. More specifically:
   a. External factors affecting farm income, such as domestic and foreign policy.
   b. Business operators' methods.
   c. Machinery technology.
   d. Buildings and structures.
   e. Range and pasture management.
   f. Marketing practices.
   g. Marketing systems.
   h. New uses of agricultural products.
   i. Consumer preference and demand.

B. Natural Resources

THE MAJOR GOAL:

The management of natural resources for maximum human satisfaction consistent with the capabilities of the resource.

1. Educational programs—for example:
   a. Wildlife—To increase understanding of the appropriate management of Montana's wildlife.
   b. Recreation—To increase the understanding and utilization of Montana's natural resources for recreational purposes.
   c. Public Affairs—To increase the understanding of public issues involved in the use of Montana's lands and natural resources.
   d. Environment—To help reduce degradation of the environment.
   e. Range—To increase the productivity of Montana ranges in line with desirable economic and ecological factors.
2. Other educational efforts are directed:
   a. To improve practices affecting natural resources (land use planning, flood plains management).
   b. To improve management and utilization of forests and forest products.
   c. To improve the physical, cultural and aesthetic environment.
   d. To insure safe use of pesticides.
   e. To reduce air, land and water pollution.

II. HUMAN RESOURCE DEVELOPMENT

A. Personal Resource Development

THE MAJOR GOAL:

To develop attitudes, knowledge and skills so that people may lead satisfying lives, with special emphasis on democratic processes.

1. Educational Programs—for example:
   a. Consumer education—To improve the quality of living of Montana people, as defined by Montana people, by improving consumers' knowledge of alternatives available to them in use of their money, time and talent.
   b. 4-H—To help develop Montana's youth as individual, family and community members so they may become responsible, productive citizens.
2. Other educational efforts are directed:
   a. To develop the surroundings of Montana homes, yards and gardens so as to provide a pleasant environment for Montana people.
3. More specifically:
   a. Improved nutrition
   b. Family economics
   c. Improved housing
   d. Physical, social and psychological aspects of clothing
   e. Interpersonal relationships
   f. Safety practices
   g. Health practices and facilities

h. Emergency preparedness
i. Helping youth develop into productive adults through acquisition of knowledge of:
   (1) Improved nutrition
   (2) Cultures of other countries
   (3) Science and technology
   (4) Personal growth and development
   (5) Responsible citizenship
   (6) Effective leadership

B. Community Development

THE MAJOR GOAL:

To incorporate economic, natural and human resources into an overall development effort that will result in better Montana communities.

1. Educational Programs—for example:
   a. Development Groups—To help Montana people form a functioning community development organization in each county or groups of counties for the purpose of achieving more constructive interaction and cooperation than people could achieve by working independently.
   b. Other educational efforts are directed:
      (1) To market analysis and feasibility
      (2) To volunteer leadership responsibilities
      (3) To improve public services and facilities
      (4) To develop marketable skills
      (5) To work with other agencies and organized groups that are committed to community improvement

*For additional details refer to *1973 Situation Statement* by the USDA Committee for Rural Development; *1972 Annual Report*, and *1973 Plan of Work of Cooperative Extension Service.*

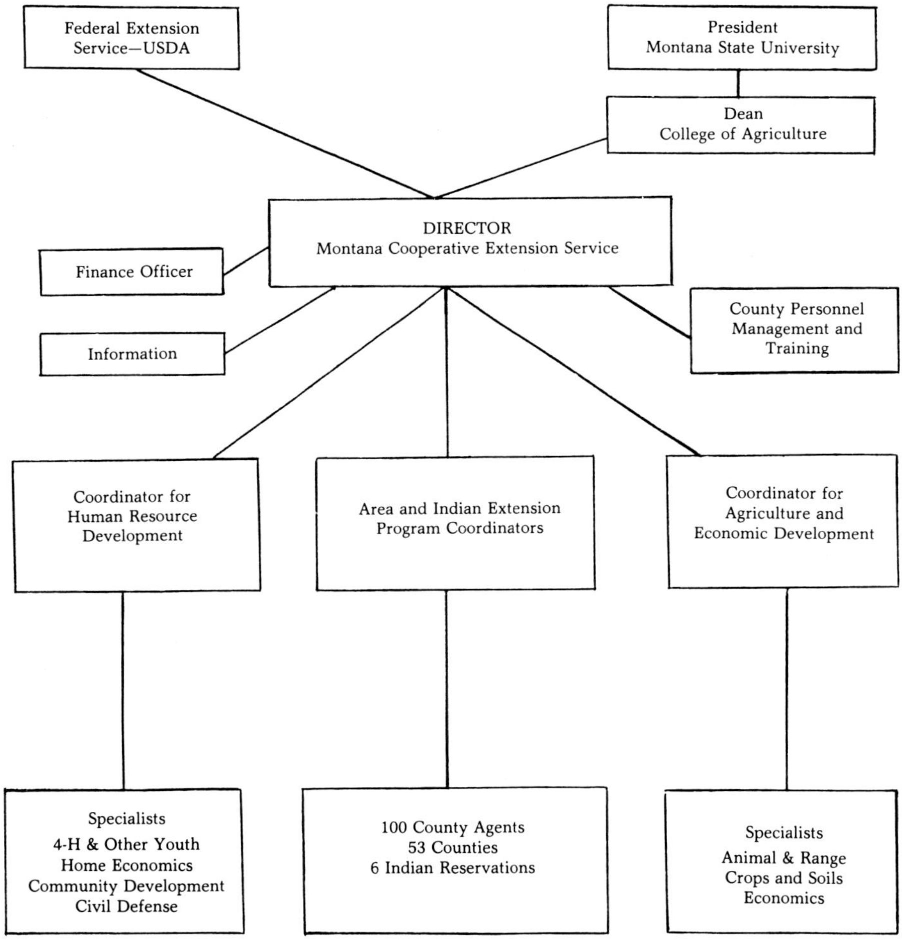

FIGURE II — Organizational Chart
Montana Cooperative Extension Service
March 1, 1973

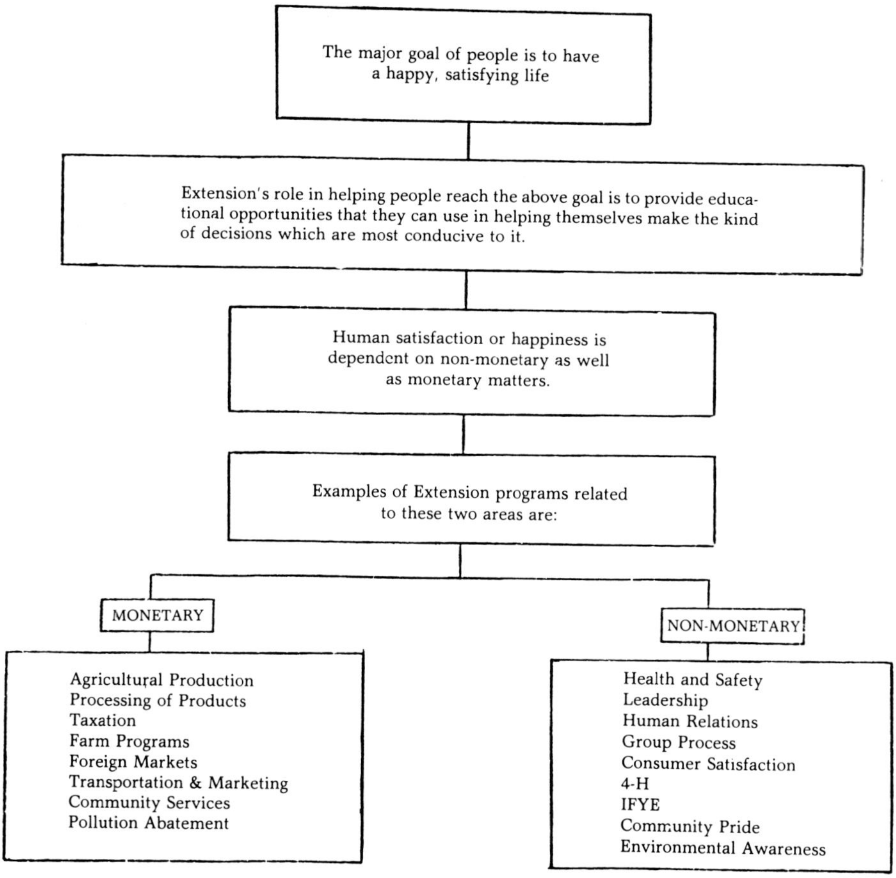

FIGURE III — A Schematic Sketch Relating Extension's Role and Activities to the Goals of People

INDEX

A

B

C

D

E

F

G

H

I

J

K

L

M

N

S

T

Y

Z

About the Authors

MERRILL G. BURLINGAME, raised on an Iowa farm, came to Montana State College as a history instructor in 1929. He served as head of the History Department from 1932 to 1965 and retired in 1967. Burlingame earned a B.A. from the University of Iowa, 1927; M.A., 1928; Ph.D., 1936, with an honorary LHD from Montana State University in 1973. He became interested in Montana history as a young man and has been involved in writing several histories of the state beginning with *The Montana Frontier* in 1942. He served on the Board of Trustees of the Montana Historical Society from 1949 until 1978. Following retirement, he wrote *A History of Montana State University* in 1968 in observance of MSU's 75th anniversary.

EDWARD J. BELL, Jr., spent his first years on a Maryland farm and then moved west with his family in 1911 to homestead in Montana near Wibaux. As a freshman in high school he worked with J. C. Taylor in organizing a county boys' corn club. Bell received a B.S. from Montana State College in 1925 and an M.S. from the University of Wisconsin in 1929. Following graduation, he worked with the Agricultural Economics Department developing a wheat marketing information system for Montana. He also was involved with federal agricultural programs, Federal Farm Board; the Agricultural Adjustment Administration; and was the first administrator of the Oregon State Wheat Commission, 1947-1953. Bell served 13 years with the U.S. Foreign Agricultural Service, spending many years in Washington, D.C., between assignments in the Philippines, Greece and South Africa. He retired in 1967 and returned to Bozeman with his wife, the former Anne Montgomery, daughter of a pioneer Chinook family. Here he has compiled a history of the Agricultural Economics Department, and wrote *Homesteading in Montana*, 1975, and *Wheat was my Life,* 1981.